The R.A.M.S. Library of Alchemy

Volume 33

The Golden Chain of Homer
by
Homerus

R.A.M.S. Publishing Company

The Golden Chain of Homer
(Aurea Catena Homeri)

by

Homerus

Translator:
Leone Muller

Editors:
Anton Joseph Kirchweger
Hans W. Nintzel
Philip N. Wheeler

Produced by

Restorers of Alchemical Manuscripts Society

R.A.M.S. Publishing Company

R.A.M.S. Publishing Company
117 Rutherford Lane
Stuarts Draft VA 24477

The Golden Chain of Homer
Copyright © 2015 R.A.M.S. Publishing Company

First Edition 2011
Second Edition 2015

ISBN-13 **978-1519337443**
ISBN-10 **1519337442**

Image Processing by Philip N. Wheeler

This book is sold for informational purposes only. Neither the publisher nor the editor shall be held accountable for the use or misuse of the information in this book.

Printed in the United States of America

Table of Contents

Dedicated to Hans W. Nintzel,

American Alchemist

and

Founder of the

Restorers of Alchemical Manuscripts Society

(R.A.M.S.)

Disclaimer

Liability: The publisher does not warrant or assume any legal liability or responsibility for the accuracy, completeness, or usefulness of any information, apparatus, product, or process disclosed. The publisher makes no representation as to the accuracy or completeness of the contents of this book and specifically disclaims any implied warranty of merchantability or fitness for a particular purpose. No warranty may be created or extended by written sales materials or sales representatives. You should obtain professional consultation where appropriate. The publisher shall not be liable for any loss of profit or other commercial or personal damages, including but not limited to special, incidental, consequential, or other damages.

ANNULUS PLATONIS

or

AUREA CATENA HOMERI

(Golden Chain of Homer)

or

a physical – chemical

EXPLANATION of NATURE

and

Its Origin, Preservation & Destruction

from

A Society of true natural researchers

revised and improved with many important

annotations given throughout

Anton Joseph Kirchweger

2nd Edition

Transcribed literally and true to the original,

rare Rosicrucian edition of 1781

To The Reader

With the production of the present work, "The GOLDEN CHAIN OF HOMER," R.A.M.S. (Restorers of Alchemical Manuscripts Society), may have accomplished their Magnum Opus. As will be discovered by the reader, many different versions of this book (Boehme et al) have been published. This includes the R.A.M.S. production of an excerpt by Dr. Sigismond Bacstrom, which was released several years ago. The present work reflects the full text of the work edited by Anton Kirchweger. This author reveals that he was a member of a secret Hermetic order. And that one of the brothers of the order, Homerus, an alchemical adept, set down the basics of this work.

Whatever the true origins of this book, no one will doubt that it is indeed a masterpiece of alchemical literature. Written in down-to-earth language, one can begin to understand alchemical principles and thus unveil the mysteries that shroud the Holy Science of Alchemy. The footnotes alone form a compendium of data that is of enormous value. It may be of interest to the reader to be aware of a question asked by the R.A.M.S. producer of one recognized as an alchemical master[1]. The question was: "If your house was on fire and you could save

---

[1] Frater Albertus.

only ONE book from your library, which one would you save"? The alchemical teacher, for so he was, smiled a secret smile and replied: "Well, it would be the Golden Chain of Homer". Pressing this point the producer then asked: "Is this then, THE most important book of alchemy, or, at least, ONE of the most important books"? The teacher looked away for a moment, reflecting and answered: "Yes. IF you can understand it!!" So, courteous reader, you have in your hands a book deemed by those who know, as being VERY important And so it is More germane, you will find it to be immensely understandable.

To give the reader some idea of the effort that went into this book, recognize that 8 people put in efforts spanning over two years. It is fitting then, that we thank those involved. First, Ms. Leone Muller who single-handedly translated almost 600 pages of archaic German text into English. We owe her much. Also, the Messrs. Kevin Masman and Daniel Dullies, two practicing alchemists in Australia. They did a tedious and masterful job of ensuring no alchemical sense was lost. Their comments and editing made a fantastic book come to life! Typists Judy Hipskind and Mary Roberts got the final version going. Jody Nintzel contributed most of the illustration while David Ham labored with the symbols and proof-reading. Final typing, editing,

symbology and production was done by Hans Nintzel
with help from some folk who wish to remain
anonymous. To all, we applaud your mighty efforts
done as a labor of love. Thank you.

Now dear reader, you can embark on a fantastic
alchemical journey. Recognize though, as Gerhard
Dorn wisely noted, the theory is dry without the
praxis. Doing the laboratory work will thoroughly
ingrain the theory and make it LIVE within and
without you. Enjoy!

Hans W. Nintzel

Introduction
Philip N. Wheeler

Given the importance of this work, I have extensively expanded it to its current state. This is a more complete and accurate edition than my previous work in 2011, and includes all of the alchemical symbols used in the original R.A.M.S. edition of 1984. In most cases I was able to verify that the symbols in the R.A.M.S. edition matched those used in the 1781 German edition; where differences were found, footnotes were added. I also added the tables that are at the start of the 1781 edition, Part 1, Chapters XXI, XXII and XXIII, which are missing from the R.A.M.S. edition.

The R.A.M.S. edition contained symbols that are often seen in alchemical manuscripts. Were these symbols used in the original German edition? Recently I was able to obtain copies of the 1757 (German), 1762 (Latin) and 1781 (German) editions. Many symbols are in the 1781 edition that were not in these earlier editions. It is unknown why these symbols were added long after the death of Dr. Kirchweger.

The footnotes present another mystery. They appear in the 1781 German edition. However, they are not in the 1757 German edition, nor in the 1762 Latin edition. In fact, each of these editions shows interesting deviations. The text in Part I Chapter I, 1757, is very short, less than a page in length, whereas it is much longer in the 1781 edition. The 1757 edition also contains an index. In the 1762 edition the paragraphs of text are numbered and it contains an index. The double-dragon image appears in the 1762 Latin edition at the very end of the book, and in the 1781 edition at the end of Part II Chapter VII, but not in the 1757 German edition. Obviously many variations were introduced after Dr. Kirchweger's death in 1746.

All additions and explanations by Hans W. Nintzel or Philip N. Wheeler are noted by the initials HWN or PNW.

The R.A.M.S. Library of Alchemy Volume 21, *Alchemical Symbols*, is highly recommended to all who wish to understand the multitude of symbols used in Alchemy.

Aurea Calena Homeri

Annulus Platonis

Superius & Interius Hermetis

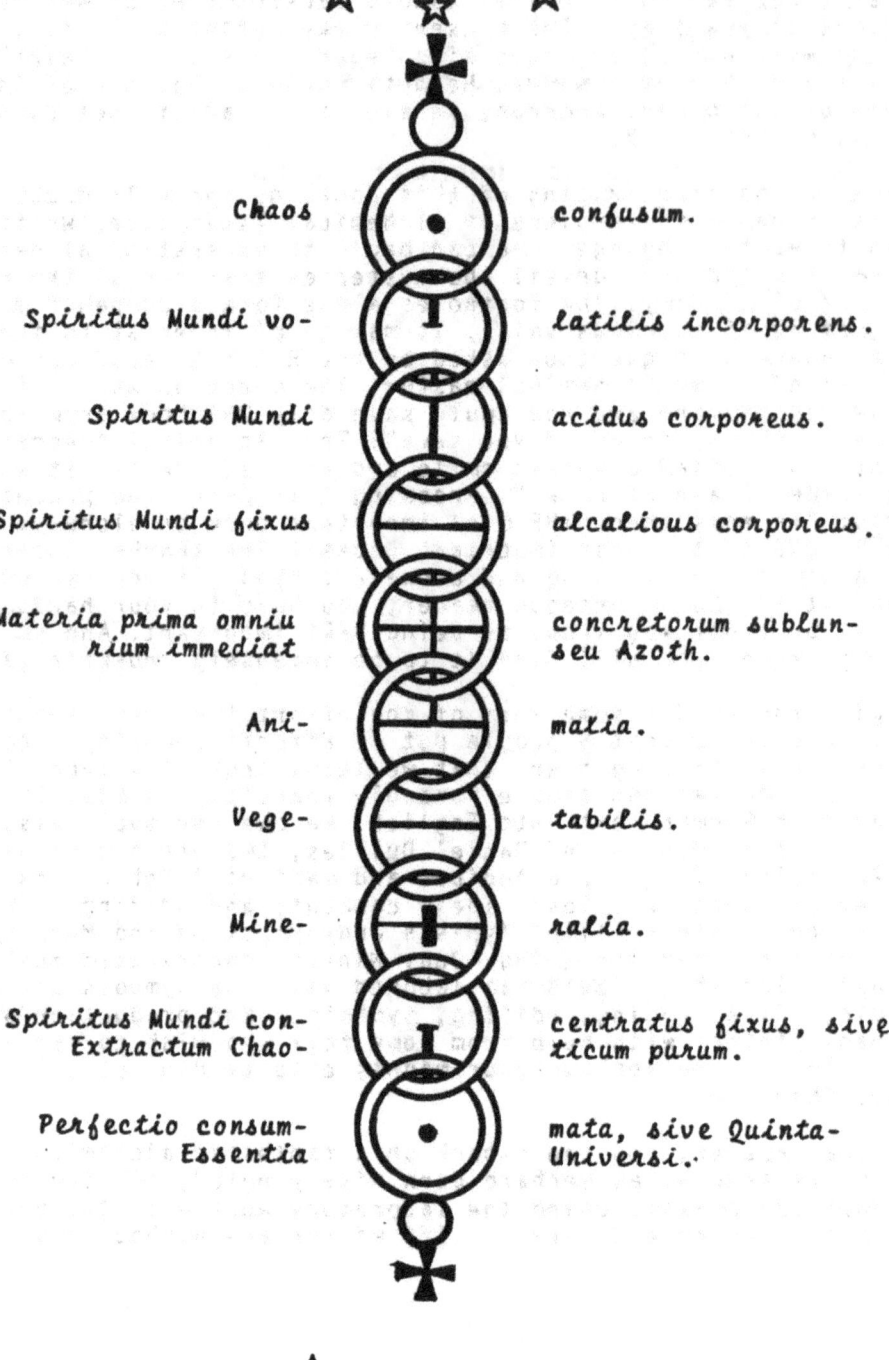

Chaos   confusum.

Spiritus Mundi vo-   latilis incorporens.

Spiritus Mundi   acidus corporeus.

Spiritus Mundi fixus   alcalious corporeus.

Materia prima omniu
rium immediat   concretorum sublun-
seu Azoth.

Ani-   malia.

Vege-   tabilis.

Mine-   ralia.

Spiritus Mundi con-
Extractum Chao-   centratus fixus, sive
ticum purum.

Perfectio consum-
Essentia   mata, sive Quinta-
Universi.

Smaragdenen Tafel (Emerald Tablet) in Phoenician.

17

A Word about R.A.M.S.

R.A.M.S., the Restorers of Alchemical Manuscripts Society, is a very loosely knit group composed of private citizens around the world who have contributed to the present work and others in one way or another. The goal of the society is to seek out and obtain copies of Alchemical manuscripts, tracts, printed works and other such materials and literature. From these, certain items are selected for 'restoration'. This consists of re-typing the material to render it readable, therefore useable. The reading is made more enjoyable by the use of charts, symbols, dictionaries, diagrams, insertion of printed illustrations and other annotations to shed additional light on the text.

R. A. M. S. is a non-profit organization with members from all walks of life. There is no 'organization' per se, just individuals desirous of performing a labor of love. It is their hope that these efforts will perhaps result in renewed interest in the science of Alchemy or even contribute to new discoveries or fields of experimentation. While this material is for all on the path of Alchemy, it is particularly for the sincere practicing Alchemist.

The transcription of this material is as verbatim as human skills permit. The only exceptions are to make the matter more readable, or understandable, with some 'modernization'. This is done ONLY where it is judged that no ambiguity will arise from a slight departure from the original and there is no danger of deleting key words or phrases where a possible use of Gematria, Temura, Notariqon or other form of written code exists. Thus, misspelled or grammatically incorrect material is to be found as well as archaic or strange words. This is in an effort to maintain the original text as far as possible. The completed material is then reproduced, sometimes bound, and offered to interested parties. The costs for these copies are used to defray reproduction costs and to obtain additional material for restoration.

The work of R.A.M.S. includes such material as "Last Will and Testament" of Basil Valentine, important selections from the invaluable Bacstrom Manuscripts such as "Golden Chain of Homer," "Lamspring," a "Process for the Lapis Sophorum," "The Chemist's Key," "The Mineral Gluten of Nitre and Sulfur", "Coelum Philosophorum" and others. Additionally, material by other writers is or will be offered, such as Geber, Kalid, Ripley, Bacon, Hazelrigg, etc.

It is highly appropriate to acknowledge the many persons who have either materially or philosophically contributed to the present effort and future ones. For some this might well be the first indication that they are considered as members of R.A.M.S. or that such a group even exists! While the list is long, it Includes C. Collins, Rick Stern, Doris Edlein, Arp Joo, D. and J. Nintzel, N. Ogle, G. Price, F. Regardie, W. van Doren, K. von Koenigseck and especially David Ham. For their labors and contributions, grateful thanks are given. Let their unselfish efforts inspire others to light the fires of Alchemy.

Hans W. Nintzel
1984

Preface

The *Aurea Catena Homeri (Goldenen Ketten Homers)* is a famous book, or better was a famous book; because today it is forgotten and buried under the dust of hundreds of years in our libraries. In the past, however, an uncountable number of people have read and studied it; even the best of that time, like Goethe, for example, gained an array of inspiring thoughts from the *Annulus Platonis* ("Ring des Plato") -- as the work is also entitled.

> From the sky it comes
> to the sky it rises
> and down again
> to earth it must come
> eternally changing.

That is the quintessence of the *Superius et Inferius Hermetis (Oberen und Unteren des Hermes)* -- as the work is also entitled thirdly. The idea that all creation, no matter what its nature, is most closely "chained together" and interconnected, that a deeply secret connection pervades all of nature, that one thing relates to the next and things depend upon each other, as well as the idea that this secret connection appears in the form of an "up" and "down," a "way-to" and "way—back," in the form of an always changing circulation —- this "possibly fantastic" double—idea was what Goethe, as he once said himself, liked best in the *Aurea Catena Homeri*. And this same grand natural philosophic idea was also the one that captivated another large circle of readers of this book for a whole century.

Today, as mentioned earlier, the book itself is long forgotten, even if its content is still lingering in all of us. Only he who concerns himself in the "art of the hermetics" will occasionally run across this "jewel"; and this "philosophical Garfunkelstein" will enlighten the hermetic darkness. And only "an artist" will not give up, but persevere through the work — once - twice - three times. "The door is always open; go into the room my friend!" *Aperta jam porta, intra in conclave, amice!* And actually, he who enters the palace of this book and gives it more than a passing glance, will finally pass "through all doors and reach the bed of the naked queen." He will also encounter the "natural king" who checks "the heart and kidneys" — *qui scrutatur corda renes*.[2] And both of them -- Queen and King -- will reveal to him the spirit of the time in which "the chain" originated. This spirit, as a "seed," will enable the comprehension of other mystical—alchemical writings.

"So I begin and present you, honorable brothers, with a work which because of its internal value and because it has been mutilated in various ways by many authors from the past, and more recently, precisely and clearly, with explanations of the theory as well as practical explanations -- a work which is considered a 'classic school—book' by our elders and which has become a requirement especially for the students of the lower grades in order to create a basis for the respective science."

That is what the preface to the "Rosenkreuzer edition" of the *Aurea Catena Homeri* of 1781 is

---

[2] Part II Chapter IX -*PNW*

saying. And even if we cannot present the complete work word by word here, we would like -- by the way without agreeing to all, especially not the details! -- to let the spirit of truth and the power of conviction come back to life, with which it has been written approximately 200 years ago. It might cause this or that follower of "hermetics" to read the original itself. He will then realize just how much enlightenment the immense progress of two centuries has contributed to the answering of specific questions, and how little in comparison to the solution of general world and human enigmas. As far as those are concerned, we are as smart as we were before. In any case we have no reason to look down disdainfully on the author of an *Aurea Catena Homeri*.

Invisible Spirit, spiritus mundi incorporeus.

The beginning of all things, the volatile seed of the cosmos, the primal seed of the universe.

The visible Steam, Fog, Smoke ; the primal steam of the universe.

The Chaotic Water.

The primeval water of the universe, the primordial Water, Mercury of the Wise, fiery Water, watery Fire, Shamajim, Azoth.

Abyssus superior seu volatilis.

The source of all separate phenomena. Hermaphroditic Mercury.
(This is composed of :

Invisible Spirit plus visible Water.
Active Spirit plus passive Body.
The ' causator ', the mover plus Instrument, receptacle.)

The great ' Nothing ', Void.

Prima Materia, primordial Matter.

CHAOS.

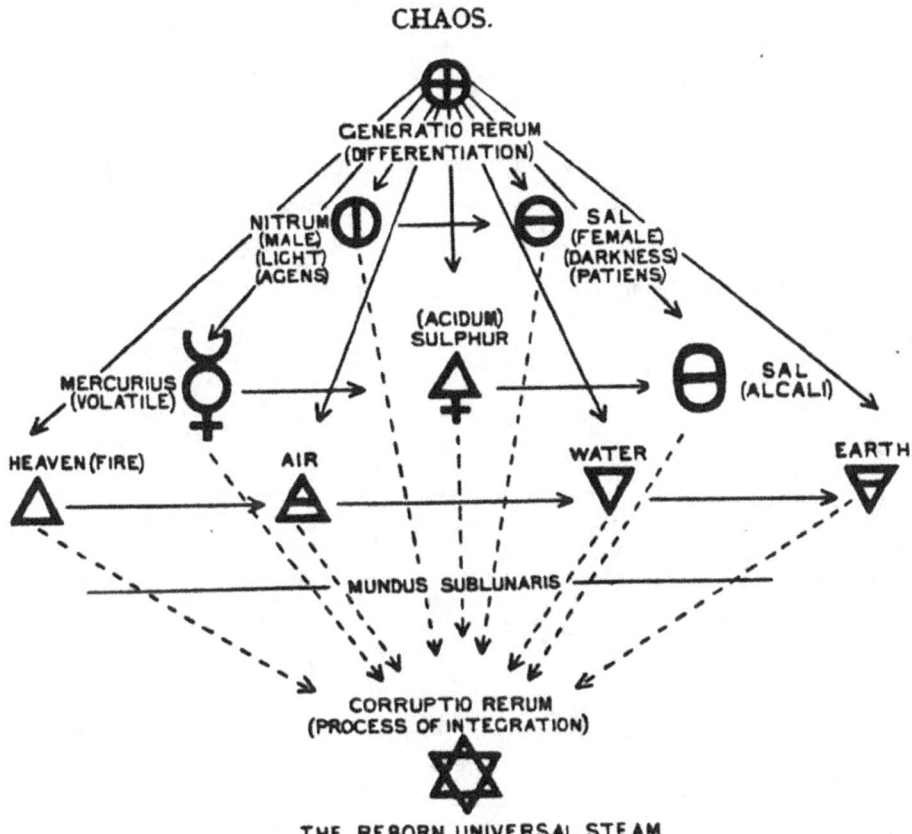

THE REBORN UNIVERSAL STEAM

INTRODUCTION: THE EDITIONS

Concerning
the several authors
and various editions of
"The Golden Chain of Homer"

The identity of the author of the *Aurea Catena Homeri* which was first printed in 1723, but which prior to that had been distributed in a handwritten version, has been unknown for a long time i.e., unknown among the "profane ones." The "informed ones" knew, however! One of these informed people was Rudolf Johann Friedrich Schmidt, born in 1702 in Celle, and who died in 1761 in Copenhagen; he was a Dr. Med., general practitioner and alchemist in Hamburg, and "Hofrat" (Councillor) of Darmstadt/Hessen. This Hofrat Schmidt, who did not receive sufficient acknowledgement from the historians of alchemy and about whom soon a major work based on some studies will be published, left all of his extensive heritage in the form of manuscripts, alchemistic and medical books to the library of the city of Hamburg in his testament. Among his books is also the *Aurea Catena Homeri*, the Leipzig edition of 1738. Schmidt was very knowledgeable of the "golden chain," which had a strong influence on him, as can be seen in his work entitled *Enchiridion Aichymico-Physicum – e Disquisitio de Menstruis Universalibus vel Liquoribus Alchahestinis Philosophorum*, etc., Jenae 1739.[3] He had the habit of making remarks on the

---

[3] This "alchemical-physical handbook or investigation of the universal solvents or the alkahest of the philosophers, etc." can be found as German translation under the title: "About the General Solvents" in the Magazine for Natural Sciences and Chemistry" (*Magasin fur die Hohere Naturwissenschaft und*

margins of his favorite books, which prove to be very valuable to us now. Also his pocket edition of the Aurea Catena Homeri was "marked" completely in this manner, and on the page of the preface it had the following remark: "Doctor Kirchweger Stirus *natione pro Autore Catenas Areae Homeri se confessus est.*"[4]

There was probably no biobibliographic researcher who could have been more pleased about this remark of Schmidt dating from 1738 or 1739 than Hermann Kopp, the deserving historian-writer of alchemy. Kopp, as a matter of fact, in his own *Aurea Catena Homeri* (Braunschweig iSSo) arrives at the same conclusion, namely that the Doctor Anton Joseph Kirchweger pg Forchenbron (dead on Feb. 8, 1746, in Gmunden in Upper Austria, Councillor and Court Physicist of the Region of Salzkammergut/Austria) was the author. Kopp concluded this from among other things a manuscript-catalogue that was published in Vienna only in 1786.

There is still another book by the same Kirchweger: *Microscopium Basilii Valentini*, etc. (Berlin 1790), in which he refers over a dozen times to "his" *Aurea Catena*; the same holds true with his third manuscript: Ars Senum seu Pandora redux." The latter, however, remained completely unknown to all hermetic literature historians.

The *Aurea Catena Homeri* consists of two parts. Later editions have had a third part added (*De transmutatione metaliorum*). But since that one is probably not by Kirchweger, we will not consider it here, even though its author assures us that we will

Chemie), Tubingen 1784.
[4] Doctor Kirchweger from Steiermark, Austria, has identified himself as the author of the *Goldenen Ketten Homers*.

find "a much brighter light" in this third part "than in the first two parts."[5]

The first edition of the *Aurea Catena Homeri* was published in 1723 anonymously under the title: *Aurea Catena Homeri*. Or: "A description of the origin of nature and natural things, how and out of what they are born and created, and how they are destroyed into their original kind, independent of what the thing is, which gives birth and destructs again, according to nature's own instructions and order, shown in the most simple way and illustrated with its most beautiful rationale and causes throughout. If you don't understand what is earthen, how do you understand what is heavenly? Frankfurt & Leipzig, Publisher Johann Georg Bohme, 1723." Of this 1723 version there are two editions, which (according to Kopp) only differ in the initials and other ornaments. Then followed these editions:

1728 Frankfurt and Leipzig

1738 Frankfurt and Leipzig

1738 Leipzig. (Unknown to Kopp; however, "Hofrat" Schmidt's pocket-size version now in the library of the City of Hamburg.)

1754 Jena (Doubted by Kopp)

1757 Jena[6]

1759 Vienna (Doubted by Kopp)

---

[5] This is not the place to investigate what reasons speak against Kirchweger as author of also the third part. I am also doubtful that he wrote "The Microscope of Basilius Valentinus." HWN

[6] German; includes alchemical symbols –PNW

1762 Frankfurt and Leipzig. (Latin version by Dr. med. Ludovicus Favrat.)

1763 Frankfurt and Leipzig. (Latin version by Favrat. Unknown to Kopp. Available at the Hamburg Library.)[7]

1771 Berlin. (Annulus Platonis. Disputed by Kopp).

1781 Berlin and Leipzig.[8] (Annulus Platonis.)

The last published (1781) Rosenkreuzer edition has this title: "*Annulus Platonis* or physical-chemical explanation of nature after its creation, maintenance and destruction, newly improved by a society of true natural scientists and amplified with many important remarks. Berlin and Leipzig, Publisher Geroge Jacob Decker, 1781."

The preface, signed "Phlebochron" is followed by the "listing of the chapters" and it is preceded by 1) an illustration plus a poetic "explanation of the figure *Abyssi Duplicatae* or the twice volatile and fixed abyss"; 2) an illustration of the catena (see page preceeding this Introduction) plus poetic "explanation of the *Annuli Platonis* or the golden chain of Homer"; 3) a "remarks." We do not show a copy of the two "Carmina" here, because even though they contain nothing different than the book itself, it would become a tremendous effort to try to make everybody understand their trenchant format rich in artificial words.

---

[7] Does not include in-line alchemical symbols –PNW
[8] German; includes alchemical symbols –PNW

Contrary to that, the "remark" says: "This Platonic Ring is clearly and completely explained in the rhymes following below, and in the whole work itself." It seems as if Plato borrowed the latter from the natural science of Pythagoras, which, except for some terrible tenets of his successors and which certainly are not those of Pythagoras, are in accordance with the mosaic and hermetic philosophy. Robert Fludd and Heinrich Cornelius Agrippa, the two very famous philosophers, except for a few crazy sentences, showed that very nicely. The latter's three books of the hidden philosophy (*de Philosophia occulta*) are very good and were considered worthy to be translated into French. They contain an explanation and an elaboration on Pythagoras' philosophy and the "Cabala." His main statement says: That in nature exists a certain connection and a common pull of hidden forces, which causes, that an upper force through its lower magnets continuously shoots off its rays extending to the farthest creature: whereas the lowest also in increasing degree swings upward through a consistent force: therefore the real "Magus" is the one who uses the things in front of the eyes, as enough magnets with which to attract the hidden forces (L.I. c. 37, 39).

The preface itself names as author a certain "Herwerd of Forchenbrunn, medical teacher in Cromau and rural doctor in Mahren, also a deserving member of our secret brotherhood, in which he took on the name Homerus," erroneously or at least inexactly.

If there were really a Rosenkreuzer with the brotherhood name of "Homerus" cannot be determined with accuracy. The *Aurea Catena*, however, did not get its name from either such a "brother" or from the known Greek poet Homer. The addition of Homeri to Aurea Catena results, however, from the fact that in Homer and since Homer the golden chain was considered the symbol of the chain that connects nature within itself. The rings of Plato constitute such a symbol.

A certain analogue to the "chain" of Homer and the "rings" of Plato are the "circles" of Scherin von Prevorst. The latter refer, however, primarily to an inner, spiritual world of looking from which they are then transferred to the outer, visible natural world. The interior world and the outside world exist not only each separately, but also each connected to the other in an orderly fashion, and in a bilateral connection, since they are basically only one world. Nothing can be recognized in an isolated form, but each thing only should be seen within its connection and its relation with other things. Lassalle once formulated this very accurately: "It is one of the most profound philosophical ideas ever expressed, that nothing is known for itself, but everything is known only within its organic relation with its 'wherefrom' and 'whereto' in the natural and spiritual universe, i.e., it may be understood only in relation with the whole or absolute true."

We would like to cite the following from the preface:

"The author shows in this work beyond doubt the real origin of all things in the so—described general spirit of the world, that the modernists hate so much, that he who does not applaud to these enlightened rules is basking in his own blindness. For the more flexible, inquisitive-minded, however, the light will come to shine to which he would be looking in vain in other philosophic literature. For miraculously as the past has taught us, most of the philosophers contradict each other in their various systems. Many modern philosophers are not concerned with the spiritual nature of things, but try to explain everything according to the laws of mechanical movements. That is why most of them stay with the artificial nature; others, however, who have realized that there is no basis for this, instead of giving up the prejudices which they have held since childhood and of becoming enhanced with the truth, prefer to stay with the most tasteless scholastic melancholy." "Each and every thing" -- says Homerus himself -- "is nothing but just this general world spirit in a more or less coagulated or firm condition." But more about this at a later time.

The "true natural science" as promulgated in the *Aurea Catena Homeri* is *nun in nuce* also contained in the famous *Tabula smaragdina*. In the preface it is stated:

"Our three—times grander Hermes (Trismegistos) in his *Smaragdenen Tafel* (Table) which we rightfully consider the most elegant of our philosophical symbolic books and at the same time as a bible of the hermetic philosophy, has presented the mentioned system of true natural science in such a small extent, however, in such a thorough and instructive manner, that he agrees even in details with those infallible rules of the most splendid of all natural philosophers who ever lived on earth; I'm referring to Moses. That is why our late author chose the aforementioned excellent, ancient monument for his "continuous theme," since as so beautifully explained in his *Annulo Platonis*, the latter may be considered to be a real, thorough commentary concerning the *Smaragdenen Tafel* (Table). In order to be able to have this theme of Ariadne always in front of your eyes, dear and honorable brothers, and so that the rules of our author may be checked in that regard...I will place it here in the German language, and for the pleasure of the scholars in the Phoenician language in which the original had been written."

If we renounce to publish the *Smaragdenen Tafel*, which had the high respect of all alchemists and natural scientists since the thirteenth century, we are doing it because it is incomprehensible in its lapidary shortness; it would take extensive explanations and the ideas anyway are expressed in the *Aurea Catena Homeri*.

Just who is the writer of the preface using the name *Phlebochron* is not known. We know, however, that the two more recent *Berliner Rosenkreuzer* Joh. Gottfr. Jugel (1707 - 1786) and Joh. Christoph Wollner (1732 - 1800) have obtained the 1781 edition. The latter was then read widely and interpreted also, especially in the Hamburg—Rosenkreuz circle.

COMMENTS

This Platonic Ring is clearly and most instructively interpreted in the verses on the following page and in the whole work itself. It would appear that Plato borrowed it from Pythagorean natural science, which, except for some basic erroneous concepts of his successors (who surely did not derive from Pythagoras), agrees with the Mosaic and Pythagorean philosophy.

Robert Fludd and Heinrich Cornelius Agrippa, aside from a few minor incorrect phrases, have proved this very well indeed. The three books on occult philosophy ("De Philosophia Occulta") of the latter are quite good and have been deemed worthy enough to warrant translation into the French language. They contain an explanation and an enlargement of Pythagorean philosophy and of the Qabala. Agrippa's principal tenet is to the effect that in nature there exists a certain connection and common characteristic in occult forces. These cause a superior force to shoot out its rays through its inferior magnets in a continuous cycle, extending to the lowest of creatures. And, vice versa, the inferior, by a corresponding attraction, can rise to the uppermost through an ascending scale. That is the reason why, he said, "He is a true Magus who uses the things that lie manifest before ones eyes as if they were so many magnets intended for the attraction of the occult forces". (L.I. c/37/39)

Note: Owing to the difficulty of translating a poem from one language to another, the rhyming has been omitted and a simple transliteration of the meaning given. This poem is given overleaf. HWN

Explanation of the Figure
ABYSSI DUPLICATAE
The Double Volatile and Fixed Abyss

One abyss (deep) calls forth another.

Together they form a hard bouquet.

The volatile must become quite fixed,

Steam and water must turn into earth.

Heaven itself must be earthly,

Or else no life will enter the earth.

The highest must become the lowest,

The Lowest again the Highest.

The fixed must become quite volatile,

Water and steam must be the earth.

The Earth must fly high up to Heaven,

Heaven must creep into the center of the Earth.

Thus, Heaven and Earth must become reversed,

If the lowest is to become the highest.

The volatile Dragon kills the fixed,

The fixed forces the volatile into Death.

Thus must stand revealed,

The quintessence and what it can do.

Of course, our modern poets would say the wonderful
things contained in these two explanations, in a
more beautiful and poetically correct manner.
Especially if a perceptive cognition would tune
their lyres up to these high truths! We have more
than one reason, however, for printing these verses
in their entire old-time format.

PHYSICO-CHEMICAL EXPLANATION OF THE ORIGIN,
PRESERVATION, AND DESTRUCTION OF NATURE by an
Association of true natural scientists, revised and
published with many important annotations by Anton
Joseph Kirchweger - Second Edition — Literal and
true-to—the-original rare Rosicrucian Edition of
1781. Illustrated. Berlin W 30, Hermann—Barsdorf,
Publ. 1921.

EXPLANATION

of   the

ANNULI PLATONIS

or: THE GOLDEN CHAIN OF HOMER

☿ The Chain of Homer is proved thus:
After the chaos pulls asunder[9]
A volatile spirit must forge it.

☉ Spiritus mundi is its name.[10]
Frost, dew, snow, rain and everything from
above
Are betrothed to it in faithful company.
Here is contained the volatile seed of the
world
From the upper realms, when it falls into the
lower.
From that it takes on a body
When it glows visibly before our eyes.

◑ Nitrum is known to the whole world.
Who is there to tell all his power?
It is he that can forge many a thing.
To him the lower realm is subject,
Neither can the upper dispense with him.
He must give birth to the whole of nature.

---

[9] The symbol ☿ in the margin comprises a great secret, of which nothing can be said at present.
[10] The symbol ☉ in the margin is here given because the World Soul, or the philosophical ☉, being the foreman of God, who forges all things above and below, therefore, also produces the meteors, has his seat chiefly in the sun, into which he was drawn and locked by Divine omnipotence on the fourth day of creation.

He is the father of all things,
Who can conquer the fortresses of the world.
His power has been given him by the Creator.
His realm is over heaven, earth and the sea.
Adam he is in all things,
Out of him Eve must also spring.
Then the goal will be reached,
when the whole earth becomes fertile,
When he becomes fixed and no longer flashes,
And Eve sits next to him.
Sun, moon, the sea, and the earth
Turn him to Eve through constant motion.
Through heat and cold, through constant
movement of the sea, With Adam rises Eve.[11]

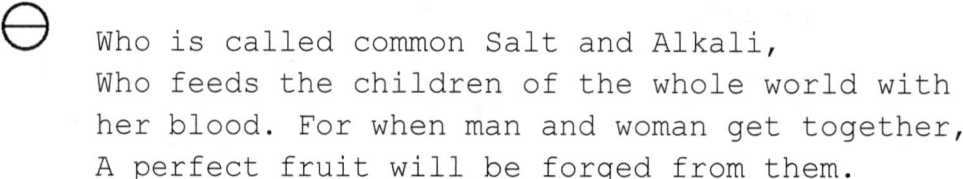

Who is called common Salt and Alkali,
Who feeds the children of the whole world with
her blood. For when man and woman get together,
A perfect fruit will be forged from them.

For the Sour and Alkali Salt Gives the fat to
every soup.[12]

---

[11] Because in our *Annulo Platonis* much is written about this
main subject, we will here only reproduce the words of
Welling, (Part I, Chap. 3, Sec. 19), they are: "Its sphere
consists of the whole world, it has the ray of the upper light
and the lower. Consequently, it consists of volatile and
alkaline fixed parts, and is a wonder salt of nature". This
author would be irreproachable in every way if he had not
tarnished himself with the error of Origen. Nevertheless, this
does not prevent us from drawing on him when his statements
are right and pertain to our subject.
[12] Here we see the character of the Philosophical Vitriol, from
which the double Mercurius is prepared, through which all
things are transformed into the pure tinctural nature.
Consequently, it gives indeed fat and good soups! Profaners
may believe this or not.

⊕ This is proved by the volatile realm of the
animals.
Not volatile, not fixed, note well.

⊕ The Vegetable Hermaphrodite also shows
Of what it is forged.

⊖ The fixed ores and stones give evidence
That they are proper (or: belong) to Niter and
Salt.
Fire and Air, Water and Earth,
Desire of it the active part.
When now the noble world—seed has been made
fixed.

☉ Steam and water have also been brought to
earth,
Then is made, and also accomplished
That which all the world esteems most highly.
Fixed must the volatile become,
Out of water and steam turn to earth.
And when it becomes a red dry blood,
It is the world's treasure and highest good.

♀ A perfect perfection
Which drives away all poverty and disease.

FOREWARD

Very worthy, dear Brothers,

Again a Rosicrucian publication! many a profane scholar will say, together with that unwelcome critic who has inserted in the *Auserlesene Bibliothek der neuesten Litteratur* (Select Library of Modern Literature, Vol. IX, N. LXVI, p. 428, f.) a very nonsensical, yet at the same time rather coarse review of the Plumenock influence on all our writings. Just, however, as this censor has been found too light on the scale of reason upon which we have been used for several thousand years to weigh the true as well as the sham scholars, to make it worth our while to reply to him, just so an identical fate will await those who are on his side.

I, therefore, come to the point and present to you, very worthy Brothers, a work which is considered by our High Superiors a classical principal textbook and has been prescribed chiefly to the disciples of the lower classes as a basis for the said science of nature. This is due to its inner worth and because it elucidates, clearly and distinctly, theoretically as well as practically, our real principles of the true science of nature which have been distorted by many writers of former and modern times.

And indeed, whoever considers how many errors have crept into this study over a period of time will himself become convinced that all of the public will have good cause for being grateful to us for trying to make books no longer available in bookstores accessible to the general public through new editions. Among these we count, with perfect justification, the present work on the true and genuine natural philosophy, tested by the most certain experiments.[13]

The author, who is resting in God, was called Herwerd von Forehenbrunn. He was a teacher of medicine at Croman and a country physician in Moravia, also a worthy member of our secret fraternity, in which he bore the name of Homerus. In this work he shows with such conclusive arguments the right origin of all things in the universal World Spirit, so much decried and so much hated by moderns that those who do not approve of these purified principles entertain themselves in their blindness. To a flexible man, eager to learn, however, that light will shine to him here which he would seek in vain in other philosophical books. For the majority of sages contradict one another in their various systems in a wonderful way. Many moderns do not wish to hear anything about spiritual essences, but wish to explain everything by the laws of mechanical motion. That is why most of them stop at the artificial nature; others, however, who recognize the unreasonableness of such a system prefer to stick to the tasteless scholastic blues instead of renouncing the prejudices they sucked in during their youth and of becoming prisoners of truth.

Yet such people should just read the first chapter of the Book of Creation with impartial eyes and should not try to be smarter than Moses, the

---

[13] The old title of this book is: *Aurea Catena Homeri*. To the last edition was added a third but false and interpolated part which does not stem from our school. We, as true brothers of the deceased author and sole legitimate heirs to this book, have deemed fit also to change the title in this new edition published by us.

natural scientist illumined by God; they should let themselves be told that which they wished he would say according to their darkened brain, and they would change their minds. They should only, I say, look at men and animals, yes, consider well all trees, flowers, herbs, stones, and minerals, know how to dissect their parts in the manner prescribed by the author and in our wisdom schools, and they would find, not without great astonishment, the universal World Spirit, which some of our philosophers, such as the first and oldest among all, know and understand very well.

In short: Everything one sees, everything that lives and grows is produced by fruitful nature; yes, the most genuine and firmest subterranean creatures are filled with spirit by the great World Spirit, or, as our Brother Homer expressed it: "Everything and each, stone, skin, and leg, whatever there is in every mineral is coagulated and fixed Spirit of the World or Life"[14] and filled with it; otherwise it would be impossible that life, rain, and motion could be in nature. For this Spirit is the direct cause and composition and augmentation in human beings, animals, plants, and ores, yes, in every single thing.[15] He can rightly be called the Spirit of the almighty Architect, the creator controlling the world, the overseer of all things, the beginning of all offspring, which proceeds and is created by the great JEHOVAH as the right F I A T,[16] the Spirit unified in itself which constitutes nature in her upper and lower levels; the right ANIMA MUNDI by which everything lives and works[17]; the right

---

[14] Part I, Chapter 23.
[15] Philalethae Anthrosoph. mag., p. 210.
[16] Zoroaster's Clay., art., p. 3.
[17] Jo. Agricola in Popp. Tract. de Mercurio, p. 757.

MERCURIUS VITAE, without which no man, animal, or plant can live[18]; the living water, into which enters the upper light with its crystalline water, by means of which the body, that is, the nethermost water, is illumined and transfigured[19]; and its previously crushed[20], as it were, dead life is resuscitated[21]. It has its seat in the upper regions of Shamajin where, in its first descent, it condensed into that nature which is called the Chaotic Water by the sages. This is the first casing and lodging of the plastic World Soul, or of that great assistant overseer of God which the old Platonists called the generating nature (*Natura genetrix*).

Now, then, this Chaotic Water, the more it sinks through the coarser ranges of the air and approaches our earthly region, the more it takes on a yet more compact degree of condensation due to its astral effluences. After it has helped the animal and vegetable realms to nourishment and growth, it proceeds to the areas of the subterranean Pluto, producing there the minerals and metals with the assistance of the fire in the center of the terrestrial globe.

---

[18] Grosser Bauer, p. 7.

[19] Regarding this, the author of *Mikrokosmieche Vorspiele* writes, p. 25: "When we enter the service of wisdom and deliver from darkness the word expressed by God, which is a true light, heaven as well as earth are engaged in nourishing and augmenting it, and giving birth to it in superperfection." Note this, as herein lies not just the ground of all tinctures but also of the great magical Stone of the age-old sages.

[20] It is contained in all created bodies of the triple realm of nature as an imprisoned, so to speak, enchanted treasure (*Thesaurus incantatus*), as a certain person calls it, crushed and locked, NB, until it is brought to fermentation by air and its acidus, and the spirit achieves a free breakthrough; or through the assistance of the artist, through general or specific expedients; but it is most wonderfully freed from its fetters by the double magical fire.

[21] Cabala chymica, which is added to the Grosse und Kieine Bauer in the edition published in the year 1753 under the title Philosophia Salomonis, p. 169 f.

However, even if the said bodies take their origin from this general procreative father, they have been marred in their infected, salty maternal places by a certain damaging, caustic corrosive which is against human nature, and which has been most strongly woven into their substance by the aforementioned central fire.

Yet, it can, by the guide to sweetening revealed by our Homer in the practical part of his work, be freed of this spot, transformed into the sweetness of sugar and made acceptable for products of the three natural realms.

Behold, very worthy, dear Brothers, the first or theoretical part of our *Annuli Platonis*, which is based on reason and experience, supported by the approbation of the divine holy scripture, and transmitted to the Egyptians by our forefathers, brought by these to all peoples and their secret fraternities, and still taught there. Among the said patriarchal forefathers our Thrice—Great Hermes is shining like the sun among the stars. In his *Emerald Tablet*, which is rightly considered by us the noblest of our philosophical symbolic books and, so to speak, the Bible of Hermetic Philosophy, he has laid down the said system of true natural science so thoroughly and instructively in such a small space that he is in complete accord with the infallible principles of the most excellent of all teachers of natural science who ever lived on earth - Moses, I mean. That is also the reason why our deceased author chose the said age-old monument as his constant guide and explained it so well in his Annuli Platonic that it can be considered a thoroughgoing commentary on the *Emerald Tablet*.

To enable our very worthy, dear Brothers to bear this guiding thread of Ariadne constantly before their mind's eye, to test the tenets of our author thereby, as well as to recognize themselves unreasonableness with which some idle talkers have endeavored to cast suspicion on this so precious document of ancient days as an interpolated product of more recent times, I am reproducing it here in German, but for the sake of scholars in the Phoenician language in which it had originally been written.

"It is true and no lie, certain and truest of all, that that which is below is like that which is above, and that which is above is like that which is below, whereby the wonder-signs of a thing can be obtained. And just as all things are created by One alone, by the will of the Only One who thought of it before; so also all things originate from the Only One Being, by appropriation. The sun is its father, the moon its mother, the wind has carried it in its belly. Its nurse is the earth. This is the father of all perfection in the whole world. His power is total when he is transformed into earth. You must separate the earth from the fire, the fire from the crude, gently, with great understanding. He ascends from the earth up to heaven, and descends again from heaven to the earth, and receives the power of the upper and the lower. When you have achieved this, you will possess the splendor of the whole world, and all darkness will fly from you. This is the strongest strength of all strength, for it overcomes all subtle and volatile things, and penetrates that which is crude. Thus the world is created. And by means of this one thing the most wonderful works are accomplished. That is why I have been called: The Thrice—Great Hermes, because I possess three parts

43

of the truth. Everything I have said about the work of the sun has been fulfilled."

That the said incomparable monument of Egyptian wisdom was originally written in the Phoenician language, the real mother tongue of the ancient Hamites, cannot be doubted if one admits, as one must admit, that its author was the second Hermes who lived at a time when no syllable was yet known of the Egyptian dialect, the oldest daughter of the said language, which those who peopled the still uninhabited Egypt under Menes brought with them into the country[22].

It is precisely this fact that Kriegsmann proves with good arguments in his commentary on the subject[23]. There is just as little reason for doubting that the said *Emerald Tablet* is the very oldest document we possess in the said language. It is quite ridiculous, therefore, that a certain J. L. ab Indagine L.M.[24] is not ashamed of contesting such high antiquity, in view of the fact that the learned Jesuit Athanasius Kircher[25], who was no friend of higher alchemy; Petrus Lambecius[26], Olaus Borrichius[27], yes, even the distilling Herr G. H. Burghart[28] admit it, and that the latter even grants: ... "This *Tabula Smargdina* is the very oldest and perhaps

---

[22] See *Compass der Weisen*, Preface, p. 30 f. If time and the limits set to this Preface would permit, this sentence taken from the learned Samuel Bochart, Canaan could be proved in greater detail.
[23] This is entitled: HERMETIS TRISMEGISTRI *Tabula Smaragdina* vindicata per W.C. Kriegsmannam, 1657, 8.
[24] In his chemico—physical *Nebenstunden*. Hof, 1780. 8. S.16 p. 12.
[25] In *Obeliec*. Painphil. 1.2.c.4.
[26] In *Prodrom*. Histor. literat. 1.i.c.i.
[27] In *Hermete Agyptiaco*. c. 4.
[28] In his *Distillierkunst*. Breslau, 1736, in 8. p. 30.

truest document derived from Hermes, be he who he may," although, according to the atomic—mechanical tenets he had absorbed from his youth, he does not perceive the universal, plastic, great Spirit of nature shown so clearly in it.

It is just as ridiculous for the aforementioned ab Inda gins[29] to ascribe this age-old work to the labor of a Latin author because, he says, the affected inscription VISITA INTERIORA TERRAE &c. is proof of it. Which is certainly a noble argument! Scillicet!

Whether it had really been engraved in an emerald big enough to hold the said writing or is only called so as a simile, I do not wish to become involved in a learned cat fight about it with him. Nevertheless, it is not so unbelievable that persons who know the art that Hermes had known, could produce such a big gem.

But that he considers the said Tablet suspect and its content confused, I cannot pardon him, especially as otherwise he produces much good in his spare time and from time to time does not show unsuitable arguments in this kind of scholarship, so that he does not seem to lack anything except that he has not studied in our academy of wisdom.

No fewer contradictions are found among scholars regarding the time at which Hermes lived. Some even consider him a chimera, asserting that he had never been on earth. Chief among these is Johannes Heinrich Ursinus[30]. The above-mentioned ab Indagine[31] would like to make out that he was Moses,

---

[29] Idem, p. 12 f.
[30] *In Diss. de Zoroactre, Hermete et Sanchoniathone*, Norimb. 1661.8.

45

although the latter lived shortly after the flood, that is, in the second millennium, while this one only lived in the fourth[32]. The most ridiculous, however, is that the author of *Famac remisaae ad Fratres Roseae et aureae Crucis* tries to aver that Hermes was a Greek and King of Athens, although not one among the true scholars agrees with this opinion. In order not to expatiate too much, however, I will only note here that our philosophical apprentices would derive great benefit from considering the Tablet of Hermes with its beautiful commentary by Hortulanus[33] together with our *Annulo Platonis*.

The second or practical part of the work we have on hand is no less worthy of consideration than the preceding theoretical. For just as in the first part the author writes with .the greatest thoroughness and clarity about the generation of things as a whole, he deals in this second part with their corruption and natural analysis. In it the true rebirth of the Chaotic. Water, and especially its astral offspring, are palpably shown, as also how to achieve it without separation of the leaven by a total volatilization of all components of these bodies. From this latter procedure arise our so useful *Menetrua radica* lie of the three kingdoms of nature, which are able to sof ten, volatilize, and again make fixed every created body of their line, and to transform it into a potent medicine for men and metals[34].

---

[31] Idem, p. 11 f.

[32] See *Compass der Weisen*, Preface, p. 30 and 41, 42.

[33] It is in the German version of Thea tro Chemico by Friedrich Rothscholz, which came of the Press in Nuremberg in 3 octavo volumes, 1728—1731.

[34] S. See *Versammiungereden*, St. VIII.

But because the creatures of the subterranean kingdom have absorbed a corrosive injurious to human nature and a caustic being, our author teaches (Chapter X) how to soften, tame, and transform this poison into a honeysweet medicine, salubrious for the animal body. If our Brother Homerus had done nothing more than show this way and teach the pertinent guides to sweetening, we would owe him our deepest gratitude even after his demise.

Finally, in the 11th and last chapter, he comes to the so much taunted Alkahest or Alka est. Some consider this famous liquor a very simple substance (*Ens simpliciseimus*), supposed to dissolve both sour and alkaline subjects, and that without any corrosive and any weakening of its virgin purity. By what Irenaeus Philaletha, Starkey, and our Brother Belmont write about it, we may infer that it must be made of urinous or urinary salt. I do not wish to go into any detail, how far its boasted virtues are founded or unfounded, or whether it can be compared to the fifth element or the volatile spirit of Mercurius of the ancient sages, which is precisely that which Basilius Valentinus and others are praising.

In one of his unpublished writings our author maintains that that the effect of the Helmontian Alkahest can be proved by other Menstrua. He calls these hermaphroditic ones, and says that they united without resistance, (*Sine Strepitu*), with both the Acidic and the Alcalicis, that they augmented thereby without precipitations and that they are a neuter between the acid and the urinary.

Regarding the most noble marcasite, various hermaphroditic solvents (*Menstrua Homogenea*) and preparations of this blessed mineral will be found

47

in our annotations, which we recommend to all pharmaceutical alchemists in true, sincere love of humanity.

We have yet to say something about the third part of this *Annuli Platonis*. There exists of it, three kinds of editions which have been added to the printing of the first two parts in three different versions. But since none of these is authentic and all have been wrongly ascribed to our Homer, we have thought it fit to print none of them and to omit the third part altogether.

Finally, the importance of this little work does not require any recommendation. It is, nonetheless, indispensable to those who wish to work intelligently in the branch covered by it. Heretofore, there have arisen complaints about all previous editions by various readers, also by some new Brothers, that is, by those who do not know the Latin Language. They are left with some doubts due to this lack of knowledge, or they must be in doubt, as Latin scraps which they do not understand occur on almost all pages, and they wonder if they have been correctly translated. All, however, are disgusted with the Moravian German manner of writing of the author. They regret that it is so tiring that many a reader must force himself not to put the book down midway, and undertake something else. This is especially as, in addition, there are obscure passages here and there, which no beginner can easily understand. That is why, for the benefit of our younger students, an effort has been made to meet these complaints, to clarify the obscure passages by explicit annotations, to accurately replace the Latin texts and single words by German terms, and in general to make the whole work clearer, easier, and more usable.

Therefore, very worthy dear Brothers, use this exceedingly serviceable tractate to the honor and praise of the lovable and adorable Creator of beautiful nature, for your own instruction, and the comforting of your needy fellow men, to which end we devoutly recommend you to the blessing Grace of Heaven with the power of the Spirit, and remain, through the sacred number, in unalterable faithfulness and brotherly love, with our usual Christian-fraternal wish,

Phlebochron[35]

---

[35] The true identity of the writer of the preface using the name *Phlebochron* is not known. We do know that two more recent *Berliner Rosenkreuzer* (Members of the Rosicrucian Society, Berlin), Johann Gottfried Jugel (1707 - 1786) and Johann Christoph Wollner (1732 - 1800), each obtained the 1781 edition. The work was widely read, especially in the Hamburg-Rosenkreuz circle. -*PNW*

PART I: THE GOLDEN CHAIN OF HOMER
On the Generation of Things

CHAPTER I
WHAT NATURE IS

NATURE is that amalgamation, which is brought together by the Creator, including the visible and invisible worlds, and containing in Itself both visible and invisible creatures, all of which function solely due to the essence (being) and presence of God.[36]

For the better understanding by men of the creation, the natural visible and the supernatural invisible realms are separated, but, in the final analysis, this is of no concern to us, because we believe that all and everything has been naturally made by God, out of the Chaos and the Great No-Thing or Void.[37]

1. The term *World* means all created beings taken all together, may they be visible or invisible to our eyes. That is why we understand it not only the whole terrestrial lump upon which we dwell, but also the sun, the moon, all the planets and the stars, together with the immeasurable space In which they exist, and all other beings which live, act, and are contained therein.

2. By the term *Nature*, on the other hand, we understand the immutably working law of motion,

---

[36] Our author's explanation of the word *Nature* is not wrong, to be sure, but formulated in such a manner so that it gives neither a clear nor a complete understanding (conception) of this subject. We will, therefore, include in this book the lecture of one of our learned brothers. It Is certainly deserving of being read.

[37] See "*Freimaurerische Versammlungsreden*" (Masonic Assembly Speeches -*PNW*) , Amsterdam, 1779, Vol. 10, and that which is also said herein.

impressed from the beginning of the world by the almighty Creator upon all these created beings by His eternally uncreated Word. This law produces both the mode of existence of the whole world in general and that of every created thing in particular.

According to the principle of the sages: *Omnis action agentis se habet as dispositionem subjecti patientis*, that is, all influence of the mover or doer is according to the structure of the object to be moved, or the sufferer, the passive recipient. The mover Himself who animates and engenders and destroys everything in all, however, is the much discussed *World Spirit*. Our great Sendivogius means Him when he uses the word Nature and calls her a "volatile spirit" that effects its work in bodies. He is one and the same in all created things. He is Universal but the Law of Motion within creatures is different in countless ways, that is, relatively different, according to the condition of the innumerable varieties of created species because of the Word of power of the Omnipotence: Let each grow, bring forth seed, and multiply after its kind (Genesis I).

That upon which precisely this law of motion is impressed is the suffering (passive) part. The universal World Spirit, on the contrary, is the active part in all creatures. Both, however, taken together as a unit, comprise what we call *Nature*.

In order to form a clear, simple and quite mechanical idea of this active and passive concept, we have only to consider paper mills, flour mills, grist mills, grinding mills, stamp mills, and innumerable other like machines. All are driven to the accomplishment of their specific action by one energy source, for instance, water, but differently

51

according to the diversity of their inner structure. Although driven differently from one another, each one, properly speaking, is in no other way driven, than in accordance with the predisposition made by the Foreman (Master Craftsman), in its inner structure, and a specific Law of Motion results. In the case of such a machine, water is like the universal *Spiritus Mundi*, or World Spirit, whereas the inner structure is like the so often mentioned immutable law of motion in creatures.

COROLLARY TEACHINGS

I. The whole of Nature and creatures, therefore, consists of two basic origins, the doer or mover, who is a spiritual being, and the moved or sufferer (passive recipient) who is a tangible or corporeal being, in regard to his moving spirit.

EXPLANATION: Air is the passive recipient (suffering) and is corporeal relative to fire, which is the subtlest and most effective element. Water is tangible and corporeal, moved and made effective by the Air, which relative to water, is an active Spiritual being. Earth is moved, impregnated, moistened and fertilized by Water, which relative to Earth, is a moving and active Spirit. But everything, as already mentioned, is originally made effective by the universal World Spirit.

II. Since the effects of Nature occur only according to the created (or: inborn) properties of natural things, while Art must faithfully follow Nature and cannot achieve anything useful without her, all practical investigation of Nature undertaken without a preliminary correct knowledge of the universal World Spirit and the created properties of those natural things in which work is done, must needs turn out as foolish and fruitless.

III. But whoever perfectly understands this active Spirit and the above-mentioned law of motion in all types of creatures in relation to the differences proper to them, will also recognize the Lawgiver in His infinite wisdom, His eternal omnipotence, His eternal justice, and His eternal mercy. He will possess natural science perfectly and is on the highest level of Hermetic philosophy.

And even if it seems impossible to some to climb to this highest level, the natural scientist eager to learn and be truly reborn in Christ, the light of grace, should not lose courage and confidently start investigating. He should not, however, take anything in hand without knowing it, or make something of it without knowing what, as the sophists are wont to do. Instead, to begin with, he must choose one thing only, not several at a time, as the object of his studies and investigations. Above all, he must get to know it inside out, and then only make of it what is possible according to its inherent law of motion. Neither must he stop working till he has completely investigated it, knows it perfectly, and has thereby obtained the final natural goal he had set for himself.

For it is not in the multiplicity that the Art consists (says one of our greatest teachers and with him the entire host of master sages). But afterwards he can proceed with the investigation of something else, in which he will already succeed more easily, because every creature is to be considered a book of Nature, as the very fine saying of our philosophers is to be understood, namely, that one book opens and explains another. In this way he can im-perceptibly reach the highest level of Hermetic wisdom, sooner than one can imagine. For the whole of Nature resembles a circular chain, the links of which hang exactly together. When then an enlightened spiritual man, by his studies and investigations, has finally come to perfectly know the first link, it will be easy for him to expand his knowledge from this to the second, from that to the third, and so on from one link to the next through the whole circle. And Nature, this faithful servant of the Lord, will herself guide an eager investigator truly reborn in the light of Grace.

CHAPTER II

OUT OF WHOM OR OUT OF WHAT EVERYTHING WAS BORN AND HOW IT ORIGINATED

The incomprehensible God has created everything out of the void or great Nothing[38] according to his will (or: as he liked). He decided and willed, and emitted from his mouth his holy Word of Power. **FIAT!** which became an immeasurable vapor, fog, mist and smoke.[39] This smoke resolved and thickened (or: condensed) into the generally known chaotic water by its perpetual increase. This water, then, is that out of which the great world with all its inhabitants has been made, and this water is that out of which God has made all supernatural and natural things, and this water is the primeval origin of all things which came before us and are to come after us[40].

---

[38] What modern natural scientists have cooked up against this statement, taken in the sense in which it is accepted in our schools, is being briefly dealt with in a discourse by one of our Brothers in the 10th of the Versamrniungsreden of the Gold -and Rosicrucians, Amsterdam 1779, 8. But because this subject is of importance, we will add yet another lecture penned by the yen. Brother who wrote the first one inserted by us in Chapter I.

[39] See aforementioned speeches, No. 9, where the fine words of our master of wisdom may be read, which throw a very bright light on this matter. You may add Camel. Drebbel's short tractate on The Nature of the Elements, which is contained in Jos. Ferd. Xleeblatt's new edition of some rare small chemical tractates, Frankfort and Leipzig 1768, 8., Chap. I. Altogether, we recommend this little work to all, and particularly to all Brothers, because it is written by a genuine and very venerable member of our blessed association, and is excellent in its kind.

[40] This is that water of which St. Peter says in his second Epistle 3:5-7: That out of such the earth arose and consisted of water by the word of God. Our old Greek Brother Thales of the Ionic School had very good knowledge of said water. He had acquired this excellent knowledge from his teachers, the Egyptian Brothers and pupils of the patriarchs, by whom he had been introduced to the temple of wisdom. Therefore, upon returning to his home country, he taught his younger Brothers that water was the first beginning of all created things. Now, to be sure, a large portion of our modern scholars believe that thereby he understood nothing but common springwater, river water, and rainwater. However, they are not quite right in this. Thales well knew how to distinguish between chaotic water as the

A LECTURE ON THE GREAT VOID OR NO-THING

FROM WHICH ALL NATURE AND CREATURES HAVE ORIGINATED

[41]

1. Here we do not teach about the primeval Nothing (void) out of which the eternal Omnipotent created the worlds of spirits before the beginning of those times, when He, in His infinite Wisdom, appointed the sun, moon and stars as measurements. We begin at Lucifer's revolt against God, and after his fall was accomplished. The Almighty had installed Lucifer as the ruler over all those hosts of the countless legions of creatures that lived in that wide and vast expanse of our planet-world. We thereby follow the infallible teaching which the most excellent Moses, that historian of creation so highly illumined by the Spirit of the Lord, has recorded and left us in his Book of Creation.

2. In the beginning, he writes, God created heaven and earth. As soon, that is, as Lucifer, the Son of Dawn, in his splendor and perfect power over the Thrones, Principalities, and inferior spirits of his whole hierarchy (desired) to rule according to his own will and thus began to resist the

---

mother and partrurient of all others and the offspring itself. Nor was it unknown to him that this offspring, that is, rainwater, had enclosed in itself a large part of those perfections which had been incorporated in the primeval chaotic water by divine omnipotence, and that it had been so much qualified by the effluences of the moon impregnated by the sun that it might bring the general Mercury hidden in it to the various births in all three kingdoms of nature as their preserver. But because our Brother Homerus is dealing with it at great length in his principal work, nothing further is said here except that the said opinion of Thales has been transmitted to us in our fraternal school as an irrefutable principle and that it will be retained as such to the end of times.

[41] This symbol appears here in the 1781 German edition but not in the R.A.M.S. edition. - PNW

irradiation of the heavenly light-waters of God, which had so gloriously illumined him in gentle stillness, perfect calm, joy and delight, and had been reflected by him upon all his Thrones and hosts of angels with great brilliance, the *Elohim*, the Judges, or the eternal justice of God, immediately withdrew this gentle irradiation of the said light-waters, thereby arousing the mighty fire, in which Lucifer together with his followers were seized, drawn together by the vehement harshness of its active strength, and (while just this fire was whirling up tremendous vapors through the entire immense space of his realm in the process of destruction, which condensed more and more and finally dissolved into water, partly adhering all around the periphery of his kingdom, in the form of a mercurial sulphur or condensed fiery water) were locked up in a dark heavy mass without water or light, for their eternal punishment.

This is the point which Moses accepted as the beginning of his history of creation and chronology. From it we understand that by his words quoted above he meant the earth to be the said dark, inert mass, and by heaven that which was outside this mass. We also understand why he immediately wrote the following after the above-cited words:

"The earth, however, was without form and void; and darkness was upon the abyss, and the Spirit of God moved upon the waters," before he mentioned the creation of this water with even one syllable. Now then, these waters, resolved out of the mighty vapors driven up by the Divine Fire of Vengeance and expanding all around the dark mass, were filled with an active spirit throughout their expansion, whereby the Almighty Word of creation brought forth all the creatures of our planetary world in the six Days of creation. It is that which we call the great Nothing, from which all nature and creatures arose.

Some ancient men gave this name void to the whole expansion and everything contained in it. They did this not in the erroneous opinion that it was really nothing in itself, but because in the beginning no form of the creatures gradually originating therein by the work of creation was real, though all were contained therein as possibilities, just as one might call for instance the ink in an ink-pot a Nothing or a void. In itself and by itself the ink contains absolutely no letters, numerals, lines, dots, words, numbers, designs, speeches, calculations, nor figures really, but only as potentialities which, however, the creative hand of a skillful, perfect writing-master can gradually all form from it and, so to speak, create.

COROLLARY TEACHINGS

I. Everything, therefore, has originated in ONE, that is, in the above-mentioned waters. According to the known tenet of all true sages: Out of which a thing becomes, into the same it can also be changed back. Therefore everything can be changed back into that ONE, into a fiery active water. This our students must henceforth deeply imprint upon their minds for further consideration.

Since Nature consists of a passive and an active essence and Lucifer together with all his angels were created by the divine flux of the Eternal Omnipotence, which he and his followers were

58

meant to use only according to the moving action of the gentle irradiation of the Divine Light-Bearers, it follows that:

II. In addition to the higher Angels, there were lower orders of Angels created, in which they could act and over which they could rule, owing to the fact that the lower order of Angels are, so to speak, tangible and are recipients in regard to the higher, according to the nature and properties of each. Consequently, the orders of Angels graduate from degree to degree passive toward an exceedingly more sublime (subtle) nature and characteristics, as is every *Agens* (agent; active) in relation to its *Patiens* (patient; passive). Since Lucifer had to act as the sole ruler over the whole hierarchy of his angels, through his Throne Princes, and had to keep *passive* only toward the above-mentioned *active movement* of the gentle irradiation of the Divine Light-Water, we may thereby conclude with what a brilliant, glorious and super-splendid nature and character the eternal Almighty had endowed this great spirit with, who should have ruled throughout all of Eternity. But since he wished to rule and act according to his *own will*, he closed off the infusion of the upper Light-Waters into himself; in respect to them, he shifted from a *natural-passive* state to a *self-acting* one. In so doing, he reversed the nature of all the orders of his hierarchy, and of his own being, by which, of course, the destruction of his realm followed as a necessity.

From this we may further conclude that Lucifer had perfectly well understood the great excellence with which he was endowed. In his inner nature, however, he either had not understood or only imperfectly; nor had he been intent on recognizing it, because otherwise he would have foreseen and avoided the reversal and destruction of his realm together with his own ruin which necessarily resulted with ghastly fright from his willful self-rule, and he would eternally have given credit to the Creator for such superior benefits. It is precisely this that teaches us.

III. That the sham-erudite children of the world, who from lack of inner self-knowledge attribute their talents not to the Creator who bestowed them, but to their own excellence; who consider, judge, and wish to deal with everything solely according to their self-conceit and imagined wisdom, are found to have a satanic nature into which they pervert everything they deal with and in whatever they operate.

A true son of wisdom, on the contrary, must especially recognize himself in his heart, according to spirit and truth, and he must let the gentle irradiation of the upper light flow and act in him. Consequently, he must steadily follow the moving light of God in nature, in true humility and annihilation of his own will, so as to receive the transformation of his own earthly nature into a heavenly nature by means of a true, rebirth through the blood of Christ, before he can think of changing earthly into tinctural bodies.

Grace of blessings, dear Brothers, which we may well ardently wish upon you according to the strength of our spirit, but which we cannot give you. It may be obtained from above, however, through steady true prayer and devoted cooperation.

To the above may be added the two incomparable speeches by our Rev. Brother Hannan, which are the second and third of the *Versammlungsreden*[42]; also, the tenth of the great Nothing. Thus you will get an adequate idea on this point.

The origin of that water[43] is thus the eternal God and his spoken word[44]. That word is a spirit full of power. That spirit changed visibly and tangibly

---

[42] Masonic Assembly Speeches -PNW

[43] This is that water of which St. Peter says in his second Epistle 3:5-7: That out of such the earth arose and consisted of water by the word of God. Our old Greek Brother Thales of the Ionic School had very good knowledge of said water. He had acquired this excellent knowledge from his teachers, the Egyptian Brothers and pupils of the patriarchs, by whom he had been introduced to the temple of wisdom. Therefore, upon returning to his home country, he taught his younger Brothers that water was the first beginning of all created things. Now, to be sure, a large portion of our modern scholars believe that thereby he understood nothing but common spring water, river water, and rainwater. However, they are not quite right in this. Thales well knew how to distinguish between chaotic water as the mother and parturient of all others and the offspring itself. Nor was it unknown to him that this offspring, that is, rainwater, had enclosed in itself a large part of those perfections which had been incorporated in the primeval chaotic water by divine omnipotence, and that it had been so much qualified by the effluences of the moon impregnated by the sun that it might bring the general Mercury hidden in it to the various births in all three kingdoms of nature as their preserver. But because our Brother Homer is dealing with it at great length in his principal work, nothing further is said here except that the said opinion of Thales has been transmitted to us in our fraternal school as an irrefutable principle and that it will be retained as such to the end of time.

[44] This is that water of which St. Peter says in his second Epistle 3:5-7: That out of such the earth arose and consisted of water by the word of God. Our old Greek Brother Thales of the Ionic School had very good knowledge of said water. He had acquired this excellent knowledge from his teachers, the Egyptian Brothers and pupils of the patriarchs, by whom he had been introduced to the temple of wisdom. Therefore, upon returning to his home country, he taught his younger Brothers that water was the first beginning of all created things. Now, to be sure, a large portion of our modern scholars believe that thereby he understood nothing but common spring water, river water, and rainwater. However, they are not quite right in this. Thales well knew how to distinguish between chaotic water as the mother and parturient of all others and the offspring itself. Nor was it unknown to him that this offspring, that is, rainwater, had enclosed in itself a large part of those perfections which had been incorporated in the primeval chaotic water by divine omnipotence, and that it had been so much qualified by the effluences of the moon impregnated by the sun that it might bring the general Mercury hidden in it to the various births in all three kingdoms of nature as their preserver. But because our Brother Homer is dealing with it at great length in his principal work, nothing further is said here except that the said opinion of Thales has been transmitted to us in our fraternal school as an irrefutable principle and that it will be retained as such to the end of time.

into a vapor and fog, and that turned tangibly into a water.

Here we have two things enclosed in one, a visible, which is water, and an invisible, which is the spirit hidden in it.

Water without the spirit is a *recolaceum*[45], or without power; and spirit without water is nothing, or without duration, because spirit must have a body if it is to effect corporeal or bodily things. For God intended it to be so that spirit should effect everything in all creatures by means of water, because water mixes easily with all things and through it spirit can soften, penetrate, give birth, and also destroy everything again.

Water is the subject, or the body, the casing, and the instrument. Spirit is the active agent (the active principle), the famous World Spirit, *anima & spiritus mundi*, the all-acting spirit and power of God, the universal seed, *sperma universi*, the true agens, the blacksmith of all natural things.

In the beginning, from its origin, this water and spirit was quite volatile and unstable, as can easily be surmised, a vapor, fog, and smoke-water[46].

From this everyone can infer what is its origin and from what beginning it stems. This can and will later be clearly put forward and explained.

That the world has arisen from vapor, and that vapor turns into water, and water into vapor, is indeed obvious. To be sure, we see nothing between

---

[45] Empty spirit. See Part II Chapter 5. -*PNW*
[46] This primeval chaotic water is called by Eug. Philaletha in *Antiquity of Magic,* p. 129, "The Second Nature of God," and "The Child of the Blessed Trinity."

heaven and earth but vapor, smoke, fog, and water which, driven by the central heat, is sublimated up from the earth-water sphere into the atmosphere or Air. And if we could see the subtle effluences or vapors of the heavens, we could also see the influences which enter from above downward into these vapors sublimated from below upward. But since we cannot do this with our dark physical eyes, we must comprehend it by analogy (reflection) touching it with our hands through the chemical praxis, "as that which can be found in the great world can also be found in the little world"[47] and that what is above is like what is below.[48]

That such vapors arise from water, we see first of all in the summer time, when the sun heats the waters, changes them into vapor, and draws them into the air; also, when it had been raining before and the sun is shining afterwards, we see how the roofs sprinkled with rain smoke and steam, and how these vapors evaporate again into the air.

When the farmer is cooking water in his saucepan, kettle or pot on his stove, he sees that the water is steaming and smoking, and if he wishes, he can boil or turn it all into pure steam.

We notice, however, that steam becomes water after fog (mist) and smoke have condensed into clouds. Such clouds thereafter condense into rain, snow and water, and fall back again to their origin. The farmer in his field, when he must work in the heat, feels with great discomfort that his whole body is breathing out and steaming out. That steam

[47] What he here calls the "little world" may be considered the human being as well as, metaphorically, the Philosopher's Stone.
[48] Read carefully the eleventh of the *Freimaurerische Versammlungsreden*. Thereby no mean light is obtained for a closer view on this subject.

settles down in his clothes and there resolves into water, so that it runs down his back in streamlets which we call perspiration.

The same is seen by those who deal with distillation, that is, that the *liquores* rise as steam into the head, there condense, trickle through the spout and run down in streamlets, and that thus the vapors turn into water.

From the above we can now conclude and be sure that the first matter of this great world, after God, is the chaotic water or vapor resolved into water. And, in our view, the latter is one and simple, but twofold in number, that is, water and spirit, visible and invisible; water is *Patiens*, spirit *Agens*. Out of these two all and everything is ceaselessly born, preserved and again destroyed, also reborn, till the end of the world.[49]

He who desires to reach the origin and fountain of the secret wisdom, let him remember this well and circle around this center in the following chapters. Then he will find that this spirit is all in all, namely, with its power it is apportioned in all subjects of the entire world; and as all things consist of this unified One, they return again to this One in their extreme dissolution, that is, the circumference returns again to the center through the natural changes. Whoever correctly understands this, do not let him feel any doubt or any scruple in the dismemberment of natural things: For he turns a volatile into a fixed, or a sweet into a sour, and vice versa; the stinking into the fragrant; poison into theriaca (treacle), because he knows that they

---

[49] This invisible, volatile, all-accomplishing spirit is called Nature by Sendivogius, and it is also described in his words, "Nature is a volatile spirit which does its work in bodies."

all stem from a single root[50] and can again be returned into that upon which he sets his most arduous desire. For they are only different because of chance, not because of their matter; because of their smaller or greater volatility or fixity; their longer or shorter digestion or maturing. That is why all highly aware philosophers say: "Our matter is in all the things of the world, in all things all around us, wherever one looks, one grasps it any moment with one's hands, one tramples on it, it flies about in front of our noses, and one often stumbles upon it." This is just said by the way.

Nevertheless, the philosophers have found a means in the selection of subjects and have directed us to where this spirit is found most concentrated, best, soonest, and fastest. Although it is all and everything, it is yet more, stronger and purer in one particular thing; otherwise, however, it is all in all[51].

The philosophers who think thus back their opinion with the known old huntsman's salute: Adam is said to have brought matter with him from paradise. Since, however, paradise must be sought

---

[50] The identity of the first beginnings *(identitas principorum)* is a tenet so clearly proven in our school of wisdom that it would be ludicrous to make a long and extensive palaver about a matter which is also not unknown to our disciples.

[51] Here the author touches upon a matter which has caused no end of confusion since olden times. Not only the sophists but also learned natural scientists considered that the Philosopher's Stone could only necessarily and inevitably be prepared from a single, quite common matter, removed from all specification. In those specified subjects there lay at most quite small particulars which, after the Fall, they said, were no longer true children of Nature, but had only received a servant's share and were covered in curses to such an extent that it was impossible to separate those completely from them. That is why most of them rejected metals as totally useless and unserviceable. Welling himself partly agreed with this opinion, as is witnessed by his words in the 12th chapter, third part, of his known work, which starts as follows: "Sal, Sulphur, and Mercurius, a wonderful Spiritus. Whoever has it, has enough; but do not look for it in the curse which, through man's great Fall, through this System everywhere has scented and crept, as it is elementary."

not on this accursed earth but in the super terrestrial pure regions, they say that it follows incontestably that the material for medicinal and transmuting tinctures must also be taken from there.

True, it cannot be denied that because of the fatal Fall the Good infused into all natural things as the Creator's blessing was swallowed up and turned into a secret. This secret consists in the true separation of the blessed matter from the cursed. And thus it follows "that this could not take place in paradise but became possible only after the Fall; and precisely this is also the reason why our sages teach that Adam brought it with him from paradise."

How to discover this secret and separate the blessed matter from accursed, in all creatures is clearly and faithfully shown in our secret schools by various methods which, however, all serve one purpose. That is, we teach this Art in order that the pupil will attain to the cognition of the Creator and His creatures.

But that the medicines, both for men and metals, are easiest and quickest found in concentrated form. In the realm of the subterranean creatures (the mineral kingdom), in that our Brother Homer is quite right, as those will be convinced who carefully heed what is extensively discussed concerning this matter in the "Compass der Weisen"[52], part II, paragraph I.

---

[52] "Compass of the Wise" - PNW

CHAPTER III
HOW EVERYTHING WAS BORN

From the above-suggested it is proven that the primeval stream, or water and spirit, is, after God, the first matter of all and each thing in this great and wide world.

That two-fold vapor turned into water by condensation, and this water was warmed, heated and made hot by the spirit implanted in it invisibly. Thus it began to work within itself, to rise, to ferment, blister, and become foul and fetid.

In the beginning this water was bright and clear, transparent, pure, without any particular taste or smell, like spring water; but through its active spirit it became turbid and gave birth to earth out of itself, giving off a foul, dead smell. It divided into different parts, into a spiritual-subtle, a half-spiritual and half-physical, and a totally physical part.

In the beginning, it was one and two. Now it is one, two and three, also four and five. It was one in the beginning as a simple water; *two*, that is a water which had hidden in it the spirit; three, as it had gone into a volatile, half-volatile, half-fixed, and a fixed status, which is, according to the teaching of the chemists, *volatile, acidum and alkali; Spiritus, Anima, Corpus; four*, as it divided into the four elements, i.e., fire or heaven, air, water, and earth; five, as it presented itself as a perfect indestructible Being aside from the four destructible elements.

After this water had become fully putrefied, the Lord gently separated one subtle thing after another leaving the coarsest, each according to its particular order and law. For by necessity the subtle rises before the coarse, and the coarse before the coarsest, thin before thick, thick before thicker, and this before the thickest. To be sure, everything goes, as children learn, according to the degrees of comparison, *positivum* (positive element), *comparativum* (comparable element), and *superlativum* (higher element).

Of the most subtle part, God made heaven and its denizens because it was the finest, clearest, purest, most spiritual, full-of-life and soul, most vivacious, fiery, fertile and mobile part.

Of the next, and one degree coarser, part, God made the Sky; subsequently, the Air followed, then Water and then the Earth. Thus God separated one after another and gave them Names and power to act, and he bid each to produce its like.

He commanded their multiplication by means of the Divine Word, which was implanted in their seed, and they perpetuated their own kind by the propagating properties of their (thus immortalized) seed. Heaven was to bring forth its inhabitants, and stars - the Air its meteors (this term means all things originating in the atmosphere including meteorites, rain, snow and hail, etc., each of which contain a virgin earth) - Water its fish, animals and plants, also rocks or stones, and minerals - the Earth its plants, animals and minerals.[53]

---

[53] All that has been said until now resembles in all respects, the Revelation, as may be seen in the first chapter of the Book of Creation. From this it is evident that no other teachings are given in our holy fraternity than those which are in perfect accord with the Divine Word. This process of Nature we must also emulate in our works. We must first rightly recognize CHAOS; also know how well to separate the four elements, how to bring out and unite according to Nature the three *principia* or beginnings, i.e., Salt, Sulphur, and Mercury, in

But God had not just given a specific power to multiply to each of these separated elemental parts, but also to each individual being, (especially those which in these separated elemental parts constitutes its own being) and each should possess its own multiplication power.[54]

God, however, willed in particular, that of all these together a universal seed and sperma should be begotten, because God saw that the primordial Chaotic Water was now divided and could never again coalesce into ONE, as it was in the beginning, without the destruction of all created things. Therefore, He commanded these four, Heaven or Fire, the Air, Water and Earth, to produce a seed from their centre, which would again combine the four into one and emanate from itself a universal seed for the birth, preservation, destruction and rebirth of all things.

To make sure that this would be done, He implanted his *Agens* and *Patiens* (active and passive principles) into each part and entity (of all kingdoms) as a whole, by means of which each entity would be kept in constant motion, warming up until it becomes hot. This heating causes each thing on its own to vaporize, exude and sweat out the superfluous matter with which it can dispense in its own body. Such sweat and vapor are commonly called the influence from above and the effluence from below. This vapor, however, turns into water in its parturient matrix during the process of birth, and this water is also a two-fold water, for it can replace the primordial Chaotic Water, precisely

---

order to stand closest to the seventh level and Solomon's throne.
[54] It would otherwise not be possible for the parts of a body separated by the Art to unite again naturally and form a third.

because it originated from it and also because it has the same shape and form, effect and quality. This will be explained in greater detail in what follows.

From this we can conclude the following: (i) that God commanded the production of a universal and general seed and *sperma totius universi* (universal seed), (ii) He so commanded the production of seeds of each of the elements, Heaven Air, Water and Earth, and (iii) he commanded the production of seeds of individual creatures, from each of the three kingdoms of Nature. This order from universal, to elemental, to particular, is the perfect (descending) order. It was so commanded that the power of multiplication and propagation should, in this fashion, be spread throughout the entire world.[55]

---

[55] By this seed or chaotic water all sublunary things are preserved, all new changes in creatures are produced, and everything is maintained in a state of efficacy. Such an admirable first beginning of all things, however, is nothing but "the pure light of Nature," (Translator's note: It is the famous "lumen naturae" of Paracelsus) which lies ever hidden in its center, so that it should preserve, nourish and refresh, the foundation and base pillar of the natural sciences. If this light is known, the whole of Nature is open, but if it is not known and obscured, the whole of Nature is dark and hidden, or it is at least covered by such dark and thick clouds that it cannot be seen. It is necessary, therefore, that we chase with a lit light this darkness which covers the light of Nature or the true science of things. But the bodies of things are shadows and darkness, by which the light of Nature is obscured, and if these hard and thick bodies are not subtilized to allow this light hidden in their center to shine and glisten, we cannot but remain in darkness and night, without seeing the rising sun. See the incomparable booklet *Geheimnis der Verwesung und Verbrennung ailer Dinge* (The Secret of the Putrefaction and Combustion of all Things), Frankfurt 1759, 8. p.19 ff.).

This light, however, cannot be discovered and separated from the dark husks in which it lies hidden except through fermentation and subsequent putrefaction, as our Brother Homer has shown throughout this chapter. As we must ever emulate wise Nature in this as in all other things, and as we wish to proceed according to Nature, we shall insert in full for reflecting readers *Welling's* fine words on fermentation, even though they are somewhat lengthy. They are contained in Part II, Chap. 4, S. 3, Page 239 ff of his known work, and are as follows:

"We find so many and varied causes of fermentation that they can hardly be told. Through *acidum* and *alkali* a fairly good proof can be had, though actually it is not yet enough. That the minera of ☿ can be totally dissolved into a red ⚴ by a sharp alkaline lye is known to all

71

That Heaven arouses a new seed is clear, because we uninterruptedly experience new influences and new effects. And so, to be brief, it follows that as the father produces a new seed, the mother must necessarily give birth to a new fruit. Therefore, Heaven as the father, causes something new in all descending, specified seeds, into the Air, Water and Earth. Consequently, it follows that all new births must proceed from below to upwards.[56]

---

crawling gold beetles. That such a solution all *three principia* (origins) of the ☿; i.e. ♀, ♃ and ⊖ are together, no reasonable man can deny. Now then, it has happened to us that such a lye with a dissolved ☿ turned with heat into an amazing fermentation of an intolerable stench, which caused not a few to wonder. Whoever understands the nature of every sulphur and alkali, will also easily understand the cause of this fermentation. Likewise, he who knows the bodily shape of every salt, will reach the goal soonest, and how a third thing has to join two, for instance, an *alkali* and an *acidum,* so that their figure can change, become bigger and swell. This motion then causes a prickly penetration and an ignition. The ignition causes a total swelling which is the fermentation. This continues until the spirit's desire to penetrate, and the body of the subject taken gets tired and therefore one is killed by the other. In this struggle of the little bodies, spirit can no longer be held back and either flees or is very easily driven away by distillation. That is why one has to be intent on getting to know the above-mentioned third thing well, *(i.e., the Spirit)* so as not to miss the right target. Whosoever understands what causes the tartness or the astringent tastes, will surely find the unerring way.

[56] What our author is here saying is beautiful and in concordance, because the fact that this seed arises in the upper light-waters and is therefore of heavenly origin, no true philosopher is likely to doubt. They call it the water of Eden and know from experience, that in it the seed-powers of all things lie hidden, just as in the waters below the firmament, being the outlet of *Pison,* the natural influences of the heavenly bodies are locked. See *Freimaureriashe Versammlungsreden* XI, p. 263. But how will our modern atom-smashers (1781 edition, p.33, Footnote D -PNW), who totally reject all astral influences and explain everything by the laws of mechanical motion, how will they make understandable the business of procreation in all three realms of Nature? (provide an understanding of). Must they not suffer just that verdict which *Plutarch* had already pronounced over their ancestral teachers, *Empedocles, Epicur, etc.,* that although they admitted some combinations and secretions in the substance of bodies, they denied all vital first beginnings and the cessation of their effects; for they asserted that generation in Nature occurred not according to the law change through inherent properties, but according to the proportion of gravity through accumulation (s. id. p. 294). Our younger Brothers are therefore advised to carefully guard against this seductive natural science and to stick unshakably to our irrefutable best-founded tenets, which are so clearly and convincingly expressed in the present work.

Again, it is axiomatic that every specific seed depends on the universal seed, and that the universal seed produces specific ones through the perpetual descent of the greater to the smaller, and similarly in the reversed order.

In order to comprehend how, and in what manner, this universal seed originates, and from this, how the universal and specific seeds originate (the *Spiritus mundi universalis & particularis*) the reader should mark well the chapters that follow.

CHAPTER IV

IN WHAT MANNER THE UNIVERSAL SEED WAS BEGOTTEN AND BORN

After God had separated the simple Chaos into a fourfold one, that is, the elements, he immediately imposed upon them the command:

Creacire & Multiplicamini! (be fruitful and multiply!) Note: Heaven and Air are the Father, the husband, the agens, the active part. Water and Earth are the Mother, the wife, the patiens, the passive part.

These four, and yet only two, must get together and again arouse a seed in the first matter, (i.e., a reborn Chaotic Water or Primordial Chaos), out of their interior or their Centre, for the birth, preservation, and destruction of all things, until God melts the mass of the great World into a stone.[57]

The number of those who are to forge this seed, or reborn Chaos, the *Spiritus Mundi*, is four, as indicated previously, Heaven (or Fire), Air, Water and Earth.

All these four appear to be quite opposed to each other and can never accomplish anything good when one extreme is considered against another, yet they accomplish everything the Creator has commanded them to do when they come together in the proper manner.

For the philosophical tenet must be and must remain true, that is: *Non transire posse ab uno*

---

[57] What our author here calls a *stone* is nothing else but that new reborn tinctural earth which the Holy evangelist John describes so splendidly in Chapter XXI of his secret Revelation, under the image of a city.

extreme ad alterus extremum ad abseque medio, (one cannot pass from one extreme (thing) to another without an intermediary). Every artist should take careful note of this. For countless thousands make mistakes and fail solely because they do not consider this point nor take heed of it.[58]

For Heaven can never become Earth without the intermediary of Air and Water, and Earth can never become Heaven without Water and Air as intermediaries between Heaven and Earth. Likewise, Heaven can hardly become water without air, and neither can earth become air without the agency of the water.

Heaven is subtle, pure, clear, thin and volatile; earth, on the contrary, is coarse, thick, dark, and highly fixed. These two are opposed to each other in consequence of their properties. For if someone wished to unite and coagulate heaven, the most volatile, with earth, the most fixed, he could never do it: For the most volatile flies back into its chaos when a little warmth is applied, and it leaves the fixed behind. This occurs in all things thru the whole of nature: The most volatile and the most fixed can never be joined or united without an intermediary. An artist should bear this in mind continually to avoid losing material, time and money.

Therefore, whoever wishes to change heaven or fire into earth must first unite them with their intermediary. Then they will immediately unite permanently in any heat,[59] while previously they

---

[58] This is a very important point, which must be very carefully observed in the fusion of the three components. Our author has very clearly explained it here both in the ascent (going from the gross to the subtle) and in the descent (going from the subtle to the gross).

[59] This is not to be understood as they become a permanent coloring tincture in the fire in

would require an eternity, so to speak, for their union.

Let heaven rise into the air as an intermediary, and they will immediately unite without a struggle, because they are both subtle. When they are united, give them water as the intermediary between air and earth, and they will again unite immediately. Then give them earth, and thus the union is accomplished in the descent of the most subtle to the next subtle level, from this to the thick, and from there to the thickest, and not from the subtlest to the thickest directly, but through the appropriate intermediate stages.

In the same way, in reverse, turn earth into water with water, change that into air through air, transform the latter into heaven with and through heaven: For they are one according to their common matter and origin. Consequently, one must be the others helper and guide and one must be prepared by means of the other.

Thus it must be, and this rule of Nature cannot be transgressed: unite heaven with air, air with water, water with earth. Or inversely: Unite earth with water, water with air, air with heaven or fire. Then heaven is turned into earth, and earth into heaven. For heaven is quite subtle, air is also subtle but coarser than heaven by one degree. Likewise, water is thicker and coarser than air by one degree, earth is thicker and coarser than water by one degree. Thus one has to proceed stepwise according to the perfect order of Nature herself.

---

such a short time, but it only means that when the primordial components of a created body are united purely and according to Nature, one cannot separate anything from them but *phlegm* (a thick, sticky, stringy substance), which does not really belong to their constituent parts.

Then a right, concentrated harmony will arise and a true operation raising them to a fifth essence. In this way, all alchemical operations must proceed or little will be accomplished.

But someone will interject: Your lecture does not seem to be quite right here. Is it right that it should be that air is only one degree subtler than Water, and the latter subtler than Earth by only one degree, since one can see that water is coarser than air by several degrees, and likewise earth is coarser than water by many degrees.

To counter this objection, the artist must understand that just as heaven, air, water, and earth are marked out in degrees of thinness and coarseness, these four elemental degrees are also marked out within elements. For the coarsest earth does not immediately mingle with the thinnest water, or the coarsest water with the subtlest air, or the thickest air with the subtlest Heaven. No! For just as one can perceive the difference in various kinds of earth as also in various kinds of water, namely that one earth is coarser and thicker than another, one can see the same in the others. That is why Nature here resumes her degrees again and mixes the subtlest Heaven with the less subtle and this with the least subtle. When they are united, they flow into subtlest air and then, after their union, into the ever thicker Air. After this, they fall into the subtle water, the latter into the thicker, and from there into the thickest water. This mixes only then with the subtle earth, till it subsequently becomes ever thicker and coarser and congeals into stones.

But let no one now understand that these degrees stand one above the other, as in a particular concentric sphere. NO! Nature has mixed the thick Earth with the thin, yes, with Water and Air, and has also forced the Fire (Heaven) into it.

In the same way Nature has also brought the coarse and subtle waters together, and likewise the air and the heaven, so that our eyes can almost notice a small difference. But it is only in the natural dismemberment that one can see how the subtle rises from the coarse and can be separated from it. In her action Nature is alternately very well ordered and intermingled in such a way that she does not throw the differing degrees together, but one subtle thing in turn after another[60] as for instance:

Take some earth from a field or meadow, whichever you like, pour water upon it and stir well together, so that water and earth be well mixed. Then let them stand for a while, and you will see that the water lets the coarse earth drop and allows only the tender earth to come out, namely, the salt. This unites immediately with the water into a virgin earth. When this earth is leached out however, the water will no longer attack the other coarse earth as it is too weak. That is the reason why you must first turn the virgin earth itself into water, with and through water. That is, you must distil it to a spiritual water, for in that way this water regains its power to separate the remaining subtler earth from the coarse and to make it like itself, so that

---

[60] Here the prime cause is explained clearly, intelligibly and in detail: why at the moment of their recombination, the three components of a natural body, clearly separated and reunited according to their natural weight, may well interpenetrate each other like water penetrating water, and appear to merge, and yet nevertheless require a long time for their real, inmost recombination and subsequent inseparableness. Our younger Brothers, therefore, must not only read this important chapter, as well all the others, more than once, and then re-read it, but they must also ponder it well, so as to deeply imprint in their minds the authentic tenets of the true teaching of Nature and the Art.

it too changes into water, and this more powerful water in turn attacks the remaining or residual earth. In this manner does Nature work in all things throughout the entire world, by dissolving and coagulating, ever through the appropriate intermediate stages.

From this an artist should now learn that Nature herself, in all her ways and doings, does not mix everything haphazardly, although it may appear so to our eyes, but by rule, measure and weight.[61]

Therefore, an artist must not see and penetrate the external appearance but the hidden secret, and he must fathom Nature by his manual work. Then he will go from one manipulation to another and grasp wisdom ever more.

---

[61] *Deus creavit omnia in numero, pondere & mesura,* that is, God has created all things according to their specific number, weight and measure. This is an old alchemical canon.

CHAPTER V

HOW THE DIVIDED AND SEPARATED CHAOTIC WATER IS REBORN AND BECOMES A GENERAL ORIGINAL (OR: PRIMORDIAL) SEED OF ALL THINGS, COMMONLY CALLED: THE WORLD SPIRIT (SPIRITUS MUNDI)

By what has been said above, a lover of the Art sees and understands how the primal and age old Chaos was divided into four parts, according to the teachings of all sages, and that it had been divided into heaven or, strictly speaking, fire, air, water, and earth by the irrevocable order of God. While God commanded them not to flow together again until a further order, he simultaneously commanded them by the Word of multiplication to regenerate unceasingly and to give birth to the divided chaotic water or the universal seed, the World Soul or the World Spirit, for the birth, preservation, destruction, and rebirth of all things.

Now this may appear strange to many a man who wonders where he will again forge the first Chaos which has been separated for so many thousands of years. Yes, it is odd. First, when a farmer hears about it, he will greatly wonder, although he sees it daily with his physical eyes and grasps it with his hands. But because it is called by a different name and he does not understand what it is and what is hidden under it. It is enough for him that his fields and meadows grow by it, by which he and his cattle are fed, and he does not speculate any further. A scholar, however, and a beginning student should heed it more, because it is clear. For if someone does not understand the main purpose and chief reason, how then will he get on in the Art? It is regrettable that so many embark upon the Art, more for shame and dishonor, since they immediately rush thoughtlessly into the processes and do not even reflect upon the origin of each thing.[62]

Now then, it is certain that the four elements, fire, air, water, earth have originated in the Chaos from one matter and foundation. That is why one of them cannot be different from another, so that one could not say to another, "I was born of another, nobler or baser matter." No! Matter is one, but the difference consists in that heaven has become subtle and volatile through putrefaction and has risen to the highest together with the air. Earth and water, however, have become fixed and have sunken into the depths with their heaviness.[63]

---

[62] He is probably partly aiming at the impatient sophists who, without having the slightest theoretical knowledge, immediately believe that they can transmute mercury into gold if they but get hold of an old musty parchment, partly at the modern democritic (not "democratic") atomists. The former will never attain the true final goal of Hermetic world wisdom, which is the cognition of the incomparably-great Creator through Nature. For "if man could acquire this secret of Nature and the Art by means of a recipe, he would never attain to the great wisdom of the true knowledge of God as also of the inner foundation of Nature, much less of himself, etc." See the booklet *Amor Proximi,* Frankfurt and Leipzig, 1746, Page 83.

These can be quite justly reproached with what *Philaletha* in the Preface of his *Anthroposoph,* Page 163, reproaches the Peripatetics, that is, that they "look upon God as an artist who builds with wood and stones, without pouring into them life; whereas the world, being God's construction, is full of moving living spirit which causes all births and increases of ores, plants, etc." One wonders how these investigators of nature accomplish anything in the true Hermetic alchemy, which deals with nothing but active spirits and in its work with nature keeps to the process which the Creator of beautiful nature kept in the separation of the Elements.

[63] What the author is here saying about the air is excellent. True, modern physicists have said many a beautiful and useful thing about the nature of this element, yes, they have even invented a special science which they have called aerometry or the art of measuring the air. They describe said air as a liquid matter, having a determined degree of weight and expanding force. They prove the weight (the compressing property of the said element) and its expanding force by the so-called air pump *(antlia pneumatica),* which *Wolf* describes most accurately and minutely in all its parts in his *Anfangrunde der Aerometrie* and in his *Versuche* (Tests). But since they also notice that it has a determined degree of heat and cold, and that the difference in the weather and the seasons depends on these properties, they have tried to determine heat and cold by the *thermometer,* its weight by the *barometer* or the weather indicator, changes in the density of the air by the *monometer* or air-measuring

But when heaven is made fixed, it is also earth; and when earth is made volatile and subtle, it is also heaven. The same applies to air and water. For it must be possible to transform one into another, or else they would not be of one matter, and no change could occur in their increase or decrease if one could not be transformed into another.

These four, then, have come into being out of the Chaos, but Chaos was nothing but water and spirit, two things comprised in one. These two then divided into four, and these four are nothing but water and spirit. From this we can therefore understand that water and spirit are distributed through all the members of the four elements, heaven, air, water, and earth, and in these four not a mote can be found where this water and spirit can't be met, be it in a liquid or dry, dissolved or coagulated condition. Thus everything in the whole world is water and spirit and a resolved and coagulated word of the eternal Creator. For the word became water, and consequently everything by it lives and is the Word.

Now it is also known that the finer and subtler a thing is, the purer, more vivacious and mobile it is: But the coarser it is, the more immobile, inert, and sleepier it is. The more mobile a thing is, the more spiritual it is, and through such motion, it becomes ever more mobile. The motion, however, produces warmth, and the more it is moved, the

---

instrument, changes in the humidity by the *hygrometer* or weather scale. These investigations are of great usefulness in ordinary life; however, when it is a question of the air's inner and physical nature, no particular advantage can be derived from this knowledge of the external effects of this element. Great profit, however, will be obtained by reading and pondering over the present and following principal piece with reflection, diligence, and deep meditation.

hotter it becomes. Therefore, it also warms and heats what is next to it.

We must admit that heaven is very subtle, and because of its subtlety it keeps moving, and through its constant, perpetual motion, it gets heated. Since steam is a Chaotic Water, and is made of Chaotic Water, this heating causes Heaven to Steam, emit vapor, and perspire. But this steam or perspiration, because it cannot rise above itself, as God has set boundaries to it, must penetrate below itself and thus comes down into the Air from above, and there the coarsest portion is retained and captured by the Air. The subtlest ascends back to Heaven again due to its affinity with it, and it continues up and down this way until it also thickens and becomes coarser by such constant course of circulation, that the Air can also retain it as well.

This then is the influence from Heaven which we enjoy by means of the Air, and the *astral seed*. Since Heaven is mobile, it also induces and causes Air, its neighbor, to be mobile. The Air moves the Water, and this the Earth, and thus the four move in perfect order. This type of motion can be observed in a clockworks where one wheel drives another, drives another, etc.[64]

---

[64] This offspring of the four elements is the aforesaid general procreator of all things, the incomprehensible Nature Spirit controlling all things, so hated by the profane natural scientists but so highly esteemed by true philosophers because of his excellence. When he could not mingle with the destructible elements in view of his delicacy, purity, most volatile quality, and indestructibility, he soared with the incredible rapidity incomprehensible to human senses, and in that way became the fifth element. That is why our philosophers say: *Quintum Elementum est Spiritus aethereus incorruptibilis,* meaning, "The Fifth Element is an ethereal, indestructible Spirit."

From these sublime dwellings he descends through the sidereal effluences into the subtlest evaporations of fire, through these into its less subtle, and with the latter into the subtlest evaporations of the air, and after that into the less subtle where he assumes, the

Just as Heaven gives its effluence, Air, Water and Earth also give their effluence. In combination, they thus forge the universal (general) seed or *Spiritus Mundi* (⏝)[65], the World Spirit.

---

first level of coagulation, and through the meteors into bodies which he engenders, sustains, nourishes, destroys, and regenerates, according to the thorough annotation of our beloved Brother Homer. He does not, however, exercise his might and power instilled into him by divine omnipotence solely above and upon the surface of our globe, but he even penetrates to its innermost center, to carry on his business there, forming subterranean creatures in the metallic workshops, and to raise them to that perfection to which creative wisdom had destined them beforehand. He is in all sublunary bodies, therefore, and none of them can exist without him or living in his way. That is why a principle of all philosophers is: "That *Mercury* (that is, of the philosophers) which is no other than our hermaphroditic *Mercurius,* can be found in all things, and that none of them can be found under the solar circle in which this divine subarchitect does not lie hidden in the more or less firm fetters of coagulation." *(Nihil sub sole existit, in quo non reperiatur argentum vivum, Hermaphroditus noster adamicus).* The Art consists in setting him free from the firm fetters of his coagulation, which Art will be greatly facilitated for our associates in the holy Fraternity through diligent reading of our present *Golden Chain.*

[65] This symbol does not appear at this point in the 1781 German edition. -PNW

CHAPTER VI
OF HEAVEN AND HIS INFLUENCES

After the separation of the Chaos, Heaven was the first and foremost, subtlest, most incomprehensible, and highest element, a very fine steam, light, pure, and volatile. That is why he went uppermost and took the highest place, which is the finest part, full of life, and became the most active element.

This is the reason why heaven is the *primum agens*, the first worker and father of all things, the male seed, the soul, or the life-giving water of life, *Nectar* and *Ambrosia*, a thinned air and water, a volatile earth.

Heaven and the air, as said above, do not have their effluence above themselves but below themselves, toward the earth and the water sphere, by well-ordered and express command of the Creator. As they flow in downward, water and earth flow out upward toward heaven and the air.

Now heaven, which is the most mobile, is heated and kindled by his ceaseless motion and begins to steam, smoke, and perspire - which he must do for his existence - quite unnoticeably and in a manner invisible to us. The exuding vapors of Heaven, with their power of effluxation, cannot rise higher but this vapor must enter the next lower sphere, the Air. Because Air is neither highly subtle nor is it too thick, Heaven is caught in it and mingles, unites and coagulates with the delicate vapors of the air. These two digest and circulate by a constant to and fro motion (the steam) till both are well merged, after which the lower effluences of water and earth are easily assimilated and can now forge the universal seed or World Spirit, i.e., the reborn chaotic water. Then, if heaven but incorporates in the air, it is subsequently easier for him to be united with his next in line, water and earth.

A lover of this art must not think, however, that heaven requires much time to unite with the air, and likewise the air with water, and water with earth. No! But as soon as they meet, the union can occur instantaneously. The longer and the more they prepare themselves for this Union through expansion and circulation or motion, so that afterwards they immediately intermix thoroughly and unite like smoke with smoke, water with water. For since the fourfold union of heaven, air, water and earth occurs in the medium of a pure subtle vapor, mist or smoke, one can easily admit that one steam mist or smoke readily enters and mingles with another, one water with another, especially as they have an absolutely natural tendency and one and the same matter and a common origin.

Now someone might ask if heaven, ever flowing out of himself, does not decrease in quantity or strength, because it appears impossible that in a natural way something should forever give of itself and not take anything back into itself without diminishing in its essence and strength. Just like a human being who due to movement and exertion is constantly and heavily perspiring, also becomes weak and weary and loses strength. Thus, one might reach a conclusion concerning a greater by faulty analogy with a lesser thing.[66]

---

[66] The essence of which Heaven consists is nothing but the primordial, etheric Light, separated on the first day of Creation from Chaos and united with the upper Waters which the Supreme Architect of the World, had divided on the second day from the Waters under

Thus one can easily understand that Heaven is not subject to the slightest change, or capable of increase or decrease. But our author is here not speaking of this enteric and indestructible heaven but of fire, and he wants this element to be properly understood under the appellation heaven. His above-quoted interpolation is also fully discussed in the sequel of his teachings and so fully cleared up that no doubt about it can remain in anyone with common sense. *Ex nihilo nihil, in nihilum nil posse reverti* (out of nothing, nothing is made) is an age-old tenet, the truth of which is confirmed by the unremitting and everlasting effects of Nature at all times. *Nam corruptic unius est generatio alterius.*[67] And since, as is well known, every destructible thing is again dissolved and separated by its destruction into precisely those parts out of which it had arisen, it may be considered just as true and evident that the four active elementary properties with their receptacles, or, what amounts to the same, the four elements, cannot experience any loss by their unceasing

---

the Firmament. (Genesis 1:6-8) This Light, the universal World Spirit, is the driving power or motion of our world system, even after His union with the Waters above Firmament, after which he had already become the universal *Mercurius,* the very purest, subtlest and simplest being of the whole visible world. He can never be affected by any adverse object, is never subject to any reaction, and also comprises within himself the whole enormous space from our atmosphere to the circle of the fixed stars, or the upper world. (*Mercurius* is the one who moves Heaven in the first place!) See *Welling,* Part II, Chapts. I,S,29 P.181. Because of his excellent teachings, this author deserves to be read carefully and frequently. Yet, he uses some paradoxical phrases which the Brothers should not be concerned about. While the influences of Heaven on our planetary system, as well as those of the latter on *ideas imaginativa,* or, they occur according to a whole system of effective ideas (Oettinger's *"Philosophic der Alten"* - Philosophy of the Ancients - Part II, P. 20), he *(Mercurius),* is also the true indestructible Fifth Element *(Quintum elementum est Spiritus aethereus incorrupti-bilis)* See Espagnet's *Enchiridion Physica Restitutae, can. 114.* The True Creator and Master Craftsman of all Things, the beginning for every offspring issuing from, and created by, the Great *Jehovah,* the great FIAT (See Zoroaster's *"Clay. Art.,"* p. 3 ff.)

[67] Corruption changes one thing into another -PNW

effects and the changes resulting from them. On the contrary, they must ever remain in just the same strength and just the same number, weight and measure as they were originally divided into out of the Chaos, until one day fire will get the upper hand, and they will melt in the heat according to the will of the Creator, and will be recast into a new heaven and a new earth (2 Peter 3).

This is easy to answer, and this 'knot' will be unraveled by what follows. For one thing is certain and can be grasped with our eyes: the so great space between heaven and earth is constantly and without ceasing full of steam, fog, smoke, clouds, and vapors, and these vapors, as soon as they condense, dissolve into rain and snow, dew, frost and hail. Soon afterwards, heat from above and below resumes, causing evaporation, so that there is no stop in the birth of those vapors. These vapors, however, we call by the common name 'Air'.

Just as everything that is steaming or breathing out desires, according to its nature, as also the command and order of the Creator, to attract its like back to itself, it is also obliged by natural constraint and command to relinquish its excess or excrement through its outlets designed for this, after it has assimilated the thing attracted and led it through all its members.

Likewise, a man who exhales and perspires profusely is forced by a natural desire or constraint and at the risk of losing his life and suffocating, to absorb again air, food and drink and assimilate it as himself, whereby he replaces the loss and revives and refreshes his life. But because this partaking or enjoying of air, food and drink is not all necessary for man's subsistence, he expels it away again through his outlets designed for that purpose, such as: the finest through perspiration, the coarser through urine, nasal secretion, and saliva, the coarsest through the stool. When this is out, he again absorbs fresh air, food and drink in a natural way, and again produces an excess or excrement to eliminate. In this cycle, the nature of air, food and drink has been completely changed inside man and so transformed into man's essence by the human *Archeus*[68] that the excrements do not give the slightest indication of their previous nature, i.e., air, food and drink. It is now altogether a different form and now of a human essence and vital spirits, saturated with volatile salt, as the art of separation (alchemy) can demonstrate in its praxis (practice).

Likewise, heaven, the air, water and earth replace their loss immediately with their likes. Heaven receives the vapors risen from below which, the longer they were on their way, the more they are thinned and prepared to the utmost degree.

They have also been drawn through all the degrees of subtlety of all of the air up to the firmament, and from there to the highest place in order to replace the loss and effluence of heaven. Of this then heaven takes as much as he requires, transforms it into his nature, and when he is satiated, he follows a natural inclination and pushes the excess or excrement away again from himself into the firmament. The air also satiates

---

[68] This word *(archaeus)* which is very often read in our scriptures and signifies the inner cause of all things in the world, is nothing but the universal World Soul or Spirit of Mercurius, which constitutes Nature in all created things. In Mankind, it is the prime origin of life which rules all of his functions. Therefore, it received its name from a Greek word meaning "I Begin," because it IS the beginning and origin of the life of all creatures.

itself and condenses all of the vapors coming from above and from below without interruption. It dissolves the excess into dew and rain and feeds them as excess to the lower sphere, the water. Water, however, discards its thick excess and hands it over to earth. The earth is overloaded by this influx, or may be satiated, and again chases the superfluous part of this water away through its implanted central heat, dissolves it into steam and vapor, fog and smoke, and drives it out into the air. By wisest regulation, the Creator has implanted this alternation of increase and decrease, absorption and expulsion into Nature and ordered her to continue till he puts an end to the world according to his will.

From this the lover of the Art clearly sees that everything must sustain itself precisely from what it casts away, but with a prior alteration. Then what we call *escrementa* or discharges, becomes our food once more. For instance: man eats bread, wine, beer, fruit. From these he produces excrements which are carried back to the fields. Seed is sown upon them, and his food grows again from his own excrements. In the same way a tree: when winter robs it of its leaves, they fall down to the roots and turn into the sap that seeps into the roots and feeds and fertilizes its own parent Tree.

Let a man observe this carefully and he will easily recognize in it the *Superius et Inferius Hermetis* (the Superior and Inferior of Hermes), the *Catena Aurea Homeri* (The Golden Chain of Homer) and the *Annulus Platonis* (The Platonic Ring). That is: one thing changes into another and through the everlasting alternation of things turns again into the same it had been before, or something similar.[69]

It is very easy to come to this conclusion, for there had indeed been one single substance[70] and from it alone everything arose, and that from which a thing arose, into that it must revert again through returning to it. Everything had been water, and everything must become water again because water had been its first beginning. Now then, let someone apply this understanding throughout the whole of the work that follows. Then, this will be no small advance in our Art. Now we shall consider the Air in this connection, as it is next in order.

---

[69] All this is so clear and can be observed daily by anyone, so that it really does not require any elucidation or further explanation.

[70] The identity of the prime origins (first beginnings) - *Identitas principiorum* - is a concept so frequently proven in our wisdom schools that only and individual could doubt it who has not yet imbibed the first principles of Hermetic philosophy. The universal primary matter from which all visible creatures have arisen the Greeks called "the All in All." According to their poetic inclination, they made a deity of it, which they sometimes mixed up with *Zeus* or *Jupiter,* that is, with the World Spirit, and they took him to be the ruler of the Chaos, which is otherwise called *Hyle* or *Sylva* by them. Our dear Brother Homerus, who reposes in God, calls this first and universal substance a chaotic water (See above, Chap. II), and he has irrefutably shown and proven that all visible things originated in and were, born out of it. This will remain the truth, no matter how Henkel (in *Flor. Saturnis.,* Chap. IV, P. 119) and other modern physicists may object to it. For, according to the well-founded annotation of Welling, (Part I, Chap. I, Sect. 16, P. 14), this water, just as all other waters now specified and fertilized by the above water, possesses the whole Heaven, even in their minutest and most incomprehensible parts, just as it does the whole, as it exists in the stars.

CHAPTER VII
OF THE AIR AND ITS INFLUENCE

The Air is the other constituent part after the separation of the great Chaos. Together with Heaven it is the smith and active male seed of all things. Heaven is the soul and life; the Air, the spirit and preserver of souls and life, and consequently the vital spirit of the microcosm. The △[71] is a delicate steam, or a water changed into vapor. It is a thicker and coarser vapor than Heaven. Therefore, owing to its thickness, it catches the subtle influence of Heaven and coagulates it in order to transform him into itself, its own airy nature and essence. Then it also absorbs the lower, still thicker, watery and earthly vapors as its equals, in view of their common origin, and copulates them with itself and with heaven.

In addition, it makes of them a unity through their motion and circulation. Finally, the subsequent vapors condense without stop from above toward the part below and dissolve into dew, rain, snow, or frost, which meteors[72] are then by their own weight precipitated to us upon water and land, to continue and complete its work.

From this we may see that the air is a first intermediate for uniting Heaven with Water and Earth, without which heaven could not become reconciled with water and earth. It is the first one which absorbs the total heavenly influence, and with this heavenly influence it converts the lower, watery and earthly discharges into a company,

---

[71] This is the symbol in the R.A.M.S. edition. In the 1781 edition it is, "Luft," aerial. -PNW
[72] Atmospheric Moisture - the virgin substance of rainwater.

connects and unites them, so as to forge in its sphere the beginning of the universal seed of all things.

For one vapor, as said above, easily mingles with another, one water with another, one earth with another. But earth does not easily attract a vapor directly, or mix with a subtle vapor. Even if Air were to retain a small part of the most fixed, most of it would fly off again. Instead, if vapor turns into Water, then the thicker the Water, the better it mingles with Earth and it also turns into Earth by its greater coagulation. On the contrary, Earth is changed into Water and Air by ever increasing refinements due to the action of Water and Air. Therefore, Nature acts through intermediates (middle courses) and not directly from one extreme to another.

For Earth and Water must become smoke and vapor, as we see daily, just as was previously explained with Heaven and Air. Adjoining elements can only combine, if they are of equal subtlety. They form a seed by their combination, which condenses into rain and dew, falls upon Water and the Earth as the centre and receptacle of all heavenly powers, out of which all animals, plants, and minerals are produced, corrupted and reborn.

As *Maria Prophetissa*[73] says: "One vapor or steam attracts or coagulates another" (of a similar nature). In the same way, the Air attracts and coagulates Heaven. Heaven, together with the Air, is condensed into Water still more powerfully. Water,

---

[73] Also known as Mary Prophetissa, or Miriam the Prophetess, a 3rd Century Alchemist. The Axiom of Maria is a precept in alchemy: "One becomes two, two becomes three, and out of the third comes the one as the fourth." It is attributed to Maria Prophetissa. -PNW

together with the Air and Heaven is absorbed and coagulated into Earth still more, yes, even to the utmost - into stone and metal. Thus Heaven becomes Earthly, corporeal, visible and tangible. Again, Water dissolves Earth, Air dissolves or thins Water and Earth into steam and vapor, Heaven dissolves and thins the Air together with the Water and the Earth, and changes them into his nature, so that one is changed into another now fixed, now volatile, in continuous alterations.

Then it is said again: *Aurea Catena Homeri, Annulus Platonis, Superius and Inferius Hermetis*, the Superior is as the Inferior, and the Inferior as the Superior. As said above, there is no difference in the matter itself, but everything occurs only by chance. Earth is a fixed heaven, heaven a volatile earth, air a rarified or thinned water, and water a condensed or thickened air.

From this one may see that one is not different from another except that one is volatile and the other fixed, one liquid or dissolved, the other thick or coagulated; and when the volatile is made fixed, the fixed made volatile, the coagulated dissolved, and the dissolved coagulated, one is changed into another and is yet what it had been before, that is, the first primordial matter of things[74].

------

[74] If this is not philosophizing thoroughly, explicitly, clearly and according to Nature, I do not know what is meant by philosophizing. Here one sees nothing of atomistic puppet shows, nothing of arbitrarily assumed hypotheses, but everything is shown as one may perceive it daily in the natural alternations, according to the laws of motion imprinted on beautiful Nature by the Creator. Blessed and doubly blessed are you, dearest Brothers, that a book is put in your hands as the basis for your natural science, whereby also those who do not reach the summit of the highest erudition can acquire a true and genuine conception of natural science, which they would seek in vain in corpuscular physics.

Finally, Nature may rightly be called the kidneys and testicles of the macrocosm, since in her especially the confluence of all the radical, essential moisture of the macrocosm takes place and the extract of the whole cosmic system collects, where the age-old Chaos, divided for so much time and so many years is daily and hourly begotten anew and reborn for the birth, preservation, destruction and rebirth of all natural things.

For what else are dew and rain - as well be further explained below - but a reborn Chaos from which all living creatures have their life and preservation? All vegetables and minerals arise and grow from it, and all this is generated and hatched in the air, as will be explained below by various examples.

CHAPTER VIII

OF WATER AND ITS INFLUENCE

Water and earth belong together as heaven and air, and these four again belong together: For earth must have water, and water must have earth; likewise, heaven must have the air, and the air heaven, and all these together must comprise one another. Else, one can neither exist nor act without the other.

Water is the third prime origin after the division of the great Chaos, and the first Patiens or passive part, the female seed or Menstruum of the great world, which must supply all sub-lunar births with nourishment, and it is, together with earth, the mother of all things.[75]

Water is a coagulated heaven, a thickened air, a liquid earth. It is steam changed into water. Water is also a middle thing between heaven, air and

---

[75] Modern philosophers do not wish to admit this. Among others, *J.G. Henkel* (in his *Flora saturnizante* (Flowers of Saturn -PNW), Chap. VIII, P. 314) attacks *Thales* and tries to contest his view by making only ONE thing or a simple something, a body-forming prime origin, while Nature, however, does not act otherwise than through male and female beginnings. Only, he does not understand the object of the query *(Status quaestionis)*. Water, in its relationship to sublunary things after the separation from the Chaos, is not a simple thing as *Henkel* believes, but it is, as our *Homerus* says very profoundly: a coagulated heaven, a thickened air, and a liquid earth - a steam turned into water. Almightiness awakened a powerful ($\triangle$) fire (our philosophers say). From this went forth an immense steam which dissolved and turned into (∇) water. (See *Versammlungsreden,* No. 14, p. 228). It is the first element of which we read in the Scripture (Genesis 1), the very oldest of the beginnings and the mother of all visible things. Without its covering the earth cannot receive any blessing, because moisture is the true cause of the mixture and the resultant fertility *(Philatetha* in *Anthroposoph,* P. 186). To this may be added *Plutarch's de Thalete,* f.l. *(de Placit Philosopher). Welling (de Sale,* c.l., Sec. 17, P. 15) says that water is the *first Qualitas fecundaria* or first element, and in it all bodies were in a primitive state and are comprised in it. The virgin earth, quicksand, however, he says, is the other element *(fecunda qualitas secundaria),* equal to the first, and from this one could recognize how this world and all its created parts had been standing in water (2 Peter 3:5-7).

earth, and it is the other intermediate by which heaven caught in the air is incorporated into earth, mixed with it, turned into earth and coagulated.

For as soon as heaven has turned into air, air into water, dew, rain or snow, they fall upon the lower, thicker water and earth, mingle with them, begin to get heated of themselves, to ferment and putrify because of the implanted primordial spirit. This is how the spirit works in the water and works one into the other until it brings their specific fruits to light and beget them through appropriate mothers (sperm, etc.).

From this, however, the artist must learn wisdom from Nature herself, that she is not content with only one intermediate, air, to turn heaven into earth, but that she uses air and water. In the same way, the artist must act in conformity with Nature and adjust his Art to the rules of Nature. How often many a man perspires to unite the elements in his work, and yet can't combine them. They stand one above the other, like oil and water, or like Water and Earth, or, both fight as violently as two fires and break the glass.[76]

He must assiduously search for a means to accomplish the conjunction, which is easy to find. Further on, the way to find it will be disclosed. For if one intermediate is inadequate, he must use

---

[76] Two opposites cannot be united without an intermediate. For when God had separated the very subtlest, clearest, driest from the first creature (that is, from the primordial Chaos), its *contrarium* was simultaneously created, that is, the coarsest, darkest, moistest and coldest, which is the coarseness of the earth and the moisture of water. They were tempered by the dryness of the air and the earth, as the coarseness and dryness of the earth were tempered by the subtlety of the air, the moisture of the water, etc. Consequently, one can see that no union can occur without an intermediate. *Cornel. Drebbel, Von der Natur der Elemente,* Chap. II, is very useful to read.

two, and if these do not produce any result, he must take three, but of the same kind and not opposites. Creatures from the subterranean kingdom and others from this same realm agree together, things from the vegetable kingdom combine easily with others from the same realm, and likewise those from the animal kingdom with others like them. For there is hardly a difference between them. While they have all arisen from a single matter, minerals are fixed vegetables, vegetables volatile minerals; likewise, vegetables are fixed animals, and animals volatile vegetables, and one can easily be changed into another. For man and cattle eat the creatures from the vegetable kingdom and make them animal by the *Archaeus*, and when man or the cattle dies, the body will be covered with earth, and plants grow again from it. Vegetables, however, feed on the mineral vapors which rise in a volatile form through the earth into their roots, and become all vegetable. But when the vegetables putrefy and turn into a nitrous-salty nature, they are dissolved by water and are carried into the sea through the crevices and fissures (cracks and crevices) of the earth, where they rise again to a mineral nature.

From this, the lover of Art can now see how one Kingdom is quite naturally changed into another. As has been so often said, it is not done indiscriminately or haphazardously - one Kingdom is made volatile and the other fixed - and so on. As one Kingdom becomes more highly fixed, or more highly volatile, it acquires properties of another kingdom, because it is the properties of fixity and volatility that determine the differences between two kingdoms in the first place.

Heaven and the Air are the father and male seed of all things, Water is the female seed and Earth is the vessel and mother in which the upper three effect all re-births and that which they have been commanded by the Eternal Creator.

CHAPTER IX
OF THE EARTH AND ITS INFLUENCE

The Earth is the fourth and last element separated out of the Chaos, and the nethermost, as heaven is the uppermost, air and water the middle parts. Heaven is the subtlest, the earth the coarsest; heaven is volatile, the earth is fixed, air and water stand in the middle, yet different from each other according to the degree of their subtlety (fineness) and volatility.

The earth is the other passive part and the female seed, the mother of all sublunary things.[77] Earth is a coagulated, fixed Heaven, a coagulated Water, a thickened Air, a vapor turned into Earth, a fixed coagulated being, the center and the vessel of all heavenly influences and of the universal seed, from which all minerals, vegetables and animals grow in the earth and through the earth.

But in order to explain briefly in what way earth and water change into steam, air, smoke, fog and vapor to achieve the generation of the universal

---

[77] The earth is the noblest seat of that womb which attracts and receives the seed from the male part of the world. It is Nature's *Aetna* or volcanic mountain, where the fire-god *Vulcanus* practices, not the lame poetic one who limps after his fall but a pure, heavenly fire that forms all things. She is the nurse and recipient of all things, for the upper natures precipitate themselves into it, so to speak. What it receives in one age, it reveals to the next and, like a faithful treasurer, it does not retain anything of what is entrusted to it. Its inherent property is cold. See *Philaletha Anthroposoph,* P. 185 ff. This cold quality is the cause of the contracting and binding power of Nature being locked in it, for we have but two prime origins of natural creatures, that is, light and darkness, or heat and cold, of which the first is active, but the other passive. The earth is the foundation of the other elements, which comprises in itself the seed and all the powers of all things, and it is precisely because of this that it is also called the universal mother of animals, vegetables and minerals. Therefore, when it is impregnated by heaven and the other elements, it produces afterwards everything out of its womb, etc. *Nuisement, Of the True Salt of the Philosophers,* which is the 10th book of the secrets of a true adept, Dresden 1757, 8., Sec. VII, P. 247 ff., which Section makes quite remarkable reading.

(general) seed or the regeneration of the Chaos, and how they soar high up into the air, yes, even into heaven, note the following:

To begin with, however, the lover of the Art must rightly understand the meaning of my words. That by "heaven" I do not understand that heaven or *Empyreum* where God dwells with his elect, which is privileged and totally exempt from any changes and natural effects, but on the contrary, that Heaven where changes DO occur. Changes in our Heaven do not influence the realm of God, by special command of God, Lord Almighty.

Having said this, I ask the reader to take note that, as I said above, "Heaven" is that which is extremely rarified (subtle) and therefore, is in a state of highest activity or constant motion. It will never stop moving as long as God is maintaining the world in its established state, and by its motion it causes what is next to it, the air, to move too, although much more slowly than heaven. The air, however, by its motion moves the water, and this in turn moves the earth, although the latter's motion gradually becomes weaker and slower.

That the air is moved by heaven can be seen by the fact that the air or the wind is constantly moving or stirring. That the air moves the water requires no proof. Sailors on small and big waters, mostly at sea, are often forced to stop work or are becalmed, while water with its tide and waves roars high up. But that water moves the earth can be seen by the fact that it is continuously carrying sand, mud and stones. They are crushed earth which the water rips off and washes off in one place and carries to another.[78] Here it erodes, and there it piles it up again mountain and valley, as befits the situation of the place.

Now then, every movement causes warmth, be it perceptible or not. For in living animals, that is, in those living on the ground, one can not only perceive warmth but also heat. In aquatic animals one does not perceive any warmth, or very little indeed, yes, so to speak, only cold. All life, however, must necessarily spring from motion and the resulting warmth, because cold extinguishes life.

From this the reader may conclude that there exists a perceptible and an imperceptible warmth. This has been inserted here because in all the elements, heat is implanted which one can feel at times, though not at other times, and which yet gives birth at all times, in one case as in the other, in all the elements, whether the sun and the subterranean central heat are present to assist or not. For everything, no matter how incredibly small it is, yes, even if one can neither see nor grasp it, because it is so tiny, still has heaven implanted in it with all the other elements.

If, therefore, this thing has implanted heaven in it, it has necessarily a motion, be it visible or

---

[78] In our Europe, this may be seen especially in places located on the German sea, such as Holland, the East and West Frisian Islands, Dithmarschen (a region in Holstein), etc. By the so-called tidal waves (spring tides) entire villages are carried off and again cast out in other places and deposited on dry land. The new land then lies untouched until the herb *Serpillum* or wild thyme grows on it, a sign that it has reached its maturity and is ready for cultivation. Then, with the approval of the sovereigns and by the granting of many privileges, it is crisscrossed with dams which they call dikes, guarded from the washing of the sea waves, and made fertile again. Yes, there are examples where in one place more whole districts have been deposited than had been carried off elsewhere. This increase undoubtedly has its origin in the fine quicksand at the bottom of the sea, which is so precisely stacked together by the strong motion and resulting mingling with the soil by means of the sulphuric-saline sea water that it constitutes such an excellent soil whose yield in fruits is more than a hundredfold, and which requires a renewal of manure only every fifteen years.

not, perceptible or not. Heaven does not rest, it must have motion, no matter from where it takes it; and even if it appears to be at rest, it yet has its invisible effluences, effects and powers. For instance, a precious stone, a root or a plant which has been torn off its mother or place of birth and is dried up, appears to be dead because it is prevented from growing. However, Heaven is within it, which does not rest but accomplishes great deeds by the imperceptible evaporation, so that such a precious stone, not just when it is worn but even when it is only touched, without loss of its power and might and without changing its size or weight, brings health or sickness to man according to its implanted nature and property.

Now the lover of the Art sees what this is and from what power each thing derives its effect, that is, from heaven and its perpetual active motion, Luke warming, warming and heating. Therefore, do not look for anything on earth, big or small, in which Heaven and all the other elements are not concentrated. Reason also indicates that everything must necessarily contain the nature and property of that from which it originated. Now all and everything has come from the Chaos, the primary matter. If it has come from that, it has indeed its properties and these properties are spirit and water. Spirit is the mover, the warming factor. Thus this spirit is distributed everywhere throughout the world, so that the smallest droplet of water and the tiniest mote of earth is also quite filled with it and with water, both in a liquid and a dry state. And just as the droplet of water is water and spirit in all its parts, the spirit being less fixed, likewise, the mote of earth is coagulated Water and spirit in all its parts, the spirit here being more fixed and coagulated.[79]

But the fact that earth and water are not as mobile as heaven is due to their thickness and coarseness, or coagulation and concentration. Make the earth volatile like heaven and it is as fast in its motion as heaven. From this can be seen, however, that the whole difference in all things depends on their volatility or fixity only, that is, fixity and volatile cause the change and varying forms in all things. And that is the whole purpose and end of Nature that heaven is to become fixed if it is to be useful and salubrious for the sublunary creatures. For it is obvious that all sublunary things are coarse and thick in comparison with heaven, therefore not as mobile either. Thus, heaven must necessarily become earthly for their usefulness. How else could they enjoy such a subtle vapor, so extremely subtle and volatile, if Heaven did not give itself to them by means of the air, water and also the earth? That is the reason why God has provided that heaven must go through all elements and transform itself into all elements as, on the contrary, the other elements must change into heaven by great subtilisation, for the salvation and usefulness, for the generation and preservation, also the destruction, and regeneration of all sublunary subjects.

In order for us to prove by what means water and earth, together with the air turn into steam,

---

[79] With this the philosophers of all times and lands agree. Therefore *Welling* says quite in accord with them *(de Sale*, Chap. I, Sec. 16, P. 14): That the intangible parts of the water also contain the whole Heaven, as it has been found and specified in the stars. In the same place he writes about quicksand, which he considers to be the virgin earth, that every particle of it, being an offspring of the water, must contain in itself the whole heavenly seed of the sun, the moon, and all the stars.

smoke and mist and how the latter change into air and heaven, let the lover of the Art keep firmly in mind that not only heaven and the other elements are everywhere mingled and present in all big and small things, but that Heaven and Air is displaying its motive power and might, in all earths as stones and bones, be it much or little, it is enough, it reveals the presence of Heaven and Air. For a subtle, thin and open thing like an animal will sooner show its power and motion than a big and immobile tree rooted in the earth, or even a seemingly lifeless stone.

Earth and water are always together: for in water there is earth, because water carries an earthy sediments. In the Earth there is water, because springs, wells and rivers flow out of it. In addition, very big lakes can be found in the earth. Now that this is known, it is also known that there is air in heaven and heaven in air, for heaven, air, water and earth are forever to-gether and in each other, and none is without having the other in all its parts. As little as man can live without a soul and spirit, just as little can one element dispense with another.

Thus, then, water and earth are filled with heaven and air. Water must moisten the earth, if it is to bear fruit. The moistening and fertilization, through the implanted heaven and air, or the mobile spirit in their mingling, together with the external sun that joins them, as well as the central heat, causes a movement, the movement a luke-warming, the luke-warming a warming, the warming a heating. This heating stirs and awakens the water into steam and vapors. The greater the heat and the more there is of water, the stronger it steams, bubbles and evaporates. When this steam breaks into the air, it is moved still more by the surrounding warmth of the sun and air and by the wind. The more it moves, the subtler it becomes, so that it rises higher as it moves and, the higher it rises, the nearer it gets to heaven; the nearer it gets to heaven, however, the nearer it gets to the origin of the motion.

This then is the reason why the longer this steam is moved the more it is subtilized and volatilized to the maximum. The more volatile it becomes, however, the more it tends toward heaven's nature, until the vapor is changed into a Heavenly nature by means of Heaven. But Heaven, the nearer it comes to the Earth, the more earthly it becomes, until changed into Earth and stone by means of Earth.

Now, an explanation has been given as to how this steam is transformed into air and heaven. We will now examine what kind of a vapor this is and what it is made of.

It is now sufficiently known that earth and water emit vapor and steam when they get heated. Whoever does not believe it may ask the farmers, they will not fail to explain to him the reason. Neither is there any scholar who would not admit that there is an inherent warmth in the earth. If there is an inherent warmth in the earth, all doubts are removed about the earth and water not emitting vapors and steam.

Yet such steam is twofold, yes, fourfold. It is twofold because it consists of water and earth; fourfold, because it consists of all four elements in regard to origin and the first matter from which these four elements, heaven, air, water and earth have arisen and, as said before, because none can exist without the other. The reason why I divide this steam into only two, into water and earth, is because in regard to heaven and air they are fixed, coagulated vapors; but when they become subtle by their motion, water and earth turn into air and heaven.

That such a steam had been water, everybody will easily believe; but that there is earth in this steam causes doubt in many, though they will not much quarrel about it after the knot of doubt has been resolved. Therefore, take note, as I said before, that one element is the other's leader and that one element dissolves and subtilizes the other. Heaven dissolves and subtilizes the air, the air water, water dissolves and softens the earth. On the other hand, Earth condenses Water, Water condenses the Air, Air the Heaven. In this way one is the other's magnet, *attrahens, solvene, coagulans, volatilisans et figens*.[80]

This, however, every artist should and must note: Just as the Chaos was divided fourfold into its parts, its grades (levels, degrees, steps), these four were in turn divided into their grades. Consequently, the heaven closest to the air is not so extremely subtle as that which touches the *Empyreum*, Heaven at its highest. In the same way, the air which borders on heaven is not as thick and coarse as that which reaches the water sphere. The uppermost water is not as thick as groundwater and the slimy, watery substance which attaches itself like gum or glue to the stones and plants that grow under the water. For not only stones and sand are

---

[80] Attract, dissolve, coagulate, make volatile and fix -PNW

earth, but there are also earthly juices, salts, pitch, resin, and wax which grow in and on the earth. These are all Earths and differs only according to *gradu*[81], that is, according to its volatility and fixity. Yes, not every earth is so extremely fixed as the stones, but there exists also a volatile earth[82] which is nevertheless on its way to becoming fixed.

Water softens, loosens and dissolves this volatile Earth, absorbs it and, prompted by heat, takes it along with it high up in the air in the form of steam, yes, by continual motion even into Heaven. It is easy to infer, and also to conclude subsequently by the test and practice of the reborn Chaos, that the thicker Heaven turns more easily into a fine (delicate) Air than the subtlest Heaven can and the finest Air is more easily changed into the thicker heaven than the thicker, coarse and

---

[81] degree -PNW

[82] This is the true elemental earth, the basis and foundation of all created things, which floats above our heads, the *Sal Astrale,* which was incorporated into all creatures at the first Creation and is daily instilled into them by the upper effluences for their multiplication and preservation; the true dust *Aphar-Min-Ha-Adamah,* out of which Adam's spiritual body was built in Eden. This salt of Nature is found in all things and in the right weight, number and measure for each matter, and it can be obtained from them.

    Its first coagulations reach us in dew, rain, snow, hail stones, shooting stars, etc. What a volatile object dew is which, as soon as it but feels the rays of the sun, is immediately attracted by them, while it contains a fixed $\triangledown$ which becomes visible during putrefaction. Rain requires a much stronger heat to make it return to its Chaos. Therefore, it also contains a greater supply of it, that is, a somewhat coarser earth, which even shows up before the putrefaction of this meteor. *D. Giese,* physician in ordinary to His Serene Electoral Highness in Cologne, Joseph Clemens, collected it in clean vessels, poured it on a porcelain plate which he put in a room where no one entered but himself and where there was no dust. He allowed it to evaporate of itself and concentrated it to such an extent that finally nothing remained-but a red earth like Kermes antimony = $\triangle\!\!\!-$. (This matches the symbol in the 1781 edition. Normally the symbol for antimony is $\overset{+}{\circ}$. -PNW) Of this he then gave to his patients only by grains, and he achieved great cures with it. It is the same with snow, hailstones and ice, which do not only consist of mere ∇ but simultaneously also of a very subtle virgin earth and a saline $\overset{\circ}{\circ}$al-∇ exist. See *Welling,* Part 3, Chap. 2, Sec. 10, P. 360.

lowest air. In the same way, the lower coarse air sooner becomes water than the upper subtle. Instead, the lower thick-slimy water sooner changes into Earth than the upper, all too thin and volatile Water; the same is true *e contra* (in reverse).

The volatile, easily soluble earth, especially the virgin saltiness, is sooner turned into water than a stone already dried out or sand. This water can sooner be changed into Air or evaporated by fire than coarse thick water. The same is true with Air and also with Heaven.

Now we have sufficiently shown the first beginning of Nature, how she was turned into water from steam by the Almighty and his word, how that water divided into two and subsequently into four parts, and how these four arose from a steam, fog, smoke, and vapor and received the command to multiply and bring forth fruit, and to produce births in the same manner in which they themselves had originally been brought forth.

For just as they arose from the primordial steam, these four must continually and in cooperation, give birth to an identical steam of exactly the same matter and nature, without any deficiency. This steam, through rebirth, is precisely to become water, that is, a Chaotic Water out of which each and all is to be born anew, preserved and destroyed, also reborn, without stop, till the end of the world.

That the four elements gave birth to such a water had been done according to the Divine Will, nor could it be otherwise in view of the natural laws of motion, for they were the children of their mother. Consequently, they have the power to produce such a seed as they receive from her, and all things produced by them, or *individua*, are an image of these four elements.

These four, when they join forces, give birth to the universal seed for the generation, preservation, destruction and regeneration of all things.

Just as these four together jointly arouse a universal seed in their union, so each of these four has in particular received the power to produce a like birth in its own sphere.

For heaven is the most subtle, most pure, most transparent, clearest above the others. Therefore, it brought forth fruit out of itself, without the collaboration of the others, that is, its stars which are full of life and light. The air brought forth its meteors, water its animals, plants and minerals; thus also the earth its animals, plants and minerals. These species, the stars and lights, the meteors, animals, plants and minerals have again been made specifically from the seed of their appropriate elemental sphere.

Just as each sphere has specifically brought forth its species from its seed, so this species, divided into its individua (parts), has again received the command to go into seed and multiply according to the image of the original matter, so that not only every star has been given a long life because of its purity and power to preserve itself for a long time, but we have experienced from century to century that various new stars have arisen, while others have been lost - which I order the astronomers to examine - and I am turning to the Air.

110

In the Air, other births of all kinds of meteors are daily hatched, so that hardly one disappears or passes away without that matter giving rise to another thing. This, however, can better be observed with our eyes and the touch of our hands in the sphere of Water and Earth.

One can indeed see that every animal and plant as it reaches its perfection generates a seed, in turn, to bring forth its kind. This power of multiplication, as it were, goes into infinity or innumerability, since as soon as one dies or decays, another or ten times more are reborn and generated.[83]

This may also be observed in the seemingly lifeless creatures of the stones and minerals. For if one were to dig ever so many stones out of the earth and used them for thousands of years for big and small buildings, one would not find an end of them, as up to now not the slightest loss or decrease has been perceived in them.[84]

---

[83] *Mors unius est vita alterius* is an age-old philosophical canon which has its perfect validity in the three realms of Nature. This is the true Pythagorean metempsychosis which cannot reasonably be explained in any other way.

[84] Much has been written and said about the daily generation of stones, but if this matter is viewed according to our reasoning, it can easily be explained, which our Brother Homer has done beautifully and clearly in the following paragraphs. Modern physicists seek the cause for the generation of stones solely in their ordinary conglomeration or juxtaposition, without the addition of a plastic and so-to-speak creative assistant foreman of God by teaching that the Air precipitates the Earthly parts existing in the Water, that is, unites Water with itself, the Air (combines the Water), lets it run off them, dries the Earth and binds small particles of this accumulated mass ever more and more firmly. (In the eighteenth century, the geologic theory was that ALL rocks were crystallized out of a watery medium). *Verbum Electri* has shown the absurdity of this system as clearly as the sun at noonday. This in the eleventh of our *Versammlungsreden*, at Pg. 293 ff. At the end of his third book, Diodor of Sicily entertains the same thoughts. Not knowing Greek, I therefore render his words in Latin: *Crystallus Lapis ex aqua oritur pura coagulata, non quidem a frigore, sed divina caloris vi, qui duritiem servet, variosque colores suscipiat,* meaning, "The crystal stone arises in pure frozen water, not from the cold, but from a truly divine power of warmth, so that it may keep its heat and take on various colors." What else is this Divine power of warmth but the etheric fire originated from the primordial-concentrated Light, the divine light of

From this, however, the reader should take it that, by what we've seen, although every elemental sphere gives birth to its *individua* and these appear to be different from each other, we may conclude here as above that the upper species of heaven as well as those of the air, water and earth differ in regard to volatility and fixity. Likewise, every *individuum* in every particular elemental sphere differs one from another in their degree of volatility and fixity.

For the fact that heaven produces illuminating creatures is due to its purity, fineness, transparency and clarity. In the same way, the lower elements produce subtle and coarse creatures according to their degree of subtlety or coarseness, and all these differ among themselves, and this difference is in proportion to their greater or lesser volatility or fixity.

For Heaven is not so wholly volatile that it does not contain fixed matter to some extent, and that is the subtle Earth. According to this content of earth, its fixity is determined. Likewise has the Air its fixity relative to its nature and Water and Earth are constituted in a similar manner.

Just as the *Volatile*, or the volatility of the Earth as measured against that of heaven is a fixity, heaven's fixity as against that of the earth is a liquidity (fluidity), or rather, a volatility. Yet it is to be understood in the following way:

Wherever there is earth - and there is some more or less of it as well as of other elements in all things - a fixity or coagulable nature exists,

---

Hippocrates, i.e., our blessed creative World Spirit.

according to which it exceeds in its *Quantuum*. On the contrary, where Heaven is present, one can assume volatility. And as an element has much or little of this or that, so one must also assess and utilize it in its practical applications.[85]

We have been speaking of the rebirth of the Chaos or universal steam. Now, however, we are going to turn this steam into water, into a true reborn Chaos as it was originally. In theory and in practice we will show its power, which it had in the beginning as well as at present and in the future, for as long as God wishes, so that the artist may grasp with his hands what he intends to do, and then establish his certainty further on.

---

[85] A very important point in Hermetic analytical chemistry is that one should become well acquainted with the degrees of volatility and fire-resistance of the bodies to be dismembered, which can provide very important advantages. In metallic bodies we find a large number of the fire-resistance class, where one can, to a considerable extent, estimate their friability or solidity by the strength of the ordinary, as well as the philosophical, solvents required to dissolve them.

CHAPTER X

DISCOVERY OF THE TRUE UNIVERSAL SEED OR REBORN CHAOS, SPIRITUS SEU ANIMAE MUNDI OR, THE FAMOUS WORLD SPIRIT

In the preceding chapters we said that heaven, air, water and earth, which originated in the Chaos-water and spirit, received the command to produce a universal seed, or to regenerate the previously existing Chaos for the multiplication, preservation, destruction and regeneration of all things.

As proven, the elements generate this seed by their evaporation. All four of them drive this steam into the air, where it is driven to and fro and circulated till it condenses from above and downward by the effect of the continually produced succeeding vapors, and finally it condenses into water due to this. We call this water, in general, dew, rain, snow, hail, frost, but fundamentally it is the real seed, the true reborn Chaos, the true *Spiritus seu Animae Mundi*, out of which all sublunary subjects are born, preserved, destroyed and reborn.

Now then, the test that this dew, rain, etc., is indeed the reborn Chaos and the universal seed or *Spiritus Mundi*, is that it must be a water from which all creatures (animal, vegetable and mineral) can arise and be born out of it which have already arisen from the Primordial Chaos. In addition, it must have the power and might to contain all four elements, Heaven, Air, Water and Earth, and if this is so, it must necessarily also contain that which the four elements comprise in themselves and everything each brings about![86]

---

[86] If this universal seed did not contain all four elements, it would be impossible for it to unite with the magnets that lie in the sublunary creatures and to operate in them.

So we now say that everything must again return into that out of which arose. The elements arose and were born out of steam and water, they also dissolve again into steam and water, which is dew and rain. From spirit and water they issued, into spirit and water they are turned again by spirit and water.[87]

That dew and rain are such a spirit and water, or such a reborn Chaos as the first, is not only proven by their daily effect, which is perhaps better known to peasants (farmers) and gardeners than to city philosophers, but their dissection also proves it, as all four elements are brought out by it (the dissection, or dismemberment, or analysis) and made manifest.

The daily effect and action of this water indeed proves that not only does every plant and grass grow and increase because of it, but that minerals and animals are also born, preserved, destroyed and reborn by means of it to the end of times.

The creatures of the animal kingdom feed on it and grow by it, (i.e., Spirit in the Air) because they take in Air constantly, and also those of the plant kingdom, grow out of this Water (spirit in the rainwater, etc.) and use it for the preservation of

---

[87] Of this spirit-water or spirit of Mercurius, *Basilius Valentinus* speaks quite beautifully in the following words: "This water (primordial universal seed) is the true Mercurius of the philosophers who have already been before me and who will come after me, without which the Stone of the Philosophers and of the Great Mystery can neither be made known universally nor particularly, nor can the metallic transmutation. And that spirit is the key to the opening of all metals as well as their locking. This spirit is also a good mixer with all metals, as long as they descend from his status and arose and were born of his blood. For he is the true *primum mobile*, sought by many thousands and not found by a single one, while the whole world is needing him, and he is sought far and is found nearby, as he is and floats before everybody's eyes." See Basilius Valentinius in *Chymical Writings*, ("*Chymische Schriften*") p. 733 ff. This should be noted.

their life. Vegetables do not require any proof, for the farmer can perceive such action in the plant Kingdom directly.

That the subterranean creatures grow out of such water and seed will be taught in its proper chapter.

Now we have partly proven theoretically that rain and dew are the universal reborn Chaos, the universal-general seed of the great world or macrocosm, the *Spiritus et Anima Mundi*, out of which and by which not only everything already born is preserved till its time arrives, but is also destroyed and reborn again by the very same Agent. And in such a cycle, this water continues to function to the end of the World, as we will elaborate further in a special chapter.

Now, however, we will examine this known universal seed or reborn Chaos through dismemberment, to see what parts it contains.

Therefore, collect some dew or rain, snow, frost and hail, whichever you wish; but you can proceed better or faster if you use rainwater; collected during a thunderstorm is best! Collect it in a clean barrel, then filter it through a felt, so that it does not contain any mist (This word, mist, has a double meaning: such as manure or dirt) from the roofs. You will find a bright, clear, transparent, crystalline water with no particular taste, it is spring-water to look at; in short, a beautiful clear water which you may drink and enjoy like other pure water.

Put this in a lukewarm place, under a roof where neither the sun nor the moon, neither wind nor rain can get at it. Cover it with a cloth or a barrel-head, so that no impurity can fall into it. Let it stand thus immobile for one month, and you will see a great change in it from its previous nature. For this water begins to get lukewarm because of the inherent spirit, although imperceptibly so; it will begin to get warm and to divide. It starts to putrefy, smell badly, and a turbidness is found in this previously clear, transparent water. A brown, spongy earth swims up on top, which increases in size more and more, gets heavier and finally sinks to the bottom. Here one may see a separation of the subtle from the coarse, the thick from the thin, by the inherent, implanted spirit *Archeus*.[88] For the earth which it separates is spongy, brown in color like fine wool, and greasy (muddy) and slippery slimy to the touch. And this is the true universal fossil dust (silicious marl or Gur, earth, guhr, or diatomaceous earth, or diatomite).[89]

---

[88] In alchemy, Archeus, or archaeus, is a term used generally to refer to the lowest and most dense aspect of the astral plane which presides over the growth and continuation of all living beings. The term was used by Paracelsus and those after him, including Jan Baptist van Helmont. -*PNW*, from *Wikipedia*

[89] On this occasion, we cannot but regret the deplorable decline of modern medicine, or the 'Art of Killing' as *Henry Cornelius Agrippa* (1486 – 1535) calls it in his first tractate *de vanitate scientiarum,* Chp. LXXXII, and that since that time the mechanical art of healing has come so much in use. I will not at all deny these gentlemen doctors the science of often judging most precisely, the nature or the true seat of an illness by the symptomatic indications or signs as laid down by their theory. Yet in such symptoms or accidents, which are proper to more than one illness, the main difficulties arise. For example, when an orgasm or boiling occurs in the blood, they immediately say: Behind this there is a hot fever, we have to do an air vesicle. Another says: Not for the life of me! A military or petechial fever is in the offing; we have to use evacuants, so that the *materia peccans* may reach the outer parts. But how, gentlemen! If the latter were the case and you had prescribed a bloodletting by mistake, or if in the first case you were to use evacuants, would not the sick person's demise be hastened? (Note: I believe the author or the German translator has mixed up the meaning of "latter" and "former.") For most assuredly, an illness will not be so complying as to become that which it should become according to their opinion. How much better would therefore young starting physicians do if, before proceeding with their medical practice, they would learn to make

In this the lover of the Art perceives two things with his eyes, water and earth, in which two (heaven and air) are concealed.

For we cannot see heaven with our dark eyes. True, we can see the air when it is flying about in our sphere as vapor, smoke or fog; but here it is dissolved into water and is contained in water as is heaven. Now there are already two elements visible to the reader, water and earth. Before, there was only a volatile water, but now earth has manifested itself visibly owing to the profitable effect of putrefaction or the lukewarm digestion. Heaven and air, however, have to be sought and ferreted out in another way.

When the rainwater has thus become turbid, stir everything well together, put it in a copper alembic, set it in an oven and heat it so that the water begins to steam. Then you will see how a steam, vapor, and smoke or fog rise out of the alembic, and that is the air which comprises heaven in itself and with itself. But if you wish to catch the air and turn it into water together with heaven,

---

good universal medicines, for these are an unknown thing in the Latin kitchen. By that they could save a great deal of discussing illness and also be certain that healing would infallibly follow as long as the date set by God had not arrived.

Assuming they had not guessed the illness: In that case these general medicines never spoil anything; they do nothing but cause the great Master of Nature, the Archaeus (Archeus) or Spirit of Life, who had fallen into disorder, to carry out his functions as the sole true physician at all times.

The dismemberment of meteors is sufficient to give them a great many of such medicines which are worth more than all pharmaceutical laboratories. How very grateful should people be to these men who labor day and night to make this wholesome use more public! Our author is giving us a good occasion to do so, and *Darnerion* has proven it theoretically and practically in his fine discourse on *astral powder*. Let us follow him, and we need no more be afraid of *Hippocrates's* judgment, which is: "The pharmaceutical art (the art of medicines or remedies) 'is the most' excellent among all sciences. But because of the ignorance of those who practice it (I am adding: due to the lack of universal medicines) it is considered the meanest of all." (Hippocrates in *Lege, ab initio*).

put a head on, add a receiver as the brandy distillers do, and the steam in the head will condense and flow like streamlets into the receiver in the form of very clear, crystalline water. Draw off one quarter of this water and you have heaven and air together, and you have separated two of four elements. You will perceive heaven by its light and shine. For this water, especially if it is rectified, is shining much brighter and clearer than before, or like a crystalline spring water. This light shows that it contains a superior power or a heavenly nature. After you have distilled heaven and air carefully, put another receiver on and continue distilling. Draw off all the water till it is thick like melted honey, and not so that it is all dry: For you would burn the tender virgin earth which has not yet reached its fixity. As to the second (drawn off) distillate, reserve this also and you have the third element, the "water of water."

What has remained in the alembic, however, is the still very moist earth. Take it out, put it in a glass dish, place it in the sun to dry up completely so that it becomes dry enough to pow-der it; then do powder it, and you have separated all the elements and have them before your eyes (this is the "earth of water").

Now, these elements must also prove to be real elements, otherwise it is very wrong what has been written about them, i.e., that all sublunary subjects are born of them. For let no one imagine that he can build heaven unless he has received a heavenly revelation as to how to make stars of this water. The same for meteors, because this water is itself a meteoric birth, and I leave it alone. Instead, we are going to see if out of this fourfold water can be born the creatures of the animal, vegetable, and mineral kingdoms, from which we have our main sustenance and with which and through which we exist and live symbiotically and in harmony.

Well then, take come earth if you wish to make minerals, moisten it a little in a retort, put it in a place where the warmth of the sun can get but not the *direct* rays of the sun. When it is dry, moisten it again with its water, (the second distillate) but not with Heaven and Air (the first distillate), and repeat this moistening ("humecting") and drying frequently. By so doing, if you wish, you can make the Earth totally mineral. By this moistening and drying you will find that the earth becomes heavy and sandy. Note well: The retort must only be closed with a paper stopper, and not firmly, to allow air to enter.

When now you see that the Earth has a sandy consistency, you will recognize it has a mineral nature: for it is not a vegetable, nor an animal; consequently a mineral. When you then have enough of that sand, take some of it and make a test, as ores are tested, and you will find a trace of silver and gold.

But if you wish to get a vegetable from the above-mentioned Earth, take the Earth referred to, sun-dried and powdered. Take two parts of its water (2nd distillate), one part of Heaven & Air, pour them together and moisten the Earth with it as gardeners do, not too wet nor too dry, place it in the air, not directly in the sun, and various little plants will grow. But if you put the seed of a plant into it (any ordinary seed) the fruit of that seed will also grow from it. In this way, we now have the vegetable birth.

If you wish to get an animal, however, take the earth dried by the sun and powdered, pour on it one part of water and two or three parts of heaven and air, such that it becomes like thinly melted honey. Put it in lukewarm air and mild warmth of the sun so that the sun does not shine on it too warm and you will see how in a few days various little animals of different kinds stir and swarm. If the water and moisture should decrease, sprinkle as before, so that it always stays at the same consistency as at first. Then you will see that in part some of the first animals disappear and others grow from them, and in part some of them will feed on it and become ever bigger. I would have liked to mention here a trick for producing all kinds of animals, whichever one desires; but I will keep it a secret, so that nobody could say that I wish to interfere in the Creator's creation. It would really be better to reason that God created everything out of the Void, and without matter. We, however, if we wish to imitate Him, must everywhere have the already created and made matter, and God has not forbidden us to delight in his creatures and his creation, but has rather commanded us to do so, and he has revealed it in secret to his saints as the Cabalistic Art, by which man attains more and more to the cognition of God.[90]

---

[90] This experience is very instructive and shows the consubstantiality of the very first beginnings and their common origin from the chaotic water so clearly that one ought to feel justly ashamed to assert the contrary. What the author says of this tender virgin earth, namely, that care has to be taken not to burn it since it had not yet attained its highest degree of fixity, is founded in Nature. But as soon as the creatures issued from her have attained their complete maturity and maturation, the imperishable plastic dot within them, which is precisely nothing but this wonderful earth or its salt-magnet, is of such an invincibility and rocklike firmness that "its power can neither be extinguished by the might of the incendiary fire nor by the cold of the dissolving water, etc." (P. J. Fabri Myrothec. Spaggr. p.m. 111) . In the XI. Versammlungsrede (p. 281 ff) Verbum Electri has had recourse to an example from the vegetable kingdom. Here now follow two examples from the animal

It is precisely the chief cause of error that neither the mob nor the vapidly disputing theologians can attain to the cognition of God. They all squabble over God, and when the squabble is over, they themselves do not know and doubt if what they have quarreled about is true. And in addition they suppress the natural sciences under the pretext that they are forbidden magic and that one blasphemes God and wishes to fathom him; but they are blasphemers themselves. That is then the beginning and origin of all idolatry and heresy that theologians as well as the mob run even more after Mammon and always preach a way to God, although they themselves do not believe and do not know what God is and who he is.

But I say that whoever wishes to have that cognition must begin with the earth and then rise to heaven from one level to another through the cognition of each of the levels. As Christ says: You who do not understand what is earthly and lies before you, how will you understand the heavenly?[91]

From this everyone can conclude that this water, or reborn Chaos, Spiritus mundi, or dew, snow or rainwater is the universal seed from which everything can be born that was born from the first

---

kingdom.

If one lets the ash of crawfish stand moistened in a humid place, or in an earthen ware vessel with some pure rain or thunder-water, one can see within twenty days innumerable small living worms, and if one squirts beef blood on them thereafter, they will gradually turn into crawfish. (Porta in his natural conjuring book, Nuremberg 1713.4. Page 173 ff.) This experiment is reliable. On the occasion of a dinner in Paris, Dygbi served a whole dish full of such crawfish made by himself, which were exceedingly large and tasty.

Paracelsus writes that if a bird is burnt in a closed glass and set thus closed in manure, a smeary moisture will arise and will finally turn again into a bird through the warmth. (Porta, ibid. p. 188).

[91] "If I have told you earthly things and you do not believe, how can you believe if I tell you heavenly things?" John 3:11-13 in the English Standard Version of the Bible - PNW

Chaos. And from this it may be seen that this water renders the earth fertile and that everything can be generated from it. The farmer or gardener sees this every day in his field and he need not be further convinced to believe it, but he sees with his own eyes that everything moistened by it grows superbly. Few there are, however, who know the core of this mystery, what it is that gives and produces fertility. True, everybody knows and will say: It is Spirit, with the help of water that makes things grow. Indeed, it is Spirit, but as a volatile spirit it can achieve little in the sublunary. For whatever wishes to be of avail in this corporeal earthly realm, in these visible bodies, must also be or become corporeal. One must also be able to grasp, touch and see it. Consequently, this volatile spirit must assume a tangible and visible body, just as the animal, vegetable and mineral seed is visible and tangible.

This is known to but a few, although people are very often handling it. Very few know the origin of this corporeal spirit or seed, even if it can be obtained in abundance. The reason is that it has another name than it should really have. For by its right origin and root it should be called *Semen macrocosmi*, seed of the great world, the fertility of the whole world. This appellation is its due because it is the concentrated, coagulated, condensed, corporeal seed and *Spiritus mundi*, in a transparent, visible body, like a crystal, a water, a dry water that does not wet hands, an earth, a watery earth, and full of heat, also full of cold, like ice, a coagulated heaven, a coagulated air which is better than all other treasures of the world.[92]

Yet in order to put this spirit palpably and bodily before your eyes and to give it into your hands, so that you may contemplate it sufficiently, do this: take the putrefied rainwater from the barrel, put it in a glass or kettle and boil it down to one-third. Then let it cool, but while it is still lukewarm, filter it clean of all sediment. Now put it in a cold cellar in a tin basin or glass dish or wooden pot, and the World Spirit will appear during the night in two forms, one a transparent and diamond-shaped crystalline matter. It attaches itself to the walls and sides of the vessel, and if some small pieces of wood are put into the vessel, it will also adhere to those. The other form, however, sinks to the bottom as a somewhat brownish colored mass.

Now you have here the spirit, the universal World Spirit and *Semen macrocosmi*, *Chaos regeneratum*, to touch bodily and see. Remove now the one attached to the sides separately and retain it pure. Carefully pour the water off from the one at the bottom and remove it also. Dry it especially well in the sun or by means of a lukewarm oven. Preserve it well. Go with these two to the lame Vulcan and he will tell you who they are and what they are called. Throw the upper seed which had adhered to the sides upon burning coal, and it will immediately tell you its name. It is called (be Silent) ⊕ **Niter.** Throw the other also upon the coal. It has a very husky, gruff voice and gnashes

---

[92] *Est in aere occultus vitae cibus, quem nos de nocte rorem, de die aquam rarefaction nominamus, cujus spiritus congelatus melior est, quam universa terra* - Sendivogius. ("There is in the air a hidden food of life, which we call dew by night and rarified water by day, whose spirit is better than the whole earth.") (Michał Sędziwój, or Michael Sendivogius (1566–1636) was a Polish alchemist, philosopher, and medical doctor. –*PNW)*

its teeth. Its name is salt, common or "rustling" alkaline salt. Now you have both names.[93]

This niter, sublimated from rainwater like every other saltpeter, has no effect different from that of any other saltpeter.

The ⊖ salt however, cracks and crackles like other common salt; it has also the same effect in all works. By this test you now see the kernel and center, the seed of all things, the *Sperma macrocosmi*, clearly visible and corporeal before your eyes, and you grasp it with your hands. These two generate, sustain, destroy and regenerate everything that exists under the moon and is visible to our eyes. In the air it is volatile and it also produces volatile meteors.[94] In water and earth it becomes corporeal and also produces corporeal things, fixed and more fixed, according to their level, also most fixed. Nothing under the moon can be found not containing these two manifesting as a

---

[93] Regarding the peculiar and wonderful effects these two have, joined to some other salts, when they have become volatile by their own liquid-volatile parts, is described in chapter VIII of *Versammlungsreden*, p. 211 *ff*. If it were worthwhile entering into a discussion about Kunckel and teachers of natural science like him, who pretend that there is no salt extant in rainwater (See *Laboratorium Chymicum*, part II, Chapter 2, page 111, *ff*.) one could write volumes! However, they need only imitate (replicate) the experiment described here in detail and faith will fall palpably into their hands.

[94] Such is the case with all *meteors* (things of atmospheric origin). The Sternschneuer or Sternpuge (shooting star), a noble meteor, and a true *Sperma Astrale*, is quite volatile before putrefaction, as every peasant is able to see. However, as soon as it requires a certain degree of fixity through putrefaction, it drops an Earth which is as fixed as is gold, which is the very (true) virgin earth sprung from the Chaotic Water and is floating over our heads, which has been extolled so highly by Philosophers since time immemorial. Concerning this subject, we strongly recommend carefully reading, several times, the incomparable discourse of our venerable Brother Damerion concerning the preparation of the astral powder, which is annexed to the known Plumenic (Plumendekischen) work. Then will there dawn a great light, not only about this, but about the dismemberment of all other meteors.

result of dissolution. All and everything consists of these two as is further proven following:

One is: ⦶.	The other is: ⊖.
One is an *Acidum*.	The other is *Alkali*.
That is *Spiritus*.	That is *Corpus*.
This is the father.	This is the mother.
Male seed.	Female seed.
Agens universal.	*Patiens universal.*
♀ *primordial.*	⊖ *primordial.*
Heaven and air.	Water and earth.
The Steel.	The Magnet.
The Hammer.	The Anvil.
The Ship.	The Existent.

To begin with, this thing was totally volatile. In order to see it we must distill the rainwater prior to putrefaction as soon as it has been collected, and it will rise over quite volatile. By putrefaction, however, it acquires a basis of fixity through the sedimentation it contains (the precipitate containing Earth).

The volatility of this water generates animals. As it grows somewhat fixed, it produces vegetables; and when it is quite fixed, it produces minerals.

Therefore, whoever wishes to generate minerals from it must take the more fixed and coarser parts, such as the water with the earth, as I said before; whoever wishes to have herbs and plants must add some heaven and air; whoever wishes to get from it various kinds of animals must add some more of the volatile, that is, more vital spirit of air and heaven, because vegetables stand in the middle between animals and minerals. From the elements, a mineral or a stone can readily be made, as well as an animal (each kingdom can be generated!) as will be shown in greater detail.

But the cause of our getting to see and touch the universal seed lies in the door and master-key of all deliverance from the natural bonds and locks, namely, putrefaction.[95] The cause of putrefaction, however, is the never resting inherent spirit which never stops but, when it has an instrument by means of which it does everything, i.e., water, it is visibly and invisibly, perceptibly and imperceptibly, incessantly at work. It causes putrefaction, turns a volatile into a fixed, and again a fixed into a volatile. It continues the alternation of this work without stop; it breaks the stones which it had itself coagulated and changes them into sand or dust. It rots the trees, decomposes the creatures of the animal kingdom and again makes a tree of the stone turned to dust, an animal of the rotten tree, a tree of the decayed animal, and a stone or mineral of the rotten tree, and this without interruption. No farmer believes this, although he must everyday watch with annoyance how the worms eat the wood of his door, and how trees and herbs grow out of his dilapidated walls,

---

[95] "Putrefaction is the *KEY* to all dissolution and separation." Basilius Valentinus in his *"Alchemical Writings,"* Hamburg 1740, page 109.

127

also that flies fly into his rooms from a decomposed ox and that fish-life generates in his ponds.

Now we have explained how the Chaos, from the beginning, had arisen from the Primordial Steam and by continually descending, became the four elements, Heaven, Air, Water and Earth, and how these had been commanded to incessantly regenerate the Primordial Steam (Vapor) and from this, the Chaotic Water.[96]

We have shown the volatile and invisible intangible seed. We have made it visible out of its invisibility, tangible out of its intangibility, so that everybody can now see it with his eyes and appreciate its powers through further investigation.

That I have said, however, that this universal seed from the reborn Chaos or rainwater, that is, *Nitrum*, is not much better (just as salt) than common *Nitrum* and salt, is due to the fact that any work can be done with one as with the other, and there is no difference in the effect unless one were more purified than the other. But if they are equally pure, one is like the other, and no artist should allow himself to be mistaken in this. If someone were to say that this is *Nitrum Vulgi* but the other *Nitrum Philosophorum*, it would be pure superstition. If common Nitrum produces the same

---

[96] Let us hear how in a very instructive way our Philosophers (with whom is God and His wisdom) express themselves, in a manner customary to them. *To wit:* "That △ causes air and vapor is known to everybody, but that this air, or vapor, or smoke (steam), when it is collected (or: gathered) condenses into a thick and a thin ▽, in which a living spirit *(i.e. archaeus - HWN),* is ceaselessly at work, until finally a separation takes place of its own accord, whereby the ▽ remains at the bottom of the vessel and above it a pure ▽ stands in which the △ and the △ lies hidden. This is all well known to all true and experienced Brothers. However, that this experiment has a likeness (similarity) with Creation may be seen by the characterization found in *Versammlungsreden,* No. IX, page 229 *u.f.* and 231. Because of this, it will not be repeated (duplicated) here.

effect as the other, it is indeed for me *Nitrum Philosophorum*.

And what scruple should one feel about this? For those little experienced laboratory workers every single thing must be doubled, one must be called *Subjectum Vulgi* and that is generally rejected but the other is called *Subjectum Philosophrum*, and that one is accepted; but when it comes to the point, they themselves do not know which is *Subjectum Vulgi* and which *Philosophorum*. They immediately say: yes, it cannot be fathomed by human intelligence.

God must always perform a miracle and reveal the subject in a dream or by an adept, while it is often a laboratory worker's own want of sense (which is the cause of his failure) because he does not pay attention to what he has in hand, to what he is doing, to what kind of result he has achieved. He does not examine the circumstances; he does not look to derive further advantage from a chance discovery through reflection.

How is this done? If I now added this or removed that, what would become of it? Instead, he lets it go, although he should keep this saying in mind: *Inventis facile est adders* - invented things are easily improved. Supposing an unlearned mason watches a house being built. He builds the house according to his simple understanding, and when he has finished building he notices in time a few faults. From this he immediately concludes: Look, if I had done thus, it would be more comfortable; here I should have built in an iron bar, and it would be stronger; or here some wood, or a big durable stone, or here a square high or low room, etc. If now he no longer wanted this house, sold it and built another, he would already have ten advantages to correct the previous mistakes. In the same way a laboratory worker or chemist should proceed when he has made a mistake. He should carefully examine it, of what there was too much or too little, what kind of an effect this has, what kind of obstacle or help that thing gave. He should investigate the kind and properties of every subject beforehand, so as not to bring together opposing things (adverse things).

So that the reader may see, however, that the universal *Nitrum* of the rain is not better than the common saltpeter, likewise with the salt; he should consider that the universal Nitrum is the generator and origin of the common saltpeter, and he should conclude that the child's blood derives from that of father and mother and that it is precisely of that primordial constitution; and if it <u>does</u> give the results of its father, it is indeed the father himself in his whole substance. In addition, as I have said and as the *Axioma* itself states: *Ex quo aliquid fit, in illud rursus solvitur* - by what something is born, into precisely the same it is again dissolved. If then the creatures of the animal, vegetable and mineral kingdoms are each one born of Niter and common salt, they must go back again and dissolve and reduce into that till their extreme prime origin. If then everything is born of this, and if everything born is again dissolved into it, there is no difference. And that everything consists of this Niter and salt and is born of it, must be clearly shown by the test: that it must necessarily be found everywhere and that it is present in everything. We shall prove this in subsequent chapters.

CHAPTER XI

A CLEAR PROOF THAT NITRUM AND SAL ARE IN THE AIR AND IN ALL THINGS IN THE WORLDS

Since we cannot ascend into heaven but must usually cognize its subjects from the inferior regions, we say: Heaven is full of light, light is an effect or offspring of fire.[97]

Saltpeter, however, is all fire. Therefore we conclude that heaven is a most volatile niter which becomes ever more corporeal and fixed in descending. Let this be enough said of the heavenly niter.[98]

Now about air. That niter and salt are contained in the air is evidently proven by lightening, thunder and hail, because here on earth we do not find any other subject that fulminates, flashes, thunders and hails like saltpeter and nitrous things.

Niter is first born volatile in heaven above but changed into something spiritual and volatile in the air; and in water and earth, into a thick visible and tangible body.

But how it happens to ignite in the air and thus hails, flashes in lightening and thunders, we will investigate theoretically with physical arguments, and then mechanically in our practice.

Niter does not fulminate unless something contrary to it is added, and it is and aroused by

---

[97] There is truth in this, in as much as the Creator had drawn the primordial light together in the sun, by which one could easily prove that light is the origin of fire, and not vice versa. But since our Brother *Homer* speaks about the separation of the elements, it does not really belong here.

[98] Here our author is saying something very beautiful, which our esteemed Brothers must well take to heart, as in these few words is contained the basis of the true natural science.

heat. The more volatile and subtle niter is, the more violently it explodes and ignites. In the same way, the finer and subtler niter's opposite is, the more violently they act upon each other.

Therefore, as we have said, in order for the light, life and fire of heaven to get caught and concentrated in the air - and then to turn into a subtle volatile niter - it must also have an opposite for it to become effective.

To give a *contrarium* to niter, it is met from below, out of the earth and water region, in the form of steam, fog and smoke, by an equally subtle, volatile, earthly body, a volatile earth or *Sal volatile*, *Sal alkalicum volatile*. When then these come together through the wind and are moved and heated by the hot rays of the sun, they affect each other, ever more heated, until they ignite, fulminate, hail and thunder and produce fierce explosions in the air, as can be sufficiently experienced on hot summer days.

On the contrary, however, when the sun does not shine too hot, they go together, the subtle niter and the volatile Alkali. They unite, but without explosions, as may be seen and observed in winter and on humid and cold days. The cause is the humidity and cold which prevent their getting thus heated and ignited, which we can clearly prove by our manual work in the following manner.

Rc. Take saltpeter, let it flow in a crucible in an open fire, add to it some volatile alkaline salt such as *sal ammoniac* or salt of urine, or another sal volatile, or a volatile earth, such as coal, sulphur, or vegetable and animal oils. Then it will ignite, fulminate, and explode like gunpowder.

132

The subtler the earth or the salt is - but in dry form - the more and the more violently it thrashes about and explodes (detonates). It does this only when receiving a dry heat, but in a damp or humid environment they readily unite.

For if such Reagentia come together in cold and dampness, they unite without exploding, because they have a third factor that does not allow any motion and ignition and prevents fulmination. Thus, if a volatile urinary salt or *sal ammoniac* is dissolved in water with niter, it will dissolve both without the least suspicion of a change. But if the moisture or water is evaporated till it is dry, coagulated on fire, and the fire is made somewhat too strong so that they begin to melt or flow together, they ignite immediately and produce the 'thunder' or explosive sound.

This may be clearly seen with *Aurum fulminante* whose primary cause of fulmination sought by many but found by few because nearly all alchymists ascribe it to the sulphur in the gold, which is wrong. This, however, is the true cause: After the gold has been dissolved in *Aqua Regis* and is precipitated with an *oleum tartari* (solution of potassium carbonate) or another alkaline salt, it falls to the bottom as a very porous calx. Though often edulcorated or washed, the fulmination cannot be removed nor can the salts that cause the gold to be heavier than before. We shall now analyze this.

Aqua Regis is made of *Aquafort* and *sal ammoniac*. *Aquafort* is made from saltpeter and vitriol; oleum tartari is a fine Alkali. When gold is dissolved in *Aqua Regis* - a volatile saltpeter - and sal ammoniac - a volatile alkaline earth - it is precipitated with *oleum tartari*, a fixed alkaline earth. The *Aquafort* is partly saturated and made fixed by the *sal tartari*, its enemy, and because it is a more open earth than gold, it precipitates the gold. The gold however, is strongly saturated and filled by the nitrous spirit of *Aquafort.* Therefore, it pulls the *Aquafort* down with itself and keeps it with itself as earth, for every dry earth is eager to absorb salt. And because these two salts, that is, the salt of *Aquafort* and that of the *sal ammoniac*, are quite subtle and volatile, they are easily excited and ignited by the least movement or heat. When they feel such warmth, they detonate or explode everything below them just as gunpowder usually explodes everything above it. This is the <u>true cause</u> of the *Fulmen* and not the *Sulphur Solis* (the sulphur of the gold). It is the volatile saltpeter and *Sal Ammoniac* both strongly interacting subjects.

The cause of the gold's exploding below, however, is in the gold itself, which is a fixed Earth, therefore having a downward inclination; while on the contrary, coal (charcoal) in gunpowder is a volatile Earth pushing upwards.

Now we also see a difference between this gold fulminate and common gunpowder in as much as this gold fulminate explodes three times as powerfully as gunpowder. The reason is that gunpowder contains a corporeal, coarse, raw saltpeter, whereas that in gold fulminate is quite spiritual, volatile and a very delicate one. The more subtle, volatile and spiritual such Reagentia are, the more violently they explode.

Gold fulminate that has been precipitated with oil of tartar (*oleum tartari*), explodes so much more powerfully than gunpowder. If we take instead of a fixed alkali, such as oil of tartar, a volatile one, such as the volatile salt of urine or ammonium carbonate, and precipitate the ☉ with it, it will detonate all the more violently. By this, the lover of the Art will see that the *Fulmen* stems from the volatile salts actually, and not from the gold. The reader will also see that if this gold were to be kept wet, it will not explode or detonate. In fact, even though it were to be standing in *Aqua Regis* for many years, it would still not explode. However, as soon as it becomes dry and some warmth is applied, it begins then to explode. Likewise gunpowder: When it is wet and damp, it will not ignite; whereas, if it is dry, it shows its effect immediately. In contradistinction, however, when this gold fulminate is dried and boiled with a fixed Alkali and water, such as *Oleum tartari* or potash, or other Alkalis or *olea salis*, it loses its fulmination at once, because the fixed *Oleum salis tartari* dissolves the volatile *Reagentia* that adhere to the gold and turns them into a third factor by means of their dissolution, and binds the reaction by its fixity, so that there can be no more explosion.

From this we can now conclude that this crackling and thundering effect in general arises from a nitric volatile substance and a delicate volatile Alkali or such a volatile earth as the sulphur of coal. The more volatile they are, the stronger they explode; the more fixed, however, the less they explode.

If now ⚇, or coal dust, arsenic, auripigment, or sulphur is added to a flowing saltpeter, one can immediately see how they drive each other out and cause a violent reaction, according to which the

Reagens is also volatile or fixed. That is to say, the degree of violence of the explosion is quite dependent on whether the Reagens is volatile or if it is fixed and to what extent they are so fixed or not fixed.

If, on the contrary, common fixed table salt or salt of tartar, another fixed Alkali or some fixed earth, such as *Terra Sigillata*, chalk or lime - which do not contain anything volatile - is added to the flowing saltpeter, one can see that they do not interact violently but unite in a friendly manner, fix each other without any change in temperature, and do not fulminate. By this we hope to have amply proven theoretically and practically that there is saltpeter and salt in the air, although volatile, and that the *fulmen* is evidence of the presence of both, which, as said above, is afterwards bodily shown in the putrefaction of rainwater.

Now we will come from the air to earth and water, and also examine their creatures, to see if saltpeter and salt can also be found in them as the generator and destroyer, sustainer and corrupter, and regenerator of all things.

CHAPTER XII

NITRUM AND SALT CAN BE FOUND IN EVERY WATER AND EARTH

That *Nitrum* and salt can be obtained from rain, snow, hoar frost, etc., is proven by the above test. But that they are also in every kind of earth and water, must be looked for in the same manner. For if we dissolve the soil - no matter which - that is on the surface of the earth, in fields, meadows, bogs, mountains and in valleys, in lime soil and red clays, filter and evaporate it to one-third, then let it shoot crystals, proceeding in everything as with rainwater, saltpeter and salt will be found, much or little according to whether the earth is strongly saturated or not. This does not require special proof. All you have to do is ask the saltpeter makers. They will tell you plenty about it, being those who know it best.

Likewise with all kinds of water and springs. For how many wells do we find that are wholly saline and nitrous? Rivers, however, show it very clearly, for they do indeed flow through the earth, dissolve the saltpeter and salt in them and carry it along into the sea through all countries.

But the cause of the sea's containing more salt than saltpeter is that it is constantly irradiated, reverberated and tossed about by the winds, so that it is ever in motion, whereby the saltpeter is reverberated. And due to such ceaseless reverberation and motion it loses the fulmen and turns into an Alkali. For if saltpeter with its unlixiviated earth is often boiled dry, and this somewhat strongly, afterwards again wetted, again boiled as before, one will find that it congeals more the longer it is boiled, etc., until it finally becomes quite fixed and alkaline. Then it never again fulminates, because salt is nothing but a reverberated and fixed saltpeter. This fixation is done faster in the dry way with quicklime or other earths, when most of the saltpeter is retained, as otherwise it detonates (fizzles out) with the coal dust and volatilizes very much and very strongly in the reaction of the opposing reagents. It congeals still faster during casting if an equal amount of common salt or another fixed Alkali is added to it. Then it gets fixed at once. If after this one lets it flow, adding sulphur or coal dust, it no longer fulminates but partly attracts the sulphur and coal to itself, rendering them stable together with itself.

CHAPTER XIII

SALTPETER AND SALT ARE TO BE FOUND IN THE CREATURES OF THE ANIMAL KINGDOM AND THEY ARE MADE OF THESE TWO AND ARE AGAIN RESOLVED INTO THEM.

Anything meant to fertilize must consist of saltpeter and salt otherwise it is not much of a manure for farmers. That all animals are saltpetric and salty is known to all right-minded chemical analysts, because in animals' anatomy we often find both a volatile and also a fixed salt and an evil-smelling inflammable oiliness.

That ⊖ Volatile is a volatile salt is indicated by its name. Fixed salt shows itself in ash. That oil is a liquid niter is proven by its ignition, because it burns, and nothing else burns except saltpeter and its addition.[99] For the fixed salt, the fixed earth, certainly does not burn. A better proof is furnished by Phosphorus made from the animal kingdom.

That the animal kingdom contains saltpeter is shown by the mechanics themselves, the saltpeter makers, who dig up the living and bedrooms of farmers, where their children urinate incessantly. That seeps into the soil and is there transformed into an excellent saltpeter.

Whoever does still not believe me, let him go to a cemetery where many people are buried. Let him take some earth from a grave, well decayed, let him lixiviate it and then investigate if the animal kingdom is not nitrous. Then he will also discover

---

[99] It is ♃, being its *Contrarium.*

that in this kingdom, a thing is reduced again to the same thing out of which it arose.

It is not cow and sheep's dung so strongly nitrous that the ⊕-makers have selected them from among all others? And if ⊕ were not an excellent spermatic food for man, God in the Old Testament would not have commanded the Jews to eat mutton and keep sheep-farms.

The farmer puts cow and sheep's dung on his field as the best manure. Even if he does not know that saltpeter promotes growth, he learns that such dung fertilizes best. He also collects their urine and waters his meadows with it, after which the grass grows very beautifully. Does he not also put human dung on his field, to render it fertile thereby? Out of that, then, out of dirt and muck grows our grain for bread and nourishment for our sustenance, food and multiplication. And as we are reflecting upon our origin, we must admit that we were not only born between muck and dirt but also out of muck and dirt, by which we are sustained, nourished and multiplied, and into which we are again dissolved - as Christ says, into dust and ashes - so that we in turn can manure and fertilize the fields, meadows and vineyards of our succeeding Adam's brothers by our dead bodies and rotten carcasses and in that way become food and drink for them.

Isn't it true that, owing to the alternation of times, many a man does not know how much he has consumed of the dead body of his grandfather and great-grandfather, father or brother, also his child, which lies perhaps buried and decomposed in his own fields and vineyards, whose corpse's juice

140

has made his wheat and his wines heavy and juicy, let alone how much he has consumed of dead cattle, also of the enemies who had invaded his homeland and died there from various diseases or had been killed, had rotted and decayed in his fields, dissolved into juice and salt. How much, I say, he has consumed thereof.[100]

By the above considerations it has been abundantly proven, and it is not necessary for any philosopher to lose many words about it, that animal creatures are not only born of saltpeter and salt and consist of them, but that they are also dissolved again into them by the Archeus or universal Life Spirit of Nature, as will be confirmed in this tractate.

---

[100] Our present bodies may well consist of the rust from the sword of a Viking, or the DNA/RNA cells of a Pharaoh or the atoms that once composed the toe of an Akkadian hog! By thermodynamic law, matter cannot be created nor destroyed, only turned into energy. - HWN

CHAPTER XIV

NITRUM AND SALT ARE TO BE FOUND IN PLANTS, AND THEY ARE MADE OF THESE TWO AND ARE AGAIN DISSOLVED INTO THEM.

That herbs and plants grow out of dew and rain, water and earth, is known to every farmer and gardener, as the evidence will also subsequently show. For we have proven in the preceding chapters that the sole essential factor, or the essence, of dew and rain etc. is saltpeter and salt. Again, all kinds of water and earth contain those as their essential substance hidden under the earthy and watery cover.

It is indeed known that the *Sperma universale*, that is, dew, rain and snow, and the saltpeter and salt hidden and dissolved in them, are that which gives and promotes growth. As said, however, these two are in all waters and in most kinds of earth. When then these two are contained in water and earth, vegetables must necessarily grow from them, because they do not grow out of mere barren earth, nor do they grow out of empty waters without essence, but out of the universal seed, which is saltpeter (⊕) and ⊖.

In a crucible, melt together two parts of salt and one part of saltpeter. Then dissolve this with ten parts of rainwater. In that, let the vegetable seed swell up. Dry it again in the sun and sow it into some soil. Likewise, take some of the same seed but which has not been soaked in that water, sow it also into the same soil, but not together with the former. Then observe the speed of growth, the beauty of the fruit and the difference in both growths.[101]

That the plants are highly nitrous, though one more than another, may be seen by their burning spirit, their acidity, their oiliness and alkaline salt. One can see how vegetables break into a strong and bright flame when they are ignited. Now, however, the inflammability, the heat and the flame are solely due to saltpeter and nothing else.

Is not the burning spirit of wine (ethyl alcohol, or ethanol) or any other *Spiritus* a very fine, yes, a heavenly Nitrum? It burns as subtly and beautifully as the stars. Oil indeed requires no proof. Farmers use the oil of many kinds of plants, as well as of animals, for their lamps. That they are a Nitrum is proven by their inflammability.

Farmers know it better than State philosophers when they gather many leaves and grass in the woods, make big heaps of them, let them rot and decay together and afterwards take the stuff to their fields to fertilize them. Enough has been said above in connection with animals about what such manure consists of.

Gardeners know very well what it means, and they are very pleased when they can get soil humus derived from a decomposing tree. They consider it too rich for common garden growths and use it to manure fine kinds of flowers and aromatic herbs, because they know that Nature makes it very subtle and has turned it into dust, mould and earth, which mould provides a very fine ⊕ and ⊖ when it is lixiviated.

---

[101] This experiment is reliable and is again recommended to each and all hardworking farmers and gardeners. But it must be done by taking into consideration that both salts must flow for a rather long time in strong heat, so that only their fixed Alkali is left.

143

By such mould and decomposition of trees we can also see that vegetables not only grow out of saltpeter and salt but also return to them and are transformed back into them as their origin, out of which in turn other products of the plant kingdom grow according to Nature. I also hope to have done justice to this kingdom and to have conferred honor upon saltpeter and salt as its origin and its direct general matter - although not yet specified or pertaining to this or that realm - which two give themselves together ⊕ to all things and generate one thing after another in accordance with the will of Nature.

CHAPTER XV

SALTPETER AND SALT ARE TO BE FOUND IN ALL CREATURES OF THE SUBTERRANEAN KINGDOM, AND THEY ARE MADE OF THEM AND AGAIN DISSOLVED INTO THEM.

The more heaven approaches earth, the more earthy and corporeal it becomes; and the more earthy it becomes, the more it becomes fixed; the more fixed it becomes, the less it burns and shines and ignites. Thus saltpeter, descended from heaven, is quite volatile and hidden in water. In putrefaction, however, it becomes manifest. The more earthy and fixed it becomes, the more it turns alkaline, and through that fixation it increasingly loses its *Fulmen*, as may be seen here with minerals. For the more it gets away from its universal nature, the more its nature is changed. Thus it takes on another nature and quality when it is specified in the animal, another in the vegetable, and yet another in the mineral kingdom. Nevertheless, it proves its fiery domination in all of them, to a greater or smaller extent, according to whether it is fixed or volatile in its grade in the animal and vegetable realms; its coarse or fine oiliness, resin, pitch, Resina, etc. in the mineral and sulphuric things, such as *Sulphur, Naphtha, Petroleum*, etc.

But since some *Mineralia* are of a stony kind, descending increasingly towards fixity, the inflammable sulphur is robbed of its inflammability by this fixation, and it acquires another grade, namely it becomes incombustible. That sulphur and similar inflammable things are saltpetric (nitrous), however, we have proven above by showing that all inflammations stem from saltpeter and its addition.

That salt, as salt, can be found in the subterranean creatures may be seen by washing the minerals with water, after they have previously been somewhat annealed. But that salt is no longer found in such quantity in the form of salt, is due to the fact that the longer it stays in the earth, the more earth it absorbs. The more earth that is absorbed, the earthier it becomes, relinquishing its salt form the longer this process continues.

For the axiom cannot be reversed, and practice shows every chemist almost daily *ex quo aliquid fit, in illur resolvitur, et per quod aliquid fit, per illum ipsum resolvitur*: Into that out of which a thing was made, it will be dissolved; and by that out of which a thing was made, it will again be dissolved. Now we do indeed see that if we are to separate the strongly interlocked minerals, we must do it with salt or salty and nitrous solvents, without which they will not open. That every *Menstruum* is salty or saltpetric, is known to every chemist. From this everyone can again conclude that, since minerals melt or dissolve in salt or salty *Menstruum*, they must be consubstantial (of the same substance) with salt, or they would not be conquered by it. Nor would the *Mineralia* melt into a *Liquor* in a salty *Menstruum* if they were not made of water or salt water and were again dissolved into it. If the remaining wateriness is drawn off to one-third, it is true that any chemist can turn it into some salt or vitriol which, by repeated distillations, can be driven over the helm. Out of this minerals were born by various preceding transformations.[102]

---

[102] See the *Der grosse Bauer* (literally The Big Farmer -PNW), p. 27, of the Augsburg (a city in the south-west of Bavaria in Germany) edition of 1753.

To summarize, minerals are born of fermented, putrefied salt and niter turned acid, which dissolves earth into itself and thereby becomes vitriolic and sulphuric, but is subsequently increasingly fixed by degrees. And just as they are born of a spiritual niter and an acidified salt, they are reconverted into their first essence by such acidified niter, as will be further related in the history of the birth of minerals progressed to their final stage.

But so as to serve the reader somewhat better, although it does not pertain to the genealogical register of minerals, we will nevertheless elaborate here a little more, so that, if it pleases him, he does not regret reading or having read.

And in order to confirm it by prior and subsequent examples, we will prove by the origin of these kingdoms and their birth that all things were generated and born of saltpeter and salt, or of a salty seed.

That animals are born of a watery ⊖-like seed and are sustained by watery ⊖-like growths and nitrous air, every philosopher clearly knows. When then they resist and decay, they turn back and dissolve into water and slime, mucus and an all watery, salty, nitrous matter and substance. That it is indeed nitrous and salty, we have already proven before and will further prove below.

So, vegetables are born of nitrous and salty rain, dew, etc., also of nitrous earth and water, and are again dissolved by fire and turned back into ▽ full of ⊕ and ⊖, their primary matter.

Likewise, minerals are born of nitrous and salty water which runs to the center of the earth through its cracks and fissures, because that salty water is strongly heated by the central heat and driven up to the circumference of the earth in the form of steam as a pure spirit. These vapors adhere to rocks because of the rebounding cold of the mountains, and condense into water. This water, however, is erosive and corrosive because it contains a spiritual salt, namely, saltpeter and muriatic acid. For if it were not corrosive, how could it attack and dissolve rocks?[103] Therefore, this water dissolves rocks and soils. Soils, however, are again coagulated by it into salt, not as before, but into a vitriolic ⊖, as much as it can contain at once.

And that which water cannot dissolve thus, it crushes into a subtle, greasy, fatty earth, generally called GUR. This Gur is ever further dissolved by the successive corrosive vapors till it is so full of corrosive that it changes into sulphur. For the more corrosive it acquires, the more sulfuric it becomes. But such a sulphur loses its combustibility owing to the lapse of time and the central heat, and it changes into arsenic. Arsenic, however, turns into marcasite, and the latter is then only the next substance in line for becoming metal, and not vitriol. That sulphur is a pure corrosive can first be noticed by its smell: More than *Aquafort, Aqua Regis, Spiritus* or *oleum vitrioli* does it infect lungs, so that a person can

---

[103] In what sense this corrosive is normally taken in our schools is shown in the *Compass der Weisen,* Part 3, Sect. I, S/5, p. 324 ff. For just as base metals are generated in the subterranean crevices, our philosophic metals are generated and produced in our artificial pits, and whoever does not imitate this process of Nature in the works of the Art, will not reach the desired goal.

hardly breathe. In addition, by its *oleum*, which is commonly distilled both under the bell from the sulphur and from common brimstone. Thirdly, because it calcines, corrodes and burns stones and bones as effectively as liquid corrosives.

One can see that *Oleum* and *Spiritus vitrioli* are dissolved sulphur by imbibing some earth with it, such as chalk or some other fixed earth, letting it smoke in an open fire, strongly, that is, and watch how it ignites and burns like sulphur. But that sulphur had been saltpeter, I have related before, tracing its origin.[104]

That they dissolve into a fermented saltpeter turned acid, or vitriol, and this in turn into the original essence, I have just been teaching. This will also be further explained below in its own chapter.

From this the reader may decide whether I understand the origin of things rightly or not. Let him go forward or backward in the analysis of minerals, and he will indeed see what he did not believe before. But if he were to believe that I wish to teach the world something else than our ancestors and invalidate them who wrote so many thousand years ago that Mercurius, Sulphur and Sal are the primal matter of all metals, I would reply to him that I do not wish to do so. That they state, however, that Mercurius, Sulphur and Sal are the original matter of metals, is well known to the modern world, but it is best known to the basic philosophers that they are to be understood as such. Whoever, nevertheless, does not follow or believe me

---

[104] Here he says something which is of far-reaching usefulness in one of the highest levels of this art.

that I only desire to proceed according to the direction of Nature, let him follow others and draw from them a better foundation. There will yet come some who will be glad that I have come to the fore and have become known to the world.

That Sulphur and Mercurius are born of saltpeter and salt, is clear indeed. The more earth absorbs saltpeter or a corrosive - an acidum - the more it becomes sulphuric. If it is rendered alkaline or salty, however, or gets into an alkaline-salty spot, which kills the corrosive or the sulphur, it turns into a Mercurius or mercurial progeny. For the time being, we have said enough of the first essence and origin of minerals, that they consist of saltpeter and salt and can again be transformed back into those. If this chapter permitted, there would be a great inducement to present the proof of it both mechanically and theoretically, but it is better to save it till later.

From what has been said so far, it is as clear as the sun that saltpeter and salt are the seed of the whole macrocosm and are volatile and fixed, depending on how they are applied. They both are father and mother, active and passive, the steel and magnet of each and all things. The visible Elements, however, Air, Water and Earth, are the casing or the dwelling place of these two essences and the matrix out of which and by which they produce and give birth to everything.

Therefore, the reader can easily draw his conclusions about all generated, corrupted and regenerated things, because he has recognized that everything is generated, sustained, destroyed and regenerated in a volatile or fixed manner, as Nature herself does.

For of the volatile ⊕ and ⊖ an animal is born instead of a mineral; of the half-fixed and half-volatile ⊕ and ⊖ a vegetable is born; and of the fixed ⊕ and ⊖, a mineral.

Therefore, after the above shown general principles, we easily derive the particular ones. For if anyone knows the origin, he also recognizes the progression and the purpose, that is, the beginning, the middle and the end.

From this we conclude that the origin of all things is the universal or watery vapor which regenerates and transforms itself into the universal chaotic water, that is, dew, rain, etc. (our primordial reborn matter). For all water turns into steam, fog and vapor through heat and fire, and all steam and vapor reverse back into water by their condensation. In this and in all waters saltpeter and salt are contained. The more delicate, volatile and spiritual the water is, the more volatile niter and salt it contains, the subtler fruits it forges. The thicker the water, the more corporeal and fixed both salts are therein, the more fixed the fruit they produce.

From these two, ⊕ and ⊖ being the first matter of all sub-lunar creatures - be they volatile or fixed - all sub-lunar creatures are born, sustained, destroyed and reborn in the animal, vegetable and mineral realms, although animals, simultaneously with vegetables and minerals and more so than they, draw the volatile ⊕ and ⊖ from the air by their breath, sustaining and nourishing themselves by them as by a special heavenly food.

Vegetables, however, increase more by dew and rain than by condensed air.

Minerals, on the other hand, are born of a thick and acidic steam and subterranean air which, due to the heat of the abyss, sublimate upwards out of the central heat into the bowels of the mountains and there become water. In all of these minerals, air and water, saltpeter and salt lie hidden as a seed

Just as the aforementioned are born and sustained by ⊕ composed of both ⊕ and ⊖, according to the difference in their volatility and fixity, they are also destroyed again by the same, according to this difference, (volatility and fixity) and also reborn, until the Creator burns everything into dust and ashes.

Now the reader will be afforded the most beautiful theory or reflection, if he considers how Nature first descends out of such an extremely volatile steam through the pertinent most beautiful levels. The more she descends, the closer she gets to those levels and through them acquires increasing fixity. For she turns the most volatile into the volatile, this into the half-volatile, this in turn into the fixed, and the fixed into the most fixed. When now she has descended from level to level, she again ascends from level to level, turning the most fixed into the fixed, this into the half-fixed, then the volatile, and this into the most volatile. As stated above, she turns heaven into air, water and earth, and of earth she makes water, air and heaven, from level to level, from one extreme to another.

She turns the most volatile heaven into the volatile air, this into the half-fixed water, and this into the fixed and most-fixed earth. Or, the most volatile heavenly niter into the volatile airy niter, this into the half-fixed corporeal and tangible watery niter, this into the earthy salt or alkali; this, in descending further, into earth, stone and mineral.

That salt is a fixed niter has been sufficiently proven above, also how and in what way it becomes alkaline and fixed. And in this the disciple of the Art can see a general description of the universal generation of natural things.

Now we will tackle the particular, which is occasionally demanded by the artists, that is, an analysis of things by which we penetrate into the center of Nature and look at her naked.

We are therefore beginning, as is right, with the main door to Nature, with the Key and originator of all generation, destruction and regeneration of everything, without which we could otherwise hardly get to the ground of Nature, and this Key and main point of alchemy is putrefaction.

CHAPTER XVI

THE MAJOR GATE AND THE KEY TO NATURE, ORIGINATOR OF EVERY GENERATION AND DISSOLUTION OF NATURAL THINGS: PUTREFACTION

Because of its tenderness and subtle purity, Heaven is not as changeable as the other Elements. But if he proceeds to the Air and from there to the Water and Earth, he putrefies together with the others, in order also to generate his like in the lower Elements which do not generate or destroy anything without putrefaction, by special disposition of God.

Therefore, without putrefaction or prior maceration, digestion or fermentation (be it quick or slow), no true dismemberment (separation) either in Universalibus (universal) or in Specificis and Individuis (specific and individual) can be hoped for.

For dew, rain, snow, hail and hoar frost putrefy without distinction, producing a separation of the fine from the coarse, and a sign of it is their giving off a smell, although only a faint one.

Like the above, animals putrefy very easily, and because of their many volatile parts or their excess of very volatile saltpeter, they stink intolerably.

Vegetables also putrefy easily because of their excessive moisture, but not as quickly as animals. Neither do they smell as badly as the above mentioned.

Minerals also putrefy and ferment but, at least most of them, do not give off such a bad smell as the preceding, except when iron is being macerated and acquires that which is consubstantial with it.

From putrefaction, then, we derive this advantage and transformation that minerals turn into vegetables, vegetables into animals, and these turn back into vegetables and minerals. Thus, Nature goes around in a ring or circle, turning the upside down and the downside up. She also turns the three realms into a general nature, not particularly belonging to any realm. As we said above, she drives the vapors from the center of the earth and water sphere, which is the realm of minerals, and vapors from the surface of the earth, which is the plant kingdom, and vapors from exhaling and decaying animals - the three living and flourishing kingdoms - into the air. There she renders them chaotic and turns them into a universality which is now neither animal nor vegetable nor mineral but common to all things, and which must be, and is, all in all.

Few budding philosophers will believe this before an explanation, even fewer of the common variety of laboratory workers. But after hearing the explanation, every farmer will see it with his own eyes before his door, even before he goes outside, as we have also spoken about.

That is why putrefaction is that wonderful blacksmith who turns earth into water, water into air, and air into fire or heaven, and again turns heaven into air, this into water, and water into earth. Putrefaction effects such transformations ceaselessly, without stopping even one minute till heaven and earth melt together into a glassy lump.

CHAPTER XVII
WHAT PUTREFACTION REALLY IS

When God created the universal steam, he implanted an active being into it, which we call Spirit. From the beginning, this Spirit has been a restless being, never standing still, at all times ceaselessly moving, acting and working without stopping. Be he fixed or volatile, he must forever be busy, bringing about one change after another in all creatures. For even when he ceases to be in one, or escapes from dead bodies, at that same moment he begins again in another, so that he does not rest one instant.

This Spirit is the mover and originator of all change, and he starts every change with putrefaction. When he has hatched this for some time, he separates the pure from the impure. After that, he binds, coagulates and congeals till the goal is attained with every individual. In the same way, he begins once more to putrefy, dissolve, and separate the coagulated body after he has attained his goal, till he has again made something else of it. This Spirit is the procreator, sustainer, destroyer and regenerator of all things in the world.

We cannot see this Spirit in his primordial essence, his steam-form, because he is then completely hidden in the steam or the water and, in addition, so spiritual that he goes up into the air in low heat. But when he descends from above into our coarser bodily elements, he is partly retained and must willy-nilly become a visible, tangible body, or rather, assume such a one. After this, he appears to us in a white, ice-cold, crystalline, transparent form, ⏀, and yet so fiery within that, if he got heated and a large amount were accumulated in the center of the earth and his enemy came toward him, he would become so enraged that he would not only blow up rocks, stones, houses and buildings, but the whole world, just as he often shows us a proof of his power in earthquakes. And if his

brother or cold wife did not exist, ⊖, his Venus, with whom he falls madly in love and on whom he becomes hooked - who alone can check and tame him - he would already have harmed the whole world long ago. But when the two clasp each other in the fiery hellish palace, his wife does not permit it. She embraces him and grabs his middle, so that he should cool and extinguish his anger in love and not cause damage through it elsewhere. And as soon as he is embraced by his Venus and is enchained in the bonds of love, he forgets himself so completely that even if his enemies approach him, he not only does not harm them but attracts them to his love, socializes with them, and as it were, forms an eternal alliance with them.

This Spirit is distributed in and through all creatures, as has been said above, so that none can live, float or exist without him. He it is that introduces all generation, destruction and regeneration into all creatures.

Therefore, putrefaction is the first key and door by means of which this double Spirit opens the palace of Nature to us and again locks it in subsequent grades.

For this Spirit is moving because he never rests, as has been said above, and due to his motion there arises a warming quality. This warmth then opens the pores of everything, so that the implanted Spirit can pass and penetrate through everywhere in order to generate or corrupt. As soon as he has permeated the members, he begins either to dissolve or to coagulate; and he continues doing this till he has completely penetrated and heated the body. Then the tender, moist, volatile parts, or the *Volatile*, begin to steam (according to whether the heat is strong or weak) and give off an odor by which one can know that the Spirit is at work and that he opens, decays, and softens the body through digestion or putrefaction, and how he continues from level to level till he has reached his intended goals.

Originally, this Spirit had been steam and vapor, and just as he himself had been water and steam in the beginning, he also makes all things out of steam and water, through and with steam and water, and without water he does nothing, because he needs water for his action, mixture and dissolution, as all things he makes are easily mixed with water.

For he makes the creatures of the animal kingdom of water, and they obviously consist of almost all soft and watery parts, and after the evaporation and extinction of the lamp of life, he turns them again into mud, slime and water through water.

In the same way, the creatures of the vegetable kingdom consist of watery, juicy and moist parts, although not as much as animals, and they are again dissolved and turned back into water with and through water.[105]

---

[105] This is also the reason why they do not decompose as fast as animals, since a moist *Vehiculum* must be used with them, especially with the dry ones, before putrefaction sets in,

Thus Nature, or this Spirit, produces minerals from water and dissolves them again into water by water, as will be explained below by several examples.

But it is not to be understood that such a water, or creatures transformed into water and coagulated out of water, is a water without power, or plain spring water, out of which the Spirit begets all Animalia, Vegetabilia and Mineralia, but a kind of water in which all four Elements are in concordance, in which there are four parts, Heaven or △, Air, ▽, and Earth, in which there are three: Corpus, Spiritus, Anima: ⊖, ☿, ⚴, *Acidum, Alkali, Volatile*; in which two are man and wife, *Agens* and *Patiens* (active and passive), *Nitrum* and *Sal* (niter and salt); out of which everything is born, destroyed and reborn; a ▽ in which Spirit is solely active and operative. Although he is called double, triple, quadruple and quintuple according to his fixity or volatility, he is nevertheless but one single Spirit and only different because of his differing effect. For when he is volatile and steam, he is called Heaven and Air, *Volatile, Agens*, the man, *Anima*. When he is half-fixed and corporeal, he is called water, *Acidum, Spiritus, Sulphur, Nitrum*; but when he is fixed, he is called Earth, *Fixum, Patiens, Alkali*, woman, magnet, *Corpus, Sal*, as said above. And this is the whole knowledge of all things, for in whatever form and shape we meet something, we soon afterwards give it a name to

---

because the latter cannot be brought about or happen without moisture and warmth. What this water is, however, out of which and by which the aforementioned generation in all three realms occurs, is very well explained by him in the following paragraphs.

distinguish it from other objects. For if there were one single name for everything, we would mistake one thing for another, as in the confusion of Babylon.

For it had originally been just one single simple water that divided in time, and in this division every part received its special name, although they all stem from one single root, one ground, one single thing, just as all separate things in the whole world can again be turned back into the single first thing of all things through reversal and dissolution, that is water.

Thus is explained what water is, namely, the implanted, moving, warming, heating, inflaming single and simple Spirit, in double and simple form, then again a fighting *Acidum* and *Alkali*, which are two in one essence, also three: *Volatile, Acidum, Alkali, Mercurius, Sulphur, Sal, Spiritus, Anima, Corpus*.

CHAPTER XVIII
WHAT THE RESULTS OF PUTREFACTION ARE AND WHAT IS ACCOMPLISHED BY IT.

In general, putrefaction results in turning a volatile into an acid, this into an Alkali, and in return, an Alkali into an acid. From this a volatile is produced, according to how the things to be transformed are constituted either naturally or artificially.

To present the true effect of putrefaction, we will take as an example rainwater, the universal regenerated chaotic water, namely: Collect rainwater, as has been said above[106] in a clean vessel, as much as you wish, the more the better, so that the generation of the universal Spirit can be observed more clearly. Let it stand covered for 14 ♂♀ or one month. As has been said above, it will begin to putrefy and break within itself; also to smell noticeably foul, so that a true separation is noticed. Then one can see a muddy, spongy, floating earth and impure water in the previously clear, pure, crystalline-transparent water, thus it is obvious that a transfor-mation has occurred.

Now, then, the cause of this breaking of the water and the putrefaction impurity is the implanted Spirit which brings some intangible warmth into the water by his continuous motion; and the longer he works, the more it gets heated and the greater the separation becomes. Then, day after day, one will find that the longer it lasts, the more impurity or earth there is, in addition to the smell or stench of putrefaction.

---

[106] See Part I Chapter X.

Now we will examine these rotten, watery bodies and inspect their parts.

We said above that water is quite volatile before putrefaction and is a pure Volatile liquid which can be drawn over completely by distillation. After putrefaction, however, it divides into three essential parts: (1) into a volatile water,[107] (2) an *Acidum*, (3) a *Nitrum* and an alkaline salt, which still leaves some earth behind after their separation, as has been reported earlier in the pertinent place. Chemists call it *Feces*.

That this decomposed water contains a spirit or active being can easily be seen and inferred. For from where would a separation or transformation come if the ▽ did not contain something active to cause it? This active factor or originator, however, we call by the name known to everyone: Spirit.

That such a Spirit is in the ▽ and warms the ▽, although imperceptibly and intangibly, can be noticed by the putrefactive or foul stench. For one never hears, or does easily hear, feel or see, that the cold causes a thing to rot, or arouses a stench. Even if in winter the whole world were covered and paved with nothing but dead apples, we could not smell anything of it. But let the weather become warm, and in one day they will begin to rot and smell foul, so strongly that no one could take it.

By this one can clearly see that it is not the cold but the heat that causes a stench and

---

[107] What he here calls volatile water is nothing but the etheric mercurial Spirit, which is easily recognized by its small droplets, the speed with which it rises over, and the fact that there is no wateriness in the head.

putrefaction, and it drives the smell, be it fragrant or foul, tangibly through the pores of the body. Consequently, putrefaction stems from the warming Spirit, that is, from warmth and not from the cold.

The stench, therefore, as well as the lovely fragrance, is caused by the warming of the volatile parts, and these are a volatile being which exhales, rises and flies off from the Spirit of the warmed rainwater, its most volatile part, in a way perceptible by the nose. This may clearly be seen in putrefied urine and its stench during distillation, when its volatile salt rises first, which has the sharpest, foulest, most penetrating stench or smell. It stinks and smells worse than its subsequent more fixed spirit and oil. Coal, or the *Caput mortuum* (dead head), burnt to coal and its Alkali have almost no smell.

This we can see with wine, especially with old wines. The longer they lie to ripen in cold cellars, the more fragrant they become. When they are distilled, the volatile strong-smelling burning ethyl alcohol (spirit of wine) goes over first and surpasses all the following parts in smell.

This we can see also with minerals. When their ores are put in the fire, immediately the volatile sulphur and the *Acidum* hit our nose, and the arsenical spirit makes our head confused. The remaining parts have practically no more smell, except for the volatile that has been congealed into them through the fire.

The *Acidum* and *Nitrum* have little or almost no smell at all, just like salt or alkali when it is separated from the putrefied rainwater, except if they were aroused again by their *contraria*.

But that this smell stems from the motion and the moving Spirit has been sufficiently proven, and that the motion causes warmth and heat, depending on whether it has a weak or a strong effect, is evident from the following.

The blacksmith sees with his own eyes and grasps with his hands, when he vigorously forges for a while, a cold iron upon a cold anvil, with his cold hammer, he sees how the iron becomes red hot by this movement.

This is what the knife grinders experience when they grind an iron on a dry grinding-stone, without water and with a rapid rotation of the grinding-stone, and it becomes so red-hot that wood can be kindled with it.

Take just a few cold stones and strike them together often, or rub one over the other and see if they do not get warm by such a movement. In addition, rub two pieces of wood together, and you will have the same experience.[108]

But how the volatile turned into an acid, and this into an alkali, then again an alkali into an acid, and this into a volatile; or how Heaven turns into Air, and Air into Water, and this into Earth, has already been reported above. Here now we will transform these three termini (limits) into each other and examine how such a transformation happens.

---

[108] This was the primitive way in which the first men kindled a fire, until in later times man discovered how to unleash it out of flint by means of steel, and finally, how to do it through concentration of the rays of the sun in a focal point.

CHAPTER XIX

HOW THE VOLATILE TURNS INTO AN ACID, THE ACID INTO AN ALKALI, AND THE REVERSE: HOW THE ALKALI TURNS INTO AN ACID AND THIS INTO A VOLATILE AGAIN.

In this chapter there follows a curious point which all artists must well take note of if they wish to progress in the Art. For in this single point thousands make mistakes, because they do not understand it in their dissolutions and coagulations, volatilizations and fixations.

One thing is certain: the whole world with all its *universalibus, specifices*, and *individuis* is so organized that one cannot be without the other, because one must be the other's leader, one must be the other's medium and link, else no union occurs and no separation.

For, as said above, the elements cannot be one without the other, because one must unite with the other and by means of the other.

Thus, animals cannot be without vegetables, and vegetables cannot exist without minerals; minerals on the other hand cannot be used either without vegetables and animals.

But just as I have sufficiently said above, that no extreme can be united with another without a medium, this statement must also here be well heeded.

For Heaven cannot become earthly without the medium of Air and Water; likewise, Earth cannot become heavenly <u>without</u> them.

The creatures of the animal kingdom cannot become mineral without the vegetable kingdom, and those of the stone kingdom cannot become animal without vegetables. Because the vegetable is the medium between animals and minerals.

Now then, just as these *universalia* and *specifica* must have their media for their union, so every individual in every kingdom must have its medium for bringing its parts together, for sustaining and conserving them.

Such a medium, however, is generally called by the newly introduced term the *Acidum* or acid, which is *universaliter, specifice* and *individualiter*, in all natural products of the world and its entire spread-out space a medium between the volatile and the alkali, between the upper and the lower (the superior and the inferior), without which the upper will not unite with the lower.[109]

For the *Volatile*, as the upper, is highly volatile, and the *Alkali*, as the lower, is highly fixed. The *Volatile* will never unite directly with the fixed, neither will the volatile with the Alkali without the *Acidum*. The *Acidum*, however, is the medium and intermediary, the Sequester, Copulator and uniter of all things: because it is neither too volatile nor too fixed, but in the middle. That is why it is a hermaphrodite, carrying on both shoulders. It is the true chymical *Janus*[110] who looks with one eye at the Volatile but with the other at

---

[109] Just as this is the process which Nature observes in all her works, so we must also emulate her in all our labors, as our wise Masters, who are the true philosophers, have so often and repeatedly recommended, and without such observance nothing good can ever be accomplished.

[110] In ancient Roman religion and mythology, *Janus* is the god of beginnings and transitions. Often he is depicted as having two heads, facing opposite directions. *-PNW*

the *Alkali*. And when he gets the Volatile, he unites with it inseparably; but if he gets the *Alkali*, he also joins with it inseparably. In the same way, when all three get together, the union is so great that rather than separating, all three stay steadfast in the fire or else they gradually vaporize, all three at the same time.[111]

But one must here understand that when the volatile, acid and fixed components of a like substance come together (and not of an unlike, although even in this case they can also combine precisely), that they become inseparable.

Rc. *Spiritum Vini, Oleum vitrioli* and *Sal fixum*, urinae together, Pour Spiritum vini (spirit of wine) and *Sal Urinae* together (salt of urine), then add *Oleum vitrioli* by drops. At first, they may well strongly resist and effervesce, but finally they calm down and combine intimately, so that if the wateriness is subsequently drawn off, no spirit of wine can be found any longer, which would congeal upon the *Alkali* with the oil of vitriol. From this, the lover of Art can see the most beautiful union of Nature even in various things of unlike substances, of which each comes from a different kingdom, and consequently is also of a nature and quality opposite to those of the others.

---

[111] Here our ingenuous Homerus has taught and disclosed more in brief words than hardly ten other philosophers have done in all their writings. From my own experience, I must give a testimonial to the blessed man: That he has thereby revealed a secret as essential as important and true. Whoever understands (and how understandably has he not revealed it!) will not encounter any difficulty in preparing, without hesitation, not only the genuine *Radical-Menstrua,* and through them the most excellent *Specifica* in all Kingdoms of Nature, for the infallible cure of the most dangerous and obstinate diseases, but also for all kinds of our so highly praised Quintessences and tinctures. We are therefore recommending to all beginners of true natural science, who are students of the Art, to take great and careful heed to this entire chapter!

To continue in order, however, and not to express ourselves too verbosely, we will explain in what way the volatile turns into an acid, and this then becomes fixed, or an Alkali. That is, how one becomes the other's magnet, for one attracts the other and changes it into itself without any interruption, that is, into whichever one of the two which has the upper hand, and exceeds the other in quantity.

Therefore hear: as soon as rainwater begins to putrefy, or as soon as the implanted spirit ($\triangle$) begins to cause the water to heat and ignite, it begins to separate and become ever more corporeal, for in its present condition it is quite volatile. Just as the volatile always aims at becoming fixed through intermediate stages, it desires, if it is fixed, to become volatile again through precisely these same stages. Therefore, this spirit becomes ever warmer and hotter in its volatility by its ceaseless motion. This heat makes it perceptibly sharp, so that it causes us to notice acidity in its taste, which we call *Acidum* after the common name. The more spirit becomes sharp, however, the more it produces earth, because the *Acidum* causes a precipitation and reveals by it a separation of the earth which has been dissolved, and made fine and subtle in the water and had been in solution in the water. This then the *Acidum* reveals, and the more acid and hotter this spirit becomes, the more it separates the earth.

But to prevent the earth from becoming too abundant, at which the *Acidum* might even eat itself to death and become alkalized, the *Acidum* takes its nourishment from the volatile, which it attracts like a magnet and transforms it into its own nature,

168

turning it into an *Acidum*. The more, however, the acid attracts the volatile, the more it becomes hot and ferments; and the more it ferments, the more it dissolves earth, in which is must subsequently act again. And the more earth it dissolves, the more the *Acidum* is alkalized and congealed. When now the earth. is saturated with the acid and the *Acidum* has dissolved as much earth as it is able to, and has magnetically attracted as much volatile as is necessary for its action, the *Acidum* is no longer as strong as before acting and precipitating but only holds a middle position. It is saturated both with the earth (the alkaline part) and the volatile. It now stands on the scale waiting to find out which side will get the upper hand. With that, it will immediately associate, and help give birth, to its like.[112]

For instance, if the Alkali or the earth gains the upper hand and is stronger and more vigorous, also as regards quantity, than the volatile, because the *Acidum* is in the middle position, the earth or alkali renders the *Acidum* totally alkaline. The *Acidum*, however, because it has been overcome by the earth, attracts the volatile and turns it completely into an Acidum. By having become an *Acidum* at the same time as the earth has gained more and more the upper hand, it makes it also alkaline and earthy, up to the most fixed kind of rock so that the volatile turns totally into an acid and subsequently totally into an alkali, earth and stone. One is the other's

---

[112] If something is to become volatile, several parts of the volatile must always be taken. In general, one reckons three or four parts of the volatile to one part of the fixed or the semi-fixed. But if something is to become <u>fixed</u>, there must be a larger quantity of the fixed, so as to enable it to change it (the matter to be transformed) gradually into its own nature and to make it fixed, as the author is here showing us thoroughly, while at the same time explaining the production of the rocklike bodies in such a clear way that no doubts can possibly remain.

strong and incessant magnet. Whichever therefore, rules and has the upper hand, transforms the others into its whole nature. On the contrary, if the volatile is too strong and there is too little earth, it transforms the Acidum into its nature and makes it volatile. The Acidum, however, turns the alkali into an Acidum, and this Acidum becomes pure volatile through and with the preponderant quantity of the volatile ⌣.[113]

It must be said, once and for all, that the earth, during the time when the acid acts thus and works in the earth, absorbs acidity and changes it into an alkali. In turn, however, the Acidum absorbs the earth, dissolves it, devours itself to death with it, and is thus alkalized and congealed. Its sharpness is thereby sweetened and blunted, so that it will not gain further ground, corrode or dissolve.

While everything acid does not dissolve nor absorb as much earth as it can at one time so that it could suddenly become fully alkali, it nevertheless absorbs enough earth to become corporeal by it, obtaining a visible and tangible form. One can see tangibly in every acid that such an earth is dissolved thereby but not completely. Pour off the part that has been dissolved and evaporate it to one third, then set it to crystallize, and the acid will shoot crystals, which would not happen otherwise should there have been too much earth in it. The remaining earth which has not been dissolved by the Acidum, is to be dried.

---

[113] One can just take an alkali, for instance quicklime, salt of tartar, or lead, in which the alkali has the upper hand, and one will soon become aware of what the author is here saying of earth, although much faster with the former than with the lead because earth still contains various acid parts.

Calcine it, dissolve it in water, and again evaporate it to one third. Thereupon, set it in the air, and it will not crystallize whatsoever, or, at best, very little of that which has remained of the Acidum. The other, however, will settle on the bottom without crystallizing, as a salt, which we call *Alkali*.

This then is the theory! We must mechanically confirm it in our praxis, because a single substantial proof is stronger than a hundred mere postulates or suppositions.

Take therefore a *Volatile* an *Acidum* and a dead earth that does not contain anything, and manipulate them in the manner as follows, and you will learn the truth.

Rc. ♈ which ignites powder - six parts; once distilled wine vinegar - four parts; ▽ or ⌒ ⊕*li* two parts. Pour the vinegar and the aqua fortis together, then pour these two upon Cologne chalk or other earth that does not contain any salt but is totally void, three parts. When you have

poured it over, now pour the ▽ upon it. Put all into an alembic, set it in *Balneo Maria*, add a head and a recipient. Let it stand thus one day and one night, or two days and nights, let it digest and dissolve in the first (or so) degree of heat. Then allow it to cool down, decant the clear liquid off from the earth that is not yet dissolved, very gently, and let the earth stay behind with as little moisture as possible. Dry this earth even further and reverberate it under a muffle (a furnace in which the subject material is isolated from the fuel

171

and all of the products of combustion including gases and flying ash). Then, leach it with distilled rainwater, filter and coagulate it, and you will find some alkaline salt which has congealed into an *Alkali* out of the acidity of the spirit of vitriol and the vinegar. The clear liquid, however, distill in B.M. (Balneum Mariae, a double cooker) to an oily consistency and the volatile will go over, although weakened, because the acid has congealed part of it in itself. Put the "oil" in a cool place and allow it to crystallize and you will obtain a *Nitrum* or nitrous salt, or another kind of saltpeter and vitriol from another acid. Now we will examine these parts, namely the *Volatile*, the *Acidum*, and the earth or *Alkali*. Thus it is proven that the volatile turns into an acid, the acid conversely into an alkali, because the earth is much stronger and more active. Consequently, it has been impossible for the earth to be overcome and totally transformed by the Acidum into its nature. Therefore the earth has the upper hand and overcomes both the above mentioned.

If, on the contrary, one takes a little earth but a lot of *Acidum* and *Volatile* so that the earth is totally dissolved, the earth will turn into an acid through the *Acidum* and *Volatile*. Add now to this acid, a volatile in sufficient quantity and the acid becomes a volatile on account of the excess weight of the volatile added. This is due to the fact that that which exceeds in quantity has the power to transform just this other into its own self or nature.

Therefore, Rc. Chalk - one part; ▽ꟻ - four to six parts; eight to twelve parts of ᐯ̃ ; take care in doing everything as indicated and you will see a proof complementary (reverse of) to that above, namely, that you have changed the earth into an *Acidum*. And if you cohobate the volatile several times, the *Acidum* is transformed into a Volatile. This is the other experiment which a lover of the Art can put into practice and can see and experience with his own eyes that how Nature acts in individual things, she also acts in general. She holds fast to this law ever and until the melting of the earth's crust. By this natural law she makes *Specifica* and *Individua* from general things, according to the proportion of their component parts or prime Essentials, as one or another of their other components dominates or is inferior according to matter and quality. One has too much *Volatile*, another too much *Alkali*, a third too much *Acidum*, in one too much, in another too little, or in some they are balanced. According to the different excess (or deficient) constituents, they assume a quality and constitute or assume a difference from the other. All creatures differ, as said already, according to the excess or deficiency of the volatile, the acid, and the *Alkali*, or according to the greater or lesser volatility and fixity.

But someone might here object and say: Why does he take for a test such very quaint and opposite things and subjects from the plant and mineral kingdoms? Why does he not make the test with rainwater, as he had begun? To this one I reply that I have given him above enough inducement and proof in connection with rainwater. The present test, however, is suggested so that the lover of the Art should get a quick assurance of the promised natural effect, for not everyone wants to put into practice the aforementioned theory on rainwater. Instead, from such a quick test everyone can quickly recognize and grasp tangibly how the volatile turns into an acid and further into an *Alkali*, and vice versa.

In addition, the beginner is not tied to the above-mentioned example. He can use all and sundry subjects of the whole of nature which have the three original substances inherent within them, viz animals and plants. With minerals, however, that test will be somewhat more difficult. But if he turns over the pages of this treatise, he will understand them easily as well.

Let someone take the volatile, its acid and *Alkali*, i.e. earth or ash, of a thing and prepare it in the above-mentioned manner, and he will also get it.

But one would like to say further: To make a quick test, it may be all right to put together several subjects. But why does he add vinegar? Is it not enough that the ⍵ as the *Volatile*, ⍵ as the *Acidum*, and chalk as a fixed earth or alkaline subject be there? Why add vinegar?

Here a very necessary and thorough discourse is not out of place, which will ignite a no mean light for those lovers of the Art who have hitherto been stuck in a maze; to correct so many thousands of mistakes they have made. This is a technique which would save many a man who would know how to apply it correctly in his alchemical work, doing it in roundabout ways, at great expense of money and time. Has it not become customary for every laboratory worker to wash his hands in ⍵ while not one in a hundred of them knows what it is? They say, the must sweeten the corrosives and purify the *Salia*, it must improve and perfect the whole work as the most noble substance.

Yes, they are right; but note well *hic latet nucleus*: "Here lies the dog buried." Tell me why it does this? If everybody looks at his work at the end, he perceives it to be so ineffective and imperfect that he wishes he had never started, although he had indeed wasted so much expensive spirit of wine and yet had not obtained the desired result. The *Spiritus vini* had been so sharp that it ignited powder, and yet it had such a poor effect. What is the reason? It is this, and the cause of the mistake is in the following: *Ab uno extreme ad alterum non datur transitus*, there is no direct passage from one extreme end to another. The spirit of wine is a most volatile substance, the aquafort, the salts, the earth are of a more fixed nature, as well as the oils and Spiritus of them, such as $\oplus$, $\ominus$, $\oplus\!\!\!\!+$, $\bigcirc$ and other salts and minerals, and these are adverse to the *Spiritus vini*, because they stand in opposition to it at the other extreme end. Therefore it will not easily unite with them and become like them without resistance. When they are poured together, one can immediately hear an adverse effect, because the spirit of wine and the aquafort, if they are strong, they at first effervesce together, so that one can hear it, since they are adverse to each other. But when the vinegar is added, being the *Acidum* to which it is entitled and whose nature is appropriate, the Spiritus vini readily unites without the least resistance. Neither does vinegar offer resistance to the aquafort, because one can see by their conjunction that they mix together like water and water, without the least resistance. The vinegar, however, absorbs the *Spiritus vini* and then unites it in the friendliest way with the aquafort, so that they lose all sharpness and rather get a sweetness therefrom.

But here someone might make a further objection and say: That, in the first place, it would be easy to make a volatile, acidum and alkali by overcharging one with the other; if one overloads it with various things and then also takes various subjects of a different kind, such as spirit of wine, and vinegar which is a plant essence; and aquafort which is mineral, as is also chalk, those contrary subjects must necessarily produce a change. First I say, proceed according to the law of Nature, as will be explained differently in its proper place in this treatise. To whomever this way of proceeding had been unknown, let him thank me for having learned through me a means of conjunction, and let him now here and everywhere take not of the most necessary and inevitable, universally known axiom in alchymical work. *Non transiri posse ab uno extreme ad alterum sine medio*, that one cannot pass from one extreme end to the other without an intermediate.

It would here lead too far afield to explain this in a greater detail. An artist, however, benefits more from physical reasons and authentic practices than by being presented with countless empty fancies and hieroglyphic ambiguities, for from a single once understood true reason and praxis, he can draw a thousand other conclusions which he had previously never observed. With such a light he can light for himself many other mental lights, whereby he is led ever nearer to the goal of truth.

Many books of processes are written, also other theories, but none or very few have been published that indicate the reason for adding this or that to this or that process, which causes fixation and volatilization, which is the coagulating or dissolving factor, and why and by what power it is done. When then an innocent apprentice stumbles over such processes, he follows them blindly until he becomes aware of his big mistake at the end. Here he is: At this moment he cannot help himself and he is at his wit's end, because he cannot fathom the reason why. But if the scholarly world would elucidate its books with true primal principles, also true and according to the praxis, no matter how bad the examples might be, we would in a short time with amazement experience a miraculous birth of our youth, who would then not have so much trouble fertilizing other fields in order to bring the Art to its highest potential.[114]

Now to return to the axiom, *non transire posse ab uno extremo ad aliud sine medio*, that one cannot pass from one extreme to the other without an intermediate. Let everyone note that no subject is found in the whole of Nature that would not have its component parts, whether they be hidden or visible, of a universal or individual nature. If ever one should lack a component part, resort to its like *Homogenous*, or the universal subjects which, according to their capability, are specified *Individua*. Consequently, they associate, and can become like all individualities, also transform themselves into them, just as the *Individua* are universals in regard to their origin and can be

---

[114] This is an irrefutable truth, grounded in our conform-to-Nature tenets, which will also cause those who do not know what we know, to recognize that the author has been a relative of our lovers of the Art.

transformed back into such by their extreme or
ultimate dissolution.

If, then, every thing has its primal Essences,
or if it should lack anything, its deficiency can be
replaced by a *Homogenous*, all have indeed a
Volatile, and *Acidum*, and an *Alkali*. Among these
three, the *Acidum* is the means in all things.

Now it is also certain that like is easily
joined to a like, such as the alkaline salts, no
matter which, which easily intermingle whether they
be in animal, plant or mineral substances. Likewise
the *Acida* and also the volatile constituents. The
Spiritus vini, or the burning spirit of plants, and
the volatile spirit of animals readily unite.
Likewise their intermediate spirit, which is the

plant vinegar and (animal) ☩, as also their

alkalis. Again, the ☩ or the ☩ *dum vegetabile*
(as plants), as a *Homogenuum*, unites without
resistance with the mineral acids, such as aquafort,
Spiritus Nitri, Salis Vitroili, Alumnis, Sulphuris,
etc., likewise their fixed salts.

On the other hand, nothing standing at one
extreme end gets along with another which is
opposite and adverse to it at the other extreme end,
e.g. the *Spiritus Vini rectificati* or the *Spiritus
animalis velatilis* will never unite with its salt or
alkali without its intermediate acid spirit.
Consequently, it does not conjoin with mineral acids
or alkalis, or if so, in a very difficult and
dangerous way, and also very slowly. But when its
medium is added to it, they are soon and in one
moment together without any later separation,
because they bind themselves together so strongly
that they can never again be separated without harm,

either by fire or by water. Add *Acetum rectificatum to Spiritus vini rectificato*, and you will immediately see its union without any resistance of these two. If you give them an alkali, it will be nicely dissolved and united, provided both or all three have their right ratio and weights. If you wished to separate the Spiritum vini or vinegar from them *per B.M.*, you would separate nothing but a pure, insipid phlegma, even if they had previously separated from the phlegma in the most stringent way, because the first substance, the essence or the volatile salt of the spirit of wine has congealed upon the alkali through the vinegar and with the astringency of the vinegar. After drawing off the wateriness, you will find a fixed salt, fluid like wax, without smoke. Consequently, the *Spiritus vini* and the *Aceti* have become so fixed by their *Sal Alkali* that they flow in an open crucible like wax, without smoke. But so as to elucidate the theory through the praxis, take note of the following.

For the same reason, I must here uncover a major mistake which nearly all common alchemists make, namely the following: When alchemists wish to sharpen the *Spiritum vini rectificatiesimi* (rectified spirit of wine), they are accustomed to do this *per Sal tartari*, or *Tartarum calcinatum*. Now I ask a practitioner if he believes that he is following the plumb-line of Nature in this way. By no means! The reason is the following, as already mentioned above: The alkali or *Sal tartari* is a fixed body at the lower extreme end. The *Spiritus vini*, on the contrary, is a most volatile essence, consequently standing at the upper extreme end and adverse to the alkali. Here they can see straight as the crow flies that an intermediate thing is missing, because the *Spiritus vini* does not mix with

the Alkali, but both stand over and against each
other as if they had never belonged together. The
Spiritus vini will never conjoin with the *Alkali* or
if so, so slowly that a man would waste too much
expense and time thereby; although by many repeated
cohobations the *Spiritus vini* either congeals upon
the *Alkali* or renders the *Alkali* volatile. But what
a terrible Herculean labor!

Now, as I see that these two do not unite or,
if so, very reluctantly, it is better, according to
the direction of Nature herself, that I look for
their proper medium which they are lacking. When I
have this medium the union takes place very nicely
and at once, yes, inseparably so in one instant.
Here everyone can see what trouble he had previously
when the *Spiritus vini* had often evaporated by so
many cohobations and that much damage had been
experienced by the worker aside from the intolerable
weariness of such tedious labor. If the *Spiritus
vini* gets together with an *Alkali* and is driven with
fire, it escapes again in the selfsame weight, and
it leaves its *Phlegma* with the *Alkali*. True, by this
the *Spiritus vini* becomes concentrated and sharper
and more fiery on account of the *Phlegma* it left
behind. This is then supposed to be a *Spiritus vini
alkalisatus* or *radicatus*. In the same way they
proceed with vinegar, as I will explain later. You
alchemists are searching for a medium, and you put
it between the *Alkali* and the volatile - the things
opposed to each other at their extreme ends. First
you allow the *Alkali* to get so intoxicated by the
medium that it can no longer go on, and is so full
of the medium that it staggers from one side to
another like a drunken peasant. When now the *Alkali*
is full of its ☨ or vinegar, let it also become

180

intoxicated with as much of its ⌒, that is its
volatile spirit, that it becomes even more drunk,
and the more ⌒ the better. When you drive it with
fire by force, the *Alkali* will rise up complete with
all its parts. Add to this again as much or twice as
much as it weighs of its volatile spirit and drive
it quickly over with the force of fire. Now the
Acidum and the *Alkali* have turned into a *Volatile*,
and this is then a radicated and alkalized *Volatile*.[115]

But so that one may understand me correctly, I
will honestly reveal it in order to benefit one who
is mistaken. Therefore let the pupil know once again
that all sublunar creatures in the animal, plant and
mineral kingdoms have their own volatile, or a
subtle volatile water, each after its own species,
be it little or much, likewise their own acidity,
which is vinegar. It goes over after the volatile
Phlegma. There is an acid, vinegar-like spirit in
each kingdom of Nature which retains the quality of
its own kingdom. Likewise, each creature has its own
alkali, which is lixiviated with water from the
residium after the first two essences are separated,
and after it has been reverberated in the fire.

If now an alchemist wishes to radicate a
volatile spirit or vinegar, let him take its own
constituents. If the individual lacks one or another
constituent, let him replace this deficiency with
the like of another, for example, a universal or
another of its own kingdom. Put part of the pure
Alkali into a retort. Pour three parts of its
vinegar upon it and distil slowly in ash or B.M.
Then the vinegar will go over quite weakly like
Phlegma, even if it had been as sharp as an aqua-

---

[115] This is the foundation of the famous *Menetrui radicalis vegetabilis,* about which one
may read in the Vilith of the *Versammlungsreden,* P. 209 ff.

fort, because the Alkali has retained the sharpness and congealed it in itself. When you have done this, pour once more three parts of its own vinegar upon it and do as before, and it will again go over weakly, and the *Alkali* will already be rather full and saturated. For the third time, pour again three parts of its fresh vinegar upon it, as before, and the more often this is repeated the better. Draw it off by degrees to an oil in the B.M. Now the *Alkali* stands dissolved and as the drunken staggering peasant. Such pouring on of vinegar is done until the *Acidum* goes over as sharply as it had been poured on. This happens at the second, third or fourth time, if the *Acidum* is sharp and concentrated, that is, not diluted with *Phlegma*. When then the Acidum is united with the *Alkali* and has become oily, two constituent parts are together. There an alchemist may see how one constituent grasps and holds the other firmly and convivially. What is alien to it, it chases away, namely, the excessive water or *Phlegma*. If now you wish to make a radicated volatile of it, proceed as follows (next paragraph). As it is, it is already a Radicated acid. Nothing has to be added to it except another six parts of fresh vinegar. With that, you drive it over through the retort into a *Liquor* after few cohobations. Then it is a radicated vinegar or acid.

If now you wish to make a radicated volatile ⌒ of it, mix it with six parts of its own volatile constituent and drive them over together. Then they will unite in quite a friendly way, without noise or discord. When this volatile spirit has been driven over, add once more fresh *Volatile rectificatissimum*, drive it also over, and repeat this three times. Then the volatile spirit is radicated according to true natural law and can rightly be called a quintessence of Nature, or a *Magisterium*, because all constituents are together in one, where the upper and the lower are united with the intermediate. It can therefore be called an empowered substance, because here the upper has entered upon a firm union with the lower. Some take one part of this *Alkali* impregnated with distilled vinegar and add to it four parts of calcined pebbles, mix them well together and drive them over in a glass retort. First they give a gentle fire for two hours. Then they increase the fire so that the flames beat around the retort, and they continue with this until the *Sal tartari* goes over with the spirit of wine vinegar in the form of a mist or spirit.

Here an alchemist may see in what way the saying becomes true: *Natura natura gaudet, natura naturam amplectitur, natura naturam vincit et superat propriam naturam*: "Nature rejoices in her own nature, Nature embraces her own nature, Nature conquers and overcomes her own nature." If now an artist did not take as much *Acidum*, but put the component parts together according to a weight contrary to the previous process, so that the fixed or *Alkali* had the upper hand, he would make a fixed salt or a fixed crystal which flows in the fire like butter and is a coagulated or congealed quintessence as the preceding one is a fluid, volatile fifth essence. Consequently, he can do just as he pleases, and can transform one into the other.

In regard to the praxis, however, how the constituents of all individual things must be separated and again combined, or how one is to be transformed into another, will follow in the other treatise on the dismemberment of things in Part II of this book, where the lover of the Art will learn everything in detail.

Now follows an instruction on how to radicate vinegar, as is commonly done: Rc. *Sal Tartari*, 1¼ lb. Pour on it *Aceti* 1 lb., draw it off again, and it is ready. Now you should examine this vinegar. Simply distilled vinegar is anyhow not sharp, and as much sharpness as it has, as much the *Sal Tartari* retains and congeals in itself. Thus a mere *Phlegma* goes over instead of true, sharpened and radicated vinegar; and if they distil the vinegar ten times over salt of tartar, it will only become ever weaker and less, and it is a futile labor. Instead, the *Sal Tartari* retains the sharpness of the vinegar in itself and thereby turns into a liquid salt which melts at a candle or light. This vinegar distilled off tartar is now supposed to be called a sharpened or radicated vinegar. What kind of extractions they can make with it, they will learn themselves. Thus they fare also when they draw off the vinegar over sal ammoniac. It will then go over also weak, leaving the sal ammoniac behind without diminishing it. In such cases they do not know how to help themselves by thinking of some remedy, which is the reason why the work then goes down, philosophy and the Art are laughed at and disgraced, decried as false and a lie.

Most artists distil the vinegar in an alembic through the head, and in that way only the subtlest *Spiritus* ☩ *volatilis* goes over together with the *Phlegma*. Some now use this for all works, while it is yet so weak that it easily proves its weakness when it is tasted on the tongue. It tastes like *Phlegma*, except that it still has a slight taste which testifies that it had retained something of

184

the vinegar. Should they drive more strongly, however, that is through the retort, they obtain a stronger vinegar but stinking of oil and of a burnt smell. They either discard it entirely or are obliged to remove its stench by frequent rectifications. If the stench is taken from the vinegar, it loses its sharpness at the same time, and then it is again of no use.

To help those to get some benefit and to obtain sharp vinegar for their works, I will show them some methods in addition to which an artist may himself think of ever better and shorter techniques. The noblest techniques by which one can distil such a vinegar immediately and in one go, sharp and clear, fragrant in its own way, and with one single distillation, are not public and do not belong in public. This one is called the gift of God, *Pandora*, whereby the whole Art is revealed. However, a good instruction can be given through examples, from which one who reflects will soon know how to help himself.

Therefore, if you wish to distil a very strong vinegar without getting a burned distillate, you must take a subject that would retain the vinegar's stinking oil and let go nothing but the vinegar's sharpness. Then the vinegar will go over clear all at once, without burning, and much can be rectified in one go, while one could previously hardly do one-fourth, and that only with great difficulty. Such subjects, however, that retain the oil are numerous, such as quicklime, the dead-head of the separated water, the *Colchotar Vitrioli*, minium (red lead), woodashes. Take therefore one of these subjects, whichever you wish, yet one is better than another, experience will teach you: Of this Rc. 1 lb., and of the strongest wine vinegar 1 or 2 quarts. Put the powdered additive into a large, proportionate retort, pour the vinegar upon it, set it in sand or ashes, and distil *per gradus* anything that will go, the last time strongly, and the acid *Spiritus* which makes the vinegar rather sharp, will also go. Thus, all at once, you get clear vinegar which has then to be sharpened with various salts. But if a tiny bit of oiliness should have gone over too, pour it all back and distil once again, and it is ready to be further sharpened. This distilled vinegar can now be made sharp in many ways, and here too one way is better than another, such as, common salt, dregs of wine, *sal tartari* and potash, *Spiritus nitri, Salis, Vitrioli* or *Sulphuris*. Whoever then wishes to sharpen it by such things can do it, and it will then have such an excellent effect as cannot be described.[116]

If someone wished to sharpen such a vinegar with wine dregs, which are full of oil, he must also give it an additive, as said above, which will

---

[116] Here someone might believe that there is some kind of contradiction in our worthy, in-God-reposing Brother Homer, since above he rejects *Sal tartari* and nevertheless orders the vinegar here to be added to alkalis. But it is not so: *Sal tartari* is a perfect higher *Alkali,* proper to wine vinegar by nature. It takes to itself the acidity of ⊹. The products indicated in this paragraph, however, are neither proper to wine vinegar nor perfect alkalis. They still contain much ▽, which absorb nothing but the oil, but do not weaken the vinegar. Instead, they rather strengthen it on account of the volatile salt they still contain. But if one were to lixiviate their salts *reverberando,* turning them into perfect alkalis, there would be a similar effect, almost as with tartar. This also explains the reason why he commands us to add a spirit of salt in connection with the strengthening of oil. But because among all these additives quicklime contains a strong *Acidum,* it is especially related to wine vinegar and is also capable of sharpening and purifying it on account of its earthy properties. True, it looks as if our author were holding back a secret, but he lists first the best *Ingrediens,* namely the *Calcem vivam,* and then he teaches the treatment without reserve; only, one has to take note that this vinegar, if it is to be used for sweetening, must not be sharpened. For extracting, dissolving and other like works, however, the ✳ or the ◠⊕ can best be used.

retain the **o o**. Then only will the volatile salt go over with the vinegar. This is the whole masterpiece that strengthens the vinegar.

Whoever wishes to take a *Sal tartari*, potash or ✳, must simply give an additive by means of which the vinegar can and must be impregnated with its *Spiritus*: such as, "glue," *Bolus*, bloodstone, red iron ore, *Minium*, tripoli, etc. He may also make a compound of such salts as Sal tartari and Sal ammonicum, or Sal, Sal tartari and Sal ammoncium together, and then draw the vinegar off quite dry.

Here I will describe another sharpened vinegar which is supposed to be better than the so-called radicated vinegar generally used for so many years and for such a long time. Rc. *Salis tartari* parts 3. *Spiritus salis* parts 2. *Aceti distillati* 6 or 10 parts. Pour the *Acetum* and *Spiritus salis* together, put the Sal tartari into a retort, pour upon it the vinegar with *Spiritus* ⊖, set it in sand and distil it over quite dry. Remove the salt of tartar from the retort, add to it two parts of calcined alum, put it back again in the retort, pour the drawn-off distillate over it, and again distill strongly *per gradus* 4. Now you have an *Acetum radicatum*, which will in one go be as effective as a common acid is in a hundred times. Remove the *Caput mortuum* from the retort, and if there is still some *Sal tartari* in it, cohobate it once or twice by pouring on and drawing off, until the *Sal tartari* has completely gone over, and you have done it very well.[117]

---

[117] The theory of this vinegar is explained above. There are several other ways to radicate and strengthen both vinegar and spirit of wine, among which the following are also described by our author. *Rc.* ⊖ *comm. punif. per* ▽ lbiv, mix them together, put it into a

That I have thrown together acid and alkaline salts, I have done right. If you know the difference between these two, it is not necessary to give you further instruction. The *Acida* are the *Subjecta* which are subtler than the alkalis, for ✝ does not have as much dissolved earth in itself as the alkalis, which makes the difference. Otherwise they are equal, and derived from one mother and one origin. The greater or lesser degree of fixation or volatilization makes the difference.

By this I wish to show that to dissolve fixed things an earthy, fixed, alkaline spirit is required. For volatile things, however, and those that are not bound so firmly, a volatile spirit is needed. Take note of the *Homogeneum*: for *simile gaudet simili*, like likes its like. Minerals and metals require a mineral homogeneous spirit, as will be said below.

Vinegar is a weak plant subject. That is why it must be strengthened to enable it to attack with double force that which is too strong for it in its

---

lines retort, pour on it 5 lb. of distilled vinegar or ♈, whichever you please. Afterwards, add thick red ○ ○ ⊕℔ij., let it stand overnight or for several hours until it no longer ferments. Then distil it over in a wide retort to complete dryness, and you have the *Acetum radicatum*. In this way one can strengthen the ⌒*vini* so that it can dissolve ☉.

Note well: If one takes ✳ or ⊖♀ instead of ⊖ comm. and ① for such a *Menstruum,* it turns into a *Menstruum Hermaphroditicum.* When it has dissolved a subject and is drawn off therefrom in B.M. to one-fourth, a ⊖ *neutrum* stays behind which unites without precipitation with an acid and an *alkali,* either separately with each, or both mixed together.

Rec I mass ⁖, add bit by bit 2 lot ○✳ and ⊖♀ one after the other. After the fermentation ceases, distil in a retort. You can also make it with ♈. These two menstruua improve opium and all poisonous plants. From opium very effective pain relievers can be extracted for diseases involving much pain. Dose is 3, 6 up to 8 grains of the quintessence, but if it is taken as a tincture in the menstruum, the dose is 1 to 2 spoonfuls.

natural state. The main reason why vinegar is used is because it softens and sweetens all corrosives that are dangerous for man's health, otherwise one could well dispense with it. Aquafort, *Spiritus Nitri, Salis, Vitrioli, Sulphuris* dissolve all and sundry subjects without vinegar. Vinegar, however, softens their sharpness and renders it pleasant for all Nature.

One can again also see that minerals may well have their volatile ⏦ but are still opposed to the other kingdoms. Neither is it as volatile. But to make them also like the other kingdoms, one gives them something volatile borrowed from the plant kingdom as their next-of-kin realm, or from a universal subject. Then the alchemist is not bound to the plant vinegar. He can take one from snow and rain, being universals, and these will do it too. But because wine vinegar is anyhow produced in great quantity, one can use that in order to avoid much fuss.

Further, we are here reporting that each kingdom carries its own solvent on its back,[118] likewise its own constituents, and if one of them was missing, one can take it in large quantity from the universals which associate with, and become assimilated to all natures such as the chaotic reborn water or rain, dew, snow, out of which one can get much volatile ⏦, in case of necessity.

Therefore, even if there were no ☦ *dum* or *Alkali* present, niter is the universal acid, salt the

---

[118] Here he says something that must be well taken note of, and in which the main point in the separation of things is comprised. For as long as one can still find the volatile and the fixed parts of a natural product in sufficient quantity, one does not need to resort to other kingdoms. But if one does not find those in adequate abundance, one proceeds as the blessed author has here very wisely written.

universal alkali. If these are distilled in *Spiritus*, they represent simultaneously an acid and an alkaline spirit, which is immediately assimilated to all creatures when it is applied.

But whoever understands (as has been sufficiently explained above), that animals, plants and minerals are not different in their centre but are essentially one, and are only different in regard to fermentation from which their greater or lesser volatility derives, has no doubts in Nature. If one thing does not please him, he takes its next like and homogeneum. By this it is proven that the *Volatile* turns into the *Acidum*, the *Acidum* into the *Alkali*; and vice versa, the *Alkali* into the acid through the *Acidum*, but the *Acidum* into a *Volatile* through the volatile. For one is the other's magnet, and one must be transformed by the other. If I take much volatile and little acid, the volatile overcomes the acid: thus the acid becomes volatile. If now I take much acid and little *Alkali*, the *Acidum* overcomes the *Alkali*, so that the *Alkali* turns into the acid. Instead, if I take much *Alkali* and little *Acidum*, the *Alkali* overcomes the *Acidum*, so that the *Acidum* turns into the *Alkali*. In the same way, if I take much acid and little volatile, the acid overcomes the volatile, so that the volatile turns into the acid, because the stronger overcomes and subjects the weaker to itself. Now we have shown theoretically and practically what putrefaction is and produces, namely that it makes the fixed volatile and the volatile fixed, that it turns the volatile into an acid, and this into an *Alkali*; and, on the contrary, an *Alkali* into an *Acidum*, and this again into a *Volatile*, destined towards the prime matter and the first origin of things. Because putrefaction has revealed to us the volatile and the fixed, the *Volatile*, *Acidum* and *Alkali*, we will examine what the *Volatile*, the *Acidum* and the *Alkali* are in general and in particular.

CHAPTER XX

WHAT IS THE UNIVERSAL AND THE PARTICULAR VOLATILE ACIDUM AND ALKALI

In the preceding chapter we outlined how the volatile turns into an acid, and this into an alkali in particulars and in universals, i.e. the reborn chaos or rainwater. Now however we will explain what these terms are.

It is known that the term volatile means a transient essence. We call it this because it is the most subtle and volatile $\bigtriangledown$ [119] or essence in all things, equally in universals, particulars and individual subjects, because in dismemberment by means of fire, it is the first obtained before its subsequent constituents. The subsequent constituents may be in dry or coagulated form. [120]

---

[119] This symbol is used in the 1781 edition. R.A.M.S. used ᒧ here. -PNW

[120] This spiritual constituent, being the originator, mover and sustainer of all created things, is called by Hermeticists, the spirit of Mercurius, of which Basilius Valentinus writes as follows in his works, page 228, "All visible and tangible things are made out of the Spiritus Mercurii, who has precedence over all things in the world, and all things are made out of him, and have their origin in him, because in him everything is found, and who can do anything the lover of the Art requires." In human beings, animals, plants and ores, yes, in every single thing, this spirit is the direct cause of their composition and multiplication. See Philalethes' *Anthroposoph*, p. 211. "The creator and craftsman of all things, the beginning of every birth that proceeds out of the great Jehovah, and is created out of the true FIAT." Zoroaster in Clav. Art. p. 3. "The one Spirit who constitutes Nature in the Superior and the Inferior." According to Sendivogious it is "the true World Soul by which everything works and lives - the right Mercurius vitae, without whom no man, animal or plant can live." See also the *Grosse Bauer*, p. 7. "This great craftsman of Nature, this true spirit of renewal and pre-servation, which penetrates naturally all creatures to their innermost core, invigorates and animates everything because he conjoins with the natural fluids of creatures like water with water, dissolves their magnet and advances it to fertility." After his separation from the chaos he took up his seat in the uppermost circles of Shamasim. That is also why he is called the "Water of Paradise" by Johann Isaacs of Holland; by Moses however (Genesis 11:11) (Referenced in the corresponding footnote, 1781 edition, p. 189. However it does not match the standard revised Bible. -PNW) he is called the outlet "Pison" which flows around the whole land "Hevila" - that is, our region and in which one finds gold - namely gold of the

The next we call the *Acidum*, because it comes after the *Volatile* and brings to our tongue and nose a simultaneous sour taste and smell, and we have proven that it is the *Nitrum* in universals, whether it be coagulated or whether a spirit has been made of it. This *Acidum* becomes elsewhere too, an intermediate, a hermaphrodite, a middle nature between the *Volatile* and the *Alkali*, between the volatile and the fixed. Therefore, because this part or constituent is ever and at all times obtained after the volatile and before the *Alkali* in universal subjects or natural products and thus stands in the middle, it also has the quality and property of the middle nature. It associates readily with the volatile and also adheres to the *Alcali* to which it is added. And without this middle nature, no volatile becomes fixed or enduring, and no alkali or fixed can become volatile without it. The *Volatile* and *Alkali* must and should necessarily be equalized and given direction, also be conjoined through a third party or an umpire, and whoever omits this, will become wise through adversity.

The next we call the *Alkali* or fixed because it is steadier in the △ than the preceding and is the third and last constituent in all things. This may also present itself to us in a coagulated or spiritual, liquid or dry form. When it has an alkaline effect it is called *Alkali* or alkaline salt, and even if it has already been driven over into a spirit, it can nevertheless immediately

philosophers, that is, their volatile Mercurius. But since the said first and excellent constituent has taken its dwelling place among all sublunary bodies - especially in saltpeter, our worthy in-God - reposing Brother Homer has also dealt with it at such length in the present chapter.

become fixed again with its like fixed. But what the volatile, acid and fixed are as chief components, which cause all effects in universal and in individual things, we will presently show.

In universal things, such as dew, rain, snow, hail, or hail-stones, the volatile is, in its dismemberment and distillation after previous putrefaction, a very subtle, bright, clear and tasteless volatile water, which is followed in continuing distillations by an ever coarser and heavier water. After this water comes the *Acidum* with its sour taste; that is followed by a thick, stinking oil which belongs to the acid, because the *Acidum* is an extended oil. The oil is a thickened *Acidum*, and the oil can also become an acid if it is mixed and distilled out of calxes or colcothar. After this, nothing else follows, but at the bottom there remains a black substance, burnt to coal, which chemists call *Caput mortuum* or the "dead head." If that is burnt to ash in fire, it is separated into two parts, into a salt called Sal alkali and into ashes. The ash, however, also belongs to the *Alkali*, because the fireproof substance is made from ashes and salt, namely glass, and ash is the most fixed part of every product of Nature. The next most fixed is salt.

With creatures of the animal kingdom, after their putrefaction, one usually obtains first during distillation a strong, volatile, stinking, very penetrating spirit and a volatile salt, and with these a *Phlegma*. Often also, if there is strong distillation, a volatile oil breaks through, which is called the volatile of animals. By continuing the distillation, these are followed by a coarser *Phlegma*, after that a strongly acid stinking spirit or animal *Acidum*, which is then followed by its stinking oil. Afterwards the coal or alkaline part stays at the bottom. From it Sal alkali and ashes are produced by calcination.

After fermentation, the creatures of the plant kingdom produce a volatile, burning spirit with their *Phlegma* and often a subtle oil. That is the volatile of plants. This is followed by a coarse *Phlegma*, after that a sour, sharp vinegar and a stinking, thick oil. This is the *Acidum*. At the bottom there stays some matter burnt to coal, which is divided into ashes and salt by reverberation, and this is the plant *Alkali*.

The creatures of the mineral kingdom, when they first come out of the mountain and are distilled, give a little phlegmatic-sweet water with an acid spirit. It is the mineral volatile. That is followed by an acid vapor, generally called oil by alchemists, (as they call the first one a spirit) and that is *Acidum*, the other constituent. Although this oil and spirit are both acid, alchemists nevertheless make a distinction in their expressions on account of their different properties. After this, some earth stays back at the bottom of various colors, according to the kind of mineral. By reverberation it is divided into two parts, that is, into earth and salt, and that is the alkaline part of minerals.

From this one can finally see of what the great world with all its parts is composed and how it originated, what original beginnings it had at the outset, and into what parts it divided subsequently, and how many there are of them, and how they are differentiated into different kingdoms, what they effect and intend, and effect to what end, and this in particular and in general. Now we will descend to particular and individual things, from the big to the small. We will consider their birth and origin, together with their middle and end, i.e. of animals, plants and minerals. We will assign to each kingdom its own chapter and research it from beginning to end.

CHAPTER XXI

WHAT IS THE BIRTH OF ANIMALS AND OF WHAT CONSTITUENTS THEY CONSIST AND INTO WHAT THEY ARE AGAIN DISSOLVED.

ARBOR GENERATIONIS ANIMALIUM.

1.	Sperma Mucilaginosum, masculine, femininum. Unimalische Gur.
2.	Formatio Infantis liquorosa crystalline.
3.	Formatio Infantis membranosa.
4.	Formatio Infantis musculosa, tendinosa.
5.	Formatio Infantis cartilaginosa.
6.	Formatio Nati ossea, Induratio Pueri, Adolescentis, Viri, Senis.

In this chapter we will only speak about perfect animals. All perfect animals are engendered through motion by means of which the seed is stimulated and allured out in the form of a viscous, watery matter, like slime or mucus. It runs into its pertinent mother's womb, where the female seed is also present in order to bring forth its like. That seed now is a thick or coagulated ▽ and can rightly be called the animal guhr (i.e. after conception). From this one can see that the animal kingdom is born out of ▽ or slimy watery substances, and that it is bred in the likewise moist mother through the juicy and watery nourishment of the blood. As soon as it is born, it feeds on moist animal and plant food, and it transforms these into its own nature, flesh, blood, skin and bones, through its *Archeus*. From that it takes its growth and the sustenance of its miserable life until its predestined end, when it dies and decomposes in or upon the earth into juice and slime, mucus and mold, and turns into a slippery watery substance. That moisture creeps into the earth to the plants and becomes for them again food and nourishment, just as before the creatures of the plant kingdom had been the animal's food. Out of that again other plants arise for the nourishment of animals, to feed them again. The animal is so to speak totally transformed into a plant by its dissolution and putrefaction and in return the growth which comes out of that is again transformed into something animal, as has been sufficiently shown above.

As soon as the seeds of the man and woman are coagulated together in the mother, they form a little skin on the exterior, within which there is a very clear, bright crystalline moisture, so clear, yes, clearer than a crystal. In this moisture a globule is coagulated, a pearl, like a small fisheye. This feeds more and more on this crystalline moisture, and it turns into a lump, a formed membranous body. After this, it gets flesh and veins, and likewise nerves. Thereupon it begins to form cartilage. When it is born, it hardens this cartilage into hard bones, and the child grows into a boy, the boy into an adolescent, this into a man, afterwards an old man, and finally a dead man.[121]

---

[121] Here I am asking all reasonable human beings to reflect whether this is not a healthier way of philosophizing than when I assert that the male *Sperma* is teaming with countless numbers of small particles resembling embryos, of which many thousands go down in the business of begetting, and to which the maternal receptacles do not give anything but quarters and warmth and instill nourishment. Certainly! One has to be very credulous or inclined to paradoxical innovations in an extraordinary way, to believe such an *Absurdura* which is grounded nowhere but in the brain of the newcomer who flutters about on the surface of things, or in the quality of the magnifying glass. Therefore, it is so true what the famous Lord *Verulam* says in the following beautiful words: *Ingenium humanum cum ad*

This is the short explanation of animal birth, destruction, transformation, and rebirth into another. In the dismemberment by fire, the Constitution of animals shows that they consist of many volatile parts and *Sal volatile*, less acid and still less *Alkali* or fixed parts.

That this is so is not only shown by anatomy but one can see with one's own eyes that animals are volatile, lively, quick, and mobile, and more mobile than plants and minerals. This is due to their extremely mobile spirit which is precisely the Volatile. In comparison with plants and minerals, animals have a great deal of it, which is proven by their agility and quick movements.[122]

If animals had more *Acidum*, which has the property of binding and contracting as may be seen in plants and subterranean creatures, animals could not move in all directions but would stay put in one place, just as plants and minerals stand immovable.

For the ☩*dum* is subtle, pushing, binding and contracting, *constringens, coagulans*, as may be seen in stiff people and hard-shelled animals, such as snails, tortoises, crustacea, which cannot carry out their gait and movements in all directions as fast as other softer animals. Therefore, because all animal creatures, one more than another, have a volatile spirit, one is also more mobile than

---

solida non sufficiat, plerumque in supervacaneis se exercere solct - Human intelligence, since it is not sufficient to contemplate very basic things, is generally used to occupy itself with superficial and insignificant ones.

[122] In the *Versammlungsreden,* Amsterdam 1779, No. 8, Page 287 (c), an incomparable passage from *Fludd* has been quoted, which explains this very wonderfully according to the tenets of Pythagorean philosophy, and which no one will regret reading over again.

another, as may be seen with birds, four-footed, and in general all animals and insects. With the latter, the difference may be noticed in regard to their creeping, crawling and walking on earth. The more volatile the spirit of an animal is, the weaker is its life, and few are durable, as may be seen in frail birds and insects which, blown at by a faint breeze lose their vital spirit. The more fixed the spirit, the longer the life, as with crows, deer, human beings and elephants.

The cause of a short life lies also in excessive wateriness and moisture, but there is little moisture but much spirit and size, there is durable life, because spirit is life and balsam, not water. Therefore, motion is advisable. It warms the whole anatomy and continuously evaporates everything superfluous, visible and invisible, tangible and intangible.

For a long life all plants are helpful to eat that are of a dry and not a moist nature. Likewise eat those animals which are constantly in motion, such as game, and better is feathered game. It has dry, pithy and therefore healthy meat. As soon now as life has gone and departed from living creatures - which is no other than a heavenly astral light that kindles the vital spirit and pushes it on to work - they fall down and die, and begin at once to decay, but with a difference. The fatter, softer, and more watery an animal is, the sooner it rots; a pithy, dry one, however, not so easily. This may be seen with fish and creatures of the water, because they have few vital spirits but more moisture. They rot quickly and return to the primary matter.

Let the reader take careful note: *It is the Spirit that acts and does, not the water.* The stronger and more plentiful the spirit is in an animal and the less water, the more sprightly and

lively we find it. But if an animal has much $\triangledown$, it drowns the spirit and the animal becomes lazy and sleepy. Spirit, however, must have water by means of which it brings everything about, and without water it can do nothing - but it must have it in the right proportion, not too much and not too little, otherwise it is excessive, which may be observed not only in the animal but also in the plant and mineral kingdoms. The beginning and origin of all things was

nothing but spirit and $\triangledown$, and spirit began to work

in $\triangledown$, to accomplish everything visible and invisible in heaven and on earth by order of his Creator.

Consequently, such an individual spirit makes everything in the animal by means of water: blood, flesh, skin and bone, and all members of the body, and it makes it hard or soft according to the directive of the Creator. Precisely this spirit, however, turns everything to manure during dissolution, and into a watery substance through the

$\triangledown$; finally, however, into water and spirit, as it was in the beginning.

By this now, the reader sees the origin out of which the animal kingdom was born and into which it returns by its destruction.

The main point to note in this chapter, however, and necessary for an artist to reflect upon, is to have ever in mind the rapidity of putrefaction in this animal kingdom, since one can see, when an animal dies, that it begins to putrefy in a few hours and days, at least in the warm season, and to smell badly and so strongly that no man can stay with it without suffering harm to his health. The cause of this putrefaction, however, is the abundant volatile spirit, the abundant volatile salt.

If an alchemist ponders this carefully, he will greatly benefit from it and be able to speed up his works, for every alchemist is aiming at becoming a dismemberer of all things. O Lord, how many thousands are not here going astray and making mistakes, who torture themselves to get their subjects into putrefaction and dissolution! What thousands of *Menstrua* and *Solventia* do they invent, and yet they do not succeed. From this all mistakes arise afterwards, expenses are incurred in vain, time and matter are lost. Here it is that they start scratching their heads, start wishing, cursing, and despising the Art together with the artist, saying it is all lies and liars. Therefore, whoever wishes to avoid such annoyance, let him study carefully in the animal kingdom; not only with his head but also with his hands must he prepare it and work upon it. "Dissect" it is said in the *Imperativo*, and one thing will come out of the other. Now it has been said that no natural anatomy can be done without putrefaction. Therefore, see and examine the reason for and cause of all putrefaction. Here in this kingdom you have the best opportunity and a wide field to work at it. If the four-footed animals and those which live on the earth rot quickly, those which live in the $\triangledown$ rot even faster. If animals

202

that live on the earth stink badly, nobody can stay around those which live in the water when they rot, because of the strong stench, as may be experienced with rotten fish and crayfish. Many an artist often takes several months, half and whole seasons to work his *putrefactiones* and dissolutions. When he has finished there is not even a first result, at least not in minerals. Therefore, if your work does not start putrefying, resort to the animal kingdom. Here you will see that animals rot in a few hours and days, and as they rot quickly they also cause other things to putrefy with them when they are used according to Nature. Here, take your eyes into your hands, for here is a basic key and cornerstone of the whole alchemical Art, a key which is able to open the strongest locks of Nature, which causes all metals and stones to fly over the highest mountains of the sages.

This is the reason why one has to ponder, and where he has otherwise spent a year, he now shortens it to a month, and what had cost him a month, will cost him a week, and the week is shortened into days and hours. But take note that this kingdom (without the plant kingdom) does not effect anything (or very little) in the mineral kingdom; yes, it even acts adversely. This animal kingdom appears so insipid without the plant or the mineral kingdoms that it does not provide any joy or alchemical sweetness. But together with the plant kingdom it effects every pleasantness in the mineral kingdom.[123]

---

[123] Ah! If only some medical doctors and *Professores Pharmaciae* (Pharmacists) took note of this, if they but studied diligently the VIlith of the *Versammlungsreden*, Amsterdam 1779, No. 8, as well as the present incomparable work, and followed its directions in the working-out of their mineral medicines. Then they would not need to push so much poison and corrosives into the body of poor patients. The manner in which this has to be done is taught so clearly below in the second part of this work, Chapter X, that anybody can understand it

Enough has been said at this point concerning animals. Now let us look into the plant kingdom.

---

who but knows the first basic tenets of alchemy.

CHAPTER XXII

WHAT IS THE BIRTH OF VEGETABLES AND OF WHAT RIME BEGINNINGS THEY CONSIST AND INTO WHAT THEY ARE AGAIN DISSOLVED.

Arbor Generationis Vegetabilium.

1.	Sperma, sive siccum semen, Inde in terra resolutum mucilagosum aquosum, die vegetabilische Gur.
2.	Formatio radices.
3.	Formatio caulis & foliorum ramosorum.
4.	Formatio florum.
5.	Formatio feminis mollis seu locustae, Quando se flores in feminificationem nodosam Glomerant.
6.	Formatio & induration feminis & perfecta ejus coagulation.

This kingdom is a realm of wonders, as is the first, and it may rightly be called the sugared and sweetened kingdom. Although in comparison with the animal kingdom it contains the bitterest *Individua*. But the bitterest things, the most harmful poisons and corrosives turn into sugar and honey sweetness in a few hours by means of this realm, but not without the animal kingdom, because one links up and joins with the other. NB. The animal kingdom must be nourished by and live off the plant realm; on the other hand, the plant kingdom is manured and fed by their excrements and dead bodies, so one is sustained by the other. Let an artist mark this as well as the preceding statements.

Furthermore, this kingdom is a true hermaphrodite and Janus, which is neither animal nor mineral but both, and it stands in the centre. With one eye it looks to the animal, with the other to the mineral, and it can become either animal or mineral, as Nature or the Art undertakes the process of transformation. It associates intimately with the first and the last, that is, with the animal and the mineral kingdoms, and is longing for it. One can see with one's own eyes that plants and trees become worms and thus become another life form. One can also see that many trees turn into stone, first those which stand and grow in water, mostly in the sea, because it is very salted.

All plant species are engendered out of their own seed or the influence of the stars, and also improperly by propagation of grafts, which are already opened and germinated.

As soon then as the seed gets into the soil, which is moist, nitrous and salty,[124] as has been proven above, it will get moist due to the water or earth, or it will be moistened by rainwater, and dissolved by the salts; it swells up and springs open and melts into a milky and slimy water, as may be seen when such a seed is soaked in a solution of saltpeter and salt. First it begins to swell, then to break open, finally to turn into slime. That mucous is the first direct element of a plant and may be called the plant ghur. This plant juice or ghur is now heated and warmed by the heat inherent in its own center and by the sun's warmth, and begins to evaporate. The most volatile vanishes into the air and the Chaos; the other, however, which is not so volatile and of a more contracting nature, coagulates into root and stem through the cold air,

---

[124] This nitrous-salty earth moisture is called *Loffas* by a special technical term. See *Versammlungsreden,* XI, P. 277 ff.

with subtle, tender and soft leaves, which is the first stage of plant generation. The more fixed part becomes the root; the not so fixed, the stem; and the more volatile, the leaves. But at the outset everything is soft, tender and young, still full of moisture and therefore weak. The root is the plant's stomach and that plant magnet[125] that draws food out of the earth and takes the approaching rain out of the air, bringing food to the plant until it has become a strong plant or tree.

Its food, however, as is proven in all dismemberments and investigations, is nothing but the earth and the water hidden in it. The earth absorbs the subterranean vapors which rise from the center of the earth, the universal realm, to the circumference and the surface of the earth, and from there to the plants. Water, however, contains the two universal seeds, salt and *Nitrum*. But there is more salt than saltpeter, because it is the magnet which must attract the nourishing moisture from below and above. Such salts, however, are uninterruptedly produced out of dew, rain, snow etc., as reported above, no less from the surrounding air filled with immeasurable atoms; but also in part where man helps Nature, from the manure which he puts on the fields and meadows, also in the vineyards and gardens, or where some cattle come and deposit their excrements.

Depending on whether the earth gets much or little, it bears much or little: fat or meager, big or small fruit. We will now leave all other nutrients and will speak only about the universal nutrient, dew and rain and the pertaining niter and

---

[125] *Versammlungsreden,* XI, P. 280 ff.

salt, because the other nutrients spring from those
original ones. In addition, they are changed back
into them through their reversal, namely into
saltpeter and salt, as has been sufficiently proven
above. The salt, however, or the fixed part of the
nutrient, is the mother and magnet which is turned
into salt and fixed precisely out of saltpeter, as
said above, by the earth and the reverberation of
the solar heat. This then attracts the nourishment
and increases through dew and rain etc., and gives
birth to saltpeter which it takes to itself out of
putrefied rain and holds on to it so that the heat
of the sun and earth can no longer chase it away. In
this way salt congeals the tender saltpeter. The
plant's root attracts those two salts dissolved by
the water, changes them into a pure spirit and
vapor, and drives that vapor into the stem and
leaves through narrow pores, where the plant then
continues its growth according to its
predestination. The salts, however, do not only
enter the plant's nature for its nourishment but
they also dissolve the earth and make it subtle and
a pure salty water. Then it can be further refined
by the root until it can serve as a nutrient.

Nature gives animals the characteristic of
dispatching food, crushed and prepared by the teeth
and tongue, into the stomach, where there is a salty
bitter juice which further refines such prepared
nutrients and turns them into a liquid substance,
then sends it into the mesentery where its best
juice is extracted. Through the natural heat and by
means of the small channels or openings which are in
all parts of the body as pores, it is then sent into
the liver and other organs and distilled, which
vapors settle in the vessels of the liver and other

organs, resolve again into water and this is dissolved through heat as vapor, sublimated or circulated in other and higher organs, and this ceaselessly until it has reached its highest perfection. Who could imagine and understand that in animals, Nature should bring the watery and juicy nutriment, which is dense, upward to the liver? It should rather sink below and flow out through the excretory organs. But if the nutrient is converted into a vapor[126] which penetrates through all pores of the body like sweat through the skin, that vapor can likewise condense into water in its particular moist places through its thickening until it is coagulated into blood, flesh, cartilage and bones through circulation.

That this is true, that Nature nourishes all creatures with vapor, we can see by the macrocosm, how it turns the ▽ from the lowest centre of the earth by force into steam by its inner heat, and drives it up into the highest heaven, and there makes it again thick and heavy, so that it turns into ▽ and falls back upon the earth by its own weight.

This may be seen in all animals, being children of the macrocosm, because the child takes indeed after father and mother. Nature drives the moisture as vapor from the innermost depth of the stomach to the outermost skin between the toes and fingers, and through its thickening it resolves and condenses to ▽ which we call perspiration.

---

[126] From where the said vapor took its origin and that it had contained the primordial constituents of all created things, has been shown in the XIth of the *Versammlungsreden*.

One can see - what is still more - that the vapors in the mines and mountains are abundant. They adhere to the bowels of the earth, and of them the ores are born. If it is so in both kingdoms, Nature will not make an exception in the plant kingdom. Since it is proven that all creatures send food as vapors into all organs, thereby obtaining their growth and preservation, they rightly imitate their origin. They have all and sundry sprung out of the universal-general vapor or Chaos, which turned into water by condensation. Therefore, minerals must rightly also follow the greater. Just as they originated in vapor, and are nourished and sustained by it, they turn into water in decomposition, and this is changed into vapor by heat. This vapor thereupon enters another natural product and again becomes corporeal after the kind of product in which it lodges.

No one must imagine that plants absorb their watery food raw, although in the form of vapor, and receive their nourishment thereby. No! If it were so that they should absorb the water which was originally vapor with all its essence, most of the plants would become all watery and soft, and would not last long, because the excessive water arouses the spirit to action, and a plant would hardly be grown up when it would begin to decompose. On the contrary, it is thus:

The roots of plants absorb only the finest volatile spirits, the clearest purest water, which quickly penetrates into the stem and leaves through the pores, is condensed there and coagulated by the air, and thus the particles (cells) of the plant grow, and are enlarged and in-creased. In addition, if plants were to absorb water with all their parts, they would draw so much nourishment out of the earth in one go that Nature would have not enough time to prepare enough food. But as there is a difference in all things, and one does not look and act like another, so it is here too. One plant has wider or narrower pores than another. Willows and elms absorb more and stronger moisture; therefore their health is not steady and they suffer from various deficiencies and always produce mold and rot. This is caused by an excess of the absorbed moisture, especially if they stand near water, rivers and humid boggy places and ditches. On the other hand, the vine, the juniper tree, the fir tree, the oak and the larch have such contracted pores that they absorb little coarse water or phlegma but only the subtlest, together with the tenderest and most abundant spirit. Therefore, they lead a durable, healthy and more fulfilled life, as one may first of all see in the fir tree, the juniper and others, that they are green in winter and summer, which virtue many creatures of the plant kingdom lose immediately and enter putrefaction: because the dryer and more spiritual a thing is, the more durable and lively it is.

But someone might say: If such plants as the fir tree etc. do not absorb moisture in abundance, how then is it possible that they grow so tall? Then there may not be so much spirit in rain, dew and the earth to make them so strong?

Well, the reader should take note that such plants generally grow in high, stony, dry places and mountains. Even if there comes rain, it flows away from the mountain on account of its slope, and simultaneously sweeps the ⊖ and ⊕, as much as it can get, off into the depressions and ditches, and carries it in torrents into the big rivers which flow on into the sea. This, although salty, penetrates back into the centre of the earth, where the ▽ is converted into pure vapor which arises into the bowels of the earth. What is heavy, attaches itself to the earth, out of which the minerals grow. The lighter such a vapor is, however, the higher it rises and reaches the roots of plants by which it is collected, and whose nourishment it becomes. But still more subtle and volatile vapors break out during the daytime. These are partly absorbed by animals through their breath, and they also feed on them. Partly, however, they rise in the air in order to again give birth to the Chaos or chaotic water.

Now mark how wonderfully the fir tree and its like must feed themselves. I have said that the general nourishment of plants is rainwater and dew, earthy ⊕ and ⊖ together with the subterranean vapors and other accidental excrements of animals, and the fallen foliage of each plant.

If the fir tree stands on stony mountains, rainwater will bring it poor food, because it runs off at once from the steep mountain. Dew, however, is too little by itself, for ⊕ and ⊖ are mostly swept away by the rainwater. Since this is known, we must admit that the fir tree and other mountain plants mostly subsist on the subterranean or mineral, uninterruptedly rising vapors and on dew, of which there is however little in comparison with the subterranean vapors. From this we conclude that the fir tree, in all its size, is mostly born, brought up and nourished out of the subterranean vapors of minerals. That is also why it is not so perishable as other juicy, marshy plants of the plains, because there, minerals decompose little or very slowly.[127] But to learn how it is that the fir tree can obtain subterranean spirits and moisture, mark the following: *Nature does not stand still for one moment but continues working ceaselessly.* We see that vapors arise from the earth continually. They turn into clouds ceaselessly, so that we do not see as many clear days as days when the air is filled with clouds. But if many vapors break into the air, many more must necessarily still be stuck in the earth, otherwise they would not accumulate so much in clouds. And because the earth is thoroughly porous, spongy and honeycombed, like the bodies of human beings and all animals that vapor breaks out everywhere like man's perspiration when there is much of it. Like the vital spirit of the great world, it goes through all wood, earth and stone, for everything has its pores, and no thing is closed to this spirit, even if our vision and intelligence do not always understand it.

Well, the fir tree[128] stands on mountains on which there is all sand, gravel and stones. They are

---

[127] Whoever would deny that the mineral vapors penetrate plants, conferring upon those that stand on mountains not only greater strength and durability but also bestow upon some of them a medicinal and on others a poisonous power, let him but take a look at the *Ipocacuancha.* This is because it grows on West Indian mountains and obtains from the rising arsenical spirits its emetic or vomit-inducing power, which is well known.

[128] The good *Homerus* takes a great deal of trouble to prove the growth of fir and other trees on high mountains, but presumably the sucking-in and reconducting vessels of the bark and leaves of trees, which *Bonner* and *Hales* have so well proven, were unknown to him. The former says: As many leaves a tree has, as many tongues it has to take its nourishment out

magnets and attracting subjects, and a coagulated
⊖ which absorbs such vapors. Through it, vapors
condense and turn into water. That water, which the
roots of the fir tree absorb and from which they
thus take their growth, is quite soft, spiritual and
strong, for as the air is full of vapors and clouds,
so is the earth. And just as such vapors turn to
rain and dew in the air, the mineral vapors adhering
to stones turn into water that plants thereafter
consume.

That vapors turn into water on stones, is
clear. One has only to dig under the earth one foot
deep, where there are stones; one will see that the
stones are always moist, although there is no well
or river in the neighborhood. This is due to the
mineral moist vapors. All one has to do is take a
heated pebble or marble, put it in a humid cellar,
and one will in a few hours see that it has got
drops on it already, as if it were perspiring: and
the longer it stays there, the moister it becomes.

Before I said that the pebble or rock is a
coagulated salt or a salt made into rock. Here some
will be surprised and say: Master, this is surely a
lie. It is easy to help those, however, through
laboratory demonstration. Let someone take some
salt, whichever he likes, melt it, pour it off,
dissolve it in water, filter it, and you will find
some gross black or grey earth. Coagulate the salt,
melt it again, pour it off, dissolve it, filter it,
and you will again find some earth, but this time
white. The more often you melt the salt, the more

---

of the air. Since, according to the author's own system, the air is full of food particles, they
can be supplied to the leaves of trees all the more without hindrance and frequently in such
high places like mountains, which are very much exposed to the open air.

earth you will find and the whiter the earth will be, like snow. Take this earth, put it into a glass vessel, melt everything together, and you will have a stone made of salt. The frequent melting causes the salt spirit to evanesce, but in part it will be congealed into this earth and be transformed.[129]

But now someone will say: Those are peculiar dealings. Where would Nature take a glass factory or crucible in the mountains? I say that myself. But Nature probably has something similar. Just as the salt had previously been vapor and has now become corporeal and fixed, by a natural change - Nature has been able to do this for a long time - so she also does the other. The more earth is added to the salt, and the more earthy and mineral salt-spirits come to the aid, the more earthy the salt becomes, with the help of water, it congeals into a thick juice which is neither volatile nor becomes volatile but ever more fixed, until it turns into a fixed, clear, transparent crystal or pebble, depending on whether the juice is pure or impure.[130] It would take

---

[129] This is an incomparable experience, which is reliable and which everyone can do himself.

[130] This has even been recognized by some modern authors, among others, *J. F. Henkel,* who rightly notes in his small mineralogical writings, Dresden 1756. 8. p. 473, that the connecting link in stones is a very tender saltiness consisting in a connecting quality *per se.* However, he again spoils everything with his little "barbed hooks," which, in his opinion, cause this connection and of which the salts are supposed to consist, without which these excellent natural scientists cannot conceive of anything else. If, for instance, I wish to define salt and I say: Salt is a dense body which consists of cubic, pyramidal, triangular, prismatic particles etc., equipped with small barbed hooks that lock into each other, embrace each other, and cause the solidity to the said bodies, no man of sound mind will understand what I wish to say but rather believe that I am insane in my brain, since this property can be attributed to all created things. But when I describe salt as a body which is supplied to all bodies by the air, after it has come from the upper spheres and been coagulated into a magnet by the air, but is chiefly engendered in the sea, from there it is conducted into the earth by certain springs and is made serviceable for all living creatures by means of certain mechanical manipulations, so that it should serve them as food, balsam and the preservation of their

too long to add such secondary matters here. Yet a natural scientist also learns thereby. We have therefore proven now how and in what manner plants grow.

Now back again to the goal. To show that there is a perfect teaching, or at least a perfect will, to be found in Nature, we will further say that after plants have used the ⊕ and ⊖ as food without interruption, how it is that such a great quantity is produced and that there is no short-coming for the purpose of growth. Therefore, mark the following:

It has been proven above that there is some ⊕ and some ⊖ in every soil, which is where plants grow, because mineral ⊕ and ⊖ also have a special abode in the soil layer. Niter and ⊖ are continually engendered from below and from above; from above by dew, rain and air, water and various accidental things and discharges made by animals and plants during putrefaction. From below, however, they are engendered by the mineral and subterranean vapors which are always breathing out toward the surface.

---

constituents, and is necessarily found in all bodies and can be extracted from them by means of the separation as the last (thing) in them and the sticky moisture by means of which the constituent parts of all created things are so to speak kept together as with a band, and linked. Then they understand.

If one says in addition *that it is of a transparent, crystalline form, soluble in water,* etc., every reasonable man can understand. Common salt is the foundation and basis of all salts in general, therefore also in particular that of *Niter,* of which *Welling,* Part I, Chapter 2, Par. 9, P. 51, writes so beautifully. Otherwise, our Brother Homer's reasoning is very thorough and of great importance, as will be seen subsequently. Take note of all words and apply them in due course, the benefit will not fail to come.

Salt is the magnet, niter the steel which is attracted by the salt and transformed into its nature, or also into salt, by the reverberation of the warm solar and central heat, because in lixiviating this earth one can generally obtain more salt than saltpeter, and there is bound to be more salt naturally as the magnet must be stronger and in greater quantity than the steel, otherwise it could not attract it.

But how the ⊕ and ⊖ are engendered by rain, dew, snow and other waters, has been proven above. There is less saltpeter in Nature than salt, however, and less of it is engendered. The reason for it can be proved experimentally. If there is more ⊕ than salt, niter converts the salt too into niter and transforms it into its own nature. Niter, however, does not attract but acts; it is the *Agens*, the *Sal* is the *Patiens*. All earth growths are attractive, because they eagerly attract Nitrum, or the universal seed, as one may see after it has rained upon the earth after long sunshine, that plants so eagerly attract the volatile saltpeter out of the rain that they often grow one inch and more in one night. Therefore, if there were more saltpeter than salt, they would attract it by force and would grow thereby in general, and they would take into themselves in one go all the sperma of the earth in a short time, so that nothing but barrenness would follow afterwards. If this were to occur, and plants had no more food, they would have to wither. And just as they grow fast, they would have to die fast, according to the axiom: *Quod cito fit, cito perit.* "What is made quickly, perishes

217

quickly." Saltpeter is a very subtle, spiritually penetrating salt which plants can quickly digest through their roots. Salt, however, is more fixed and gross, which they must digest more slowly and subtly. In addition, salt and its spirit are balsamic, which sustain everything. Saltpeter, on the contrary, is a volatile, corrupting, corrosive, putrefactive salt, which may be seen with one's eyes. Take some pure saltpeter that has no salt; dissolve it in some rainwater. Sprinkle this on an apple or pear tree frequently, and it will bear the most beautiful fruit that same year, so much that you will be surprised. On the other hand, if you wait for the fruit the following year, you will hardly get any. Yes, if the tree does not stand in good earth, it will gradually start withering. Instead, however, as has been reported above, melt 1 Part. of saltpeter and 2 Part. of common salt together, dissolve this in rainwater, pour it upon the tree or soak some seed in it, you will obtain good, magnificent fruit in large quantity without any damage, and that every year, provided the tree is watered twice or three times in the spring.[131]

The reason for the quick fertility has been indicated above, namely, that plants absorb saltpeter very eagerly and in great quantity. Salt, however, they cannot absorb so fast on account of its fixity. Therefore, because the salt together with the earth have the upper hand over saltpeter, the salt turns the *Volatile* in the rain or dew into saltpeter by its attractive power. Plants absorb this in part, but in part the salt congeals it into salt by the earthy central and solar heat. It thereby increases and rejuvenates the excess of its

---

[131] *Damerion,* about *Astral powder* in *Plumenoek,* p. 251.

magnetic attractiveness and its quality, so that it incurs constant increasing and diminishing. What the plant has absorbed, is again replaced from below and above.

But so that not all saltpeter turns into salt, Nature frequently sends down the volatile dew and rain, first after long sunshine when the earthy salt or *Sperma* is already reverberated too strongly. Then there comes a good amount of the volatile, namely rain, out of which the salt very eagerly absorbs the volatile saltpeter and congeals it. But because the plants have been very much dried out by the sun, they are also eager to attract this ⊕ and thus tear the ⊕ forcefully away from the ⊖. Thereby the salt is partly increased but also partly robbed of it, and thus it goes without stop in a perpetual cycle, until the Creator changes his Law.

As soon as the *Alkali* or salt would dominate - which is the right primordial constituent of all minerals in view of their fixity - it would produce nothing but minerals, stones, sand and sterility instead of plants. To prevent this from happening, the volatile has been put in opposition to it.

Someone will say: He contradicts himself, because he says that ⊖ makes ⊕ fixed and turns it into salt; then comes the volatile or rain, and it turns the Alkali into saltpeter. Above he had said that one extreme does not act upon another without an intermediate, and here he contradicts himself.

Answer: The earth is never without Niter, and although it is congealed by salt, it is not

congealed all at once. Therefore it retains its
intermediate status, so that the volatile of the
Niter adheres to the corporeal Niter, and this
adheres to the *Alkali* or salt, and one is the
other's magnet, as I said above.

From this the reader can now see the plant
birth and its growth, as much of it as can be said
in this treatise. But whoever would wish to get a
more perfect explanation, let him look for it with
scholars and Messrs. *Botanicis*. They will tell him
about it in general and in particular. Consequently,
the seed of the creatures of the plant kingdom is a
coagulated water, but a slimy one in dissolution,
and therefore a plant ghur. By this one can further
see that everything is born out of water and is
again returned into water. Everything gets its
growth and sustenance from water and at the same
time its death and dying, as is clearly described in
the other treatise *de Anatomia*. This is only a short
report on plant birth. During dismemberment, it is
found that they are constituted of much volatile and
little acid, still less *Alkali*. Yet their whole
constitution, the *Volatile, Acidum* and *Alkali*, is
more acid or stringent than the constitution of
animals, which may be felt and noticed by their
volatile burning spirit which has always got a
little stringent effect. Their acid or vinegar,
however, does not require any proof, *because it is
self-evidently* contracting, while the *Alkali* is
almost like animal *Alkali*.

That this is so, is obvious. They must have
more volatile than acid, or else they could not grow
tall so quickly, high and big, which is then their
motion. The volatile must overbalance the acid, but

the volatile is also (in form) acidic. If the astringent acid were to be in excess, plants could not thus grow up and would stay closer to the earth or even become minerals, since the mineral kingdom is often provided with strong acidity.[132] Plant acidity is somewhat less sour but strongly astringent and vigorous, which may be seen by the fact that it contracts and coagulates many trees and plants so strongly, and also makes them so hard wooded and tough that one can often hardly subdue them with iron and fire.

One can also notice that they have a strong *Acidum* because they are attached so firmly and immobile to the earth, for if their volatility were to overcome their acidity, as with animals, they would be much more mobile, as may be seen with zoophites. In them the volatile has much the upper hand and is already not as contractive as with the immobile plants, which have a strongly astringent *Acidum*.

But that they do have motion may be seen by the fact that they grow from day to day, from week to week, in length, thickness and size. Increasing and growing is motion, although different from animal motion by many degrees.

Plant *Alkali* is fixed, not contractive, as in animals. This is shown in dismemberment. These are the more noble parts of every plant. Although they have still others, these others are counted with

---

[132] This is again a hard nut to crack for the corpuscular physicists and those who do not wish to admit that the constituents of one kingdom can be transformed into those of another kingdom, although our author indeed proves the contrary in many places of this Golden Chain. (1781 edition, footnote on p. 226 -*PNW*)

these three, that is, the subtle *Phlegma* together
with the volatile, the grosser phlegm and the oil
together with the acid; and the third part, however,
or *Caput mortuum*, and ashes together with the
Alkali.

Further, the reader should take note that one
plant has more essential and nobler parts than
another, just as animals have each more or less
Volatile, more or less *Acidum* and *Alkali*, according
to their destiny.

Again, the reader should also note that he can
change plants and animals altogether into a
Volatile, or into a pure *Acidum* or *Alkali*, depending
on what processing he carries out. If he distills
them without previous fermentation, he obtains
almost all *Phlegma* which has only a very fleeting
smell, according to the natural quality of the
subject. After this, much *Acidum*; the *Alkali* stays
in the *Caput mortuum*. But if it is fermented and
putrefied, there will be the more volatile the more
it is processed. Again, the reader can see by this
that the difference lies in the volatile and the
fixed, the acid and the *Alkali*, or between the
volatile and the fixed. The volatile can be quite
fixed, and the fixed quite volatile. Therefore these
constituents are not essentially different but only
accidentally so. If it is all too volatile, it is
called the *Volatile*; if it is a little more fixed,
it is called *Acidum*; if it is quite fixed, it is
called *Alkali*. Everything derives from one root and
stem, namely from the volatile chaotic water and the
indwelling volatile spirit which is transformed by
putrefaction and fermentation into many thousand

forms, like a *Proteus*, according to which it is also called by different names.

This chapter is getting fairly long, because I always mix in other works with my subject, although they are not without usefulness: for if they do not help this man, they help that, and many are glad when they find a doubt resolved. But so as to reach my goal, I will relate some virtues of this kingdom, and everyone is to take note of the following as a principal point: namely, many alchemists have endeavored, and quarreled about it for a long time, to make the mineral realm homogeneous with the animal kingdom, so that it might be absorbed by the latter pleasantly, nicely and sweetly, without any corrosive, for its food and sustenance, to cure and heal its infirmities. They see that the mineral kingdom, after its dismemberment by fire, becomes quite sharp, pungent, corrosive, and poisonous. Consequently, they see that it is directly opposite and alien, also highly harmful to the animal kingdom. To change this now into pleasantness, they have always kept to the burning and alkalized spirits which they digested over it, circulated, decanted, distilled, etc. and much similar work. Yet there has been endless tedious effort and great expense. Now, however, so as to open my heart and to make my loyalty to my neighbor felt,[133] I will give

---

[133] The whole Aurea Catena of our blessed Brother Homer is an irrefutable proof of his kindheartedness. It is a fertile fruit of the true Imitation of Christ. What a glorious light this excellent book would not have shed among God-loving and Art-loving truth seekers in all nations, if the children of darkness, less clever than they think they are, were not blind with seeing eyes and deaf with hearing ears! They consider the author too patriarchal, verbose, incomprehensible, and therefore also - quite naturally - too tiresome. His praxis is too laborious, too tedious, and unusual. In short, the whole work is diametrically opposed to the taste of the greatest scholars of our so very enlightened century. We will here not examine how far our age is more or less enlightened than others. But, you fool! Your proud self-conceit, or your own imagined wisdom, does not change the just, immutable, truthful and

here my toil and sweat both theoretically and practically, so that everyone may obtain at least one hundred times more pleasure than he had before.

To make a greater distinction, however, I will first deal with the usual and common practice of alchemists - how they generally sweeten the corrosives, whereby they believe that they have reached their goal: namely, the most frequently used sweetening, improvement and edulcoration of all corrosives has until today been done with *Spiritus vini rectificatissimo sive alkalisato*. They digested and circulated it for a long time over the corrosives or corrosive precipitation, distilled it several times, or else they rectified it 6, 7, or 9 times. Then it was said that it was now sweetened and improved etc. But the effect has proven that they gave this improved medicine to people with trembling and awareness of the danger. Now, however, I will here put down my method of sweetening, with the reasons for doing so and the proof why the Spiritus vini can never properly sweeten a single corrosive without an intermediate. Namely, I have in this treatise often taught and explained the *Axioma: Non transiri posse ab uno extreme ad alterum sine medio*: "that one cannot pass from one extreme end to the other without an intermediate." Indeed, let

---

infallible ways of Nature. These our unforgettable Homerus has ingenuously taught, not counting on our thanks, with the honorable intention of dealing sincerely, without roguery, with the talent entrusted to him by the Heavenly Father, according to the sweet law of pure love of Jesus Christ. But as the ways of Nature are as old as the world itself, his teachings must also be patriarchal and cannot be after your taste, that is, modern. Consequently, the cause of the disgust with which you look upon them is not their substance or the style of the author - for who would not rather leave the most beautiful avenues, even if they smiled nothing but order, joy and pleasure toward the eye, for a quite natural sidewalk on which he knows he can discover essential goods and treasures? But in your own, alas!, corrupted taste which resists the holy truth which is not as deeply hidden from you and all worldly-wise ones as it is hated.

every alchemist heed this point well and let him
ponder it day and night, if he wishes to achieve
something useful in alchemical works.

All *Philosophi Baccalaurci* etc. probably know
this *Axioma* by heart, but in practice they
nevertheless do not know what is a medium. In this
the world is full of error, which is yet so easy to
recognize and also easy to find.

Every artist should be well acquainted with the
nature and characteristic of everything, and it is
also easy for him to see if something is fixed or
volatile. The volatile (first the *Spiritus vini*)
flies away in weak fire through the alembic, over
the head, which is proof of its highest volatility.
At such a degree of the fire no corrosive will rise
with it, although they have already been turned into
a spirit and driven over in a volatile form, such as
Aquafort, Spiritus Nitri, Salis, Vitrioli, Sulphuris
or their ° °. These do not rise through such a high
alembic, or if so, only with great difficulty. The
same with strong △ through a low alembic or a
retort. From this an artist should again learn that
these spirits are of a more fixed kind in comparison
with the highest volatility of the *Spiritus vini*.
Consequently, they are adverse to the *Spiritus vini*
and stand at its opposite end. From this one can see
and conclude that a medium is lacking, which the
artist should seek. It is indeed easy to find if one
considers the homogeneous natures more carefully.

See how slowly and in heavy drops a corrosive
rises over and, on the contrary, how quickly, indeed
in streamlets, the *Spiritus vini* runs into the
recipient. Well, an artist must perceive that there

is a big and mighty difference between these two, as
practice proves.

Rc. Some well dephlegmatized aquafort, *Spiritum
Nitri, Salis, Vitrioli, Sulphuris*, etc. or their
oils. Take one of these, and pour on it some spirit
of wine rectified to the highest degree, or
alkalized but gently, so that you do not expose
yourself to danger. For two wondrous fires come
together, especially the *Spiritus Vini* and the *Oleum
Vitrioli*. You will see how the *Spiritus Vini* will
not at all conjoin, but the two will stand one above
the other like water and oil, whistle and sing
together like adders and weasels. And if they should
conjoin, they must be digested and circulated for a
long time in a tedious way, and yet the corrosive

will not readily accept the ⚛ . This anyone can
experience by the said experiment.

Then let a man see, and look himself at the
Spiritus of wine or wine dregs, what a contrary
"personality" it has assumed! Because for such a
fixed acid a like Acidum must be taken, and not

immediately the most potent ⚛ . Now distil the
burning spirit of wine together with all gross
Phlegma to the thickness of honey. Drive this *per
Retortam*, and you will obtain a very sharp vinegar
of *Acidum* which is already more fixed than its
preceding Spiritus. This *Acidum* now pour upon a
corrosive spirit and then watch their quick
conjunction. Afterwards, pour the *Spiritus vini* on
it and watch again their lovely union.

Since not everybody wants to risk taking wine
to make vinegar of it, and getting only a small

amount, I will teach, as a favor to him, how to make good vinegar quickly and in quantity, which is anyhow also described in the plant chapter, namely: Rc. In the fall or somewhat earlier, take some unripe grapes with stalks and everything, pound them to juice in a stone mortar. Put that juice into a glass bowl or a glazed vessel, set it in the sun or a warm stove, and let it become quite dead and dry, so that it is absolutely dry. You may make as much of this juice as you wish. But you must not discard the stalks but dry the juice together with the stalks. Upon that dried up juice pour the following wine:

Rc. Take the worst, sourest wine. Pour it into a burning-kettle and distil all its $\mathbb{V}$ off. Pour the rest upon the dried grapes in a cask in the following way:

Have a small cask made of oak or birch, containing ten or twenty quarts. Fill it completely with the dried grapes, or at least half. Pour the distilled wine upon them, set the cask in a warm spot or in the sun, and when the wine smells quite sour in a few days, let it run off through the bung, and pour more such distilled wine upon it. You may do this kind of pouring on and pouring off for ever, as long as you live, and you need no other new vineyards. They form a mother and become an eternal ferment, and soon turn the wine into vinegar.[134] Many

---

[134] This is a beautiful and useful way of making vinegar. If it is afterwards concentrated over quicklime according to the method described above in a footnote to Chapter XIX, beautiful, sweetening vinegar is obtained. The metallic and mineral subjects can also be sweetened in another way, if the first (metals) are turned into ⊕ and the ☿ put into an alembic and made into a tincture by the following *Menstruo*.

people make vinegar, but they do not know the cause of the manufacture of vinegar. The cause, however, of the wine's turning into vinegar is in part that its *Spiritus volatilis* escapes with the heat, and in part that it becomes congealed and ⏝ed. As long as it remains unchanged, no vinegar results. Therefore, to enable you to make vinegar quickly and to obtain a double benefit, I have taught you to distil the *Spiritum Vini*, which would otherwise be lost. Then you can also use it. Therefore, give thanks for the Art: because a good trick can help many. Now to the point. From the above, the lover of the Art sees that vinegar or ☩ of wine is an intermediate nature between the spirit of wine and the corrosive, which very few are yet reflecting upon. Neither have I read it, nor seen it, nor heard it anywhere. When then the corrosive is first united to the intermediate thing and subsequently to the ♒, the lover of the Art will immediately notice a sweetness and mildness, so that it is already much more agreeable to human nature than before, and with it there remains a liquid, volatile and pleasant spirit which can be distilled quite nicely, and by

---

Rc. ⊖♇tartar salt P.I. ℞tiss P. viij., pour the ♒ upon the *Sal tartari,* then pour upon the vinegar. It will not effervesce much, and when fermentation is finished, gradually add to it ▽ P. iv. Let it digest, then draw it over *per Retortam,* and you will have a medicine and somewhat sweet *Menstruum,* to be put into the alembic with the ♁nii. But with the said *Menstruo* a little ⊖y *Caput mortuum* stays behind. Dissolve it with the *Menstruo* and digest it for several ♃♀. Then distil, and everything will go over. Now you have a radical and excellent *Menstruum,* which experience will prove. This is also confirmed by our author, and he reasons about it thus: "Distilled vinegar is just the right medium by which ♒ has entry to the tincture of ♁ii, whereby it is very nicely dissolved. The whole Art consists in this—that ☿ (and all mineral products) are brought to ⊖ or ♃ by means of the volatile salt and distilled acid. Then it can be turned into a liquid essence and a tincture by ♒ and not otherwise."

distilling it, it conjoins and is more and more
sweetened and refined.

Although there is still another method to kill
the corrosives, to make them lose their corroding
ways, it is not as sweet and nice and by far not as
good as the above-mentioned manner. I will also
describe it here to show the difference.

Rc. *Alkali vini*, that is, the salt of wine
lixiviated out of the dead head, such as ⊕ or
another *Alkali* that is pure, white and clear. Put it
into an alembic, pour on ♏tissimum, three or six
times as much. Then pour into it by drops a
corrosive, whichever you wish, and the two will
effervesce together. Do this until there is no more
effervescence. Then draw off all moisture *per B.M.*

It is all a tasteless *Phlegma*, because the ♏
has congealed. At the bottom, however, you will find
some salt which has killed and congealed the
corrosive, so that it can be swallowed without harm,
but this method is by far not so good as the above-
mentioned spiritual way. Here the reader again sees
a medium for joining the corrosives and acids
sweetened by the alkalis. But it is somewhat
violent, as may be seen by their great
effervescence, and not as nice as the previous one,
when they unite like water with water quite
deliciously. For ☩ is basically related to the
spirit of wine, likewise to the corrosive, because
its sharpness and acidity prove that it has a
mineral homogeneity and acidity. This is the medium

which follows in the wake of the ⟨symbol⟩ in separation,
for we count the phlegmatic parts as superfluous,
because the spirit uses them only as a tool for
action and does not absorb more of them than is
necessary for its assistance, as may be seen in the
rectification of the parts.

   Therefore mark: Rc. Sharp and strongly
distilled wine-vinegar partes iii. The sharper it
is, the sooner and better it will sweeten. *Corrosive*
part j. pour together, then carefully pour into this
four or six part. *Spiritus vini rectificatissimi*,
and you will see a noble conjunction when they get
together very peacefully. In this way one can also
sweeten all corrosive ⟨symbol⟩ates and calcined

products. First, pour two-thirds of ⟨symbol⟩ upon it,
draw it off from it twice or three times, later only

the ⟨symbol⟩. Draw off from that as well. And if the

corrosive were not killed sufficiently, and the ⟨symbol⟩

or ⟨symbol⟩ had been too weak, pour a fresh one upon it,
and repeat this until it is enough.
   Mark this well. The sharper and stronger the

⟨symbol⟩ and ⟨symbol⟩ are, the sooner and faster they
sweeten. Indeed, they do this but not as perfectly;
far, far from it, as when they enter on a friendship
and union with the animal kingdom, as will
truthfully be revealed in the second book of my
Praxi de Corruptions rerum seu Anatomia Naturae for
the sake of my neighbor and the poor sick people.

I must here make another point, and cannot avoid it as I see that all physicians are in the habit of using the *Mercurium dulcem* (i.e. Calomel!) as a great medicine in almost all desperate diseases, which can become exceedingly dangerous at times. Here, however, I will give them an excellent improvement, upon which they can rely. Namely: take this ⚛ and *Spiritum*, specially prepared, the *dulcification* of which I will teach in a special chapter of another book. With this dissolve the *Mercurium dulcem* completely, filter it and draw it off in B.M., very slowly, as much as you can. Pour once more three parts of vinegar upon it, dissolve, filter and coagulate it each time again in B.M. into an ⚬ ⚬ ⚬. After that, take the ♉ indicated in the same chapter, pour four parts upon it, draw it off very gently in B.M., again pour on four parts of fresh, draw it off again, and do this for the third time. If you wish, you may leave it as oil, or coagulate it down to a sweet salt or powder, of which, one grain will work better and is safer to use than the former ten. Success will teach you this.[135]

---

[135] This is quite an incomparable way to improve, to sweeten and to render ☿ dulc. quite harmless, so that it can be used safely and to great advantage. What dreadful poison is ☿ ♎ but *Basilius Valentinus* nevertheless makes an excellent balsam for wounds with this horrible corrosive and with ANTIMONY. Yet, hidden in the *Butrym* there is still a far greater power and medicine for the severest internal and external diseases, and it is the right ☿ *vitae* of *Paracelsus,* so that this medicine alone justifies the *Triumphal Chariot.* But this is how it is done:

Rc. ♁ ibj, ☿ ♎ lbiij, each ground to a powder separately, then mixed well together and put into a well-lined, wide necked retort, and moistened with ⚛ or ⚭ ☉ tri. Distill it in ⁂ with a recipient, through all four degrees (of fire) until everything has gone over. Finally, you must also put some burning coal on the retort. When everything has gone over and has

231

cooled off, gradually pour āā of the strongest 🝩⦶, so that everything dissolves into a beautiful red and is turned into a blood-red juice. This must then be circulated to a nice sweetness with sufficient 🜹 *urinosus* by frequently pouring on fresh 🝩 and drawing it off again. If the *urinosus* goes over weakly and the sweetness does not come on properly, take fresh 🝩. When it is quite sweet. everything in the retort must be driven over *per se* two or three times with a subtle degree of fire, in sand, very gently and slowly, in a wide recipient, so that everything can be and all marcasites and metallic ores that have not yet experienced 🜂 they all give off an acid 🜖*ic* ⦶*rous* vapor and *Acidum* in a strong fire. Since it is a true statement that, according to Nature, everything must be dissolved into that in which it originated and from which it had sprung, we can easily understand why minerals yield so readily to acid solvents such as 🝩⦶, 🜆, and ∴ 🜔*lis,* 🜖 *li,* 🜍*ris, etc.,* because they can easily be brought by them into a salty, vitriolic, nitrous-watery state.

But suspect and poisonous are all these strong waters (so the unwise bellow with all their might), because they are all corrosive and the worst enemies of human nature. This is the reason why many alchemists look for nothing but insipid solvents with they wish to deal with metals. But our author, in his unpublished writings, calls those, very politely, the true idols of EBRON and their subordinate gods, who allure the seekers away from SILO until they are forced to return to the right source of Nature. It is indeed well known to us that ☉, by means of its mere amalgamation with common ☿ - provided it is often repeated - can be dissolved to such an extent that it can be transformed into volatile 🜔in distilled vinegar and 🜹, and that it can almost be completely driven over into a volatile spirit by means of a retort. This process is clearly described in the German *Keslero Redivivo,* Frankfurt 1713, in 8., P.36 *ff.* It is a true fact that *Rupescissa* thinks it is the true solution of gold. It is therefore not wrong to ask how this effect comes about since even *Glauber* calls it his white potable gold! Surely, from nothing else but the mineral acid, covered in ☿ more than in all other with little earth, which is very subtle and is introduced by ☿ in a penetrating way.

But to return to the corrosives. First, one has to know what a corrosive is, or any *Acidum.* It is a substance, a nitrous essence mingled with ∴ or greasiness: a dissolved ⦶which has been turned into a liquid or moisture, or the specified universal juice of salt. In a word: the true, universal *World Spirit,* made visible and tangible in all three kingdoms, the animal, the plant and the mineral. Just consider the vast immeasurable space between heaven and earth, and investigate to see if it does not contain salt; which can be proven with certain magnets. Thus, what need is there for great elaboration? In Hungary, around Lake Neustadt on the Austrian border, not far from Wienerisch Neustadt, there are villages there, where the farmers, early in the morning, or at sunrise, gather pecks (a peck = 3.44 liters) of 🜔 congealed by the air and the earth. They call it *wild saltpeter* and sell it quite cheaply. This is a true air-salt, which is daily made corporeal and presents itself visibly by means of an earthy convenient magnet. When this is distilled, it turns into a 🜂ry corrosive spirit. Consider meteors (rain, snow, dew and hail, etc). Doesn't one find a fine sulphureous salt during their dismemberment, after their putrefaction? It is 🜔 *tenerrimum* ⦶ *tri* which, however, is surrounded by and wrapped in much, excessively well joined together. In this way, one obtains one of the most excellent medicines and the true Mercury of Life with all

In conclusion, I advise every true and earnestly (honestly) seeking artist who would like to learn my secret directive, to be on his guard, as much as he loves his life and soul, honor, name and fame, his temporal and eternal salvation, against the great and rich godless of this world, who do not take note of the simple and lowly, but are like unto bees, everywhere searching to suck out the honey, while they are actually trying to put poison into the heart, who promise vast golden mountains, so long and so much, until they have allured the sweat out of an honest but persecuted man. And when they have got it, they no longer hold in esteem the one who acted in good faith with them. That is why it serves them right that they get cheated in a multitude of ways, and also learn to understand how

---

its powers as PARACELSUS describes it in the *ARCHIDOXES* and in his *Clavi* of the tenth book. Its dose is: 1, 2, 3, 6, 8, or 10 drops in water, or any other *vehiculo* This medicine passes through the entire structure of the body and dissolves all. *(i.e.* burns off all the dross! - HWN)

♈ *urinosus: Rc.* ♈ one quart; pour it over two Lots (one Lot or *Lath* is equal to a half ounce) of ✳ and as much *Sal Tartari*. Draw it over just to the point of dryness and it will be ready.

There is still another Spirit of mine which also dissolves quite well, it is made as follows: *Rc.* ⊖♀ and ♀ *crud* in a̅a̅ two Lots, ♈ one quart, rectify this two or three times, but not quite to the point of dryness.

With just the corrosive ♎☿ and *auripigment* (aripigment) he also makes a *Menstruum* to which he ascribes very great virtues. This is to be made as follows:

Take *auripigment* and ♎☿ a̅a̅ , lbj ✳ ci lb ℥. Each is to be made separately into a ☿ . Mix them together rapidly and place into a retort to which connect a recipient. With half a quart of the best ♈ distill it *ex arena* very gently until all the o o ☿rii has gone over. This is an excellent *Menstruum*, especially if it is strengthened with ♎ ⊖ lis., about one fourth. Then it will dissolve common ☉ into o o and when it is drawn off by *B.M.* and some ♎*vini* is poured upon it, it dissolves the ☉ to a high red and sweet-to-the-tongue substance. Likewise △ ♂ and all fine metallic ♀.

233

much effort and sweat a man deeply in love with the Art has to suffer. They should perceive it and feel it, and avarice and envy should eat their hearts out when they see that man a peasant, or a man considered a simpleton in their wanton eyes, knows and understands more of the Art than such high and mighty, vain and yapping braggart, who imagines that the unmoving forests and mountains should bow down to him and abase themselves in his presence. Therefore, you who have gleaned a technique in these writings, laugh up your sleeve and enjoy it in peace and quietness in the fear of God and for the service of your neighbor, and let the big blowhards go - so that they may bravely get blackened in coal until they recognize that the peasant as well as they themselves have sprung from **ONE** God. For the greatest vain-glory must be intertwined with the greatest misery, by troubles and worries as their medium, so that vain gloriousness must recognize what is the medium and what is forced humility. Therefore, let an artist suffer in every way: *Post nubila phoebus*, "after the rain, the sun." God Himself will provide enough means to enable the artist to enjoy the Lord's blessings. And if he does not do so during the day, he will surely receive whilst sleeping. To whomsoever God wishes to give, he gives it in his sleep. That is what happened to me, a poor persecuted husband, man and peasant.

But back to our purpose. I have also promised to show how a poor worried artist should search for a medium when he gets stuck in his work. So I will add the following to conclude this chapter. Therefore, if I were to put two things together and saw that they do not mix or conjoin readily, I would look and consider what kind of subjects I have in hand. If they are creatures of the animal kingdom, I look in that kingdom for its own medium. For instance, if I have a subject lacking a *Volatile* such as bones, horns and claws, but it has *Acidum* and *Alkali* I would therefore like to supply it with a homogeneous Volatile. But from where am I to take this from? Well, ask the subject, bone, horn or claw, from what kind of animal it has been taken! If you know the animal and can obtain it, take its urine or meat, excrement or fat, putrefy and distill its *Volatile* in B.M. and you have already replaced the medium or the missing part. But if you cannot get such an animal, look around to see in which animal there is the same virtue and quality. If it cannot be found, take the subject in which all animal virtue and power are together concentrated as one, that is, in man, who has hidden in his centre the power of all animals, whose urine and excrement can be of help everywhere if you are lacking in a *Volatile*, an *Acidum*, or an *Alkali*. But if you still do not have sufficient of these, return to universality, where all animal, plant and mineral powers are concentrated, and whose subjects join up with all and sundry creatures in a homogeneous way, such as rain, dew and snow. They have a *Volatile*, *Acidum*, and *Alkali*. With that, you can replace every deficiency. Then, putrefy the rainwater, distill off all moisture with a distillation train, rectify it of its *Phlegma* in B.M. through a high alembic as usual. From the remainder, draw off all Phlegma to a consistency similar to honey. From that thickness, distill a ☩ and out of the *Caput Mortuum* extract yet some salt or *Alcali*.

As it is with animals, so it is with plants. In this kingdom, when one comes across an obstacle, one takes wine and its constituents, because in that all plant powers are contained! In the end, however, one can resort to the universals as indicated above.

Similarly with minerals: In alum for instance, all white minerals are concentrated, and in ⊕ all red ones and *astrae*. But if these are not sufficient or are inadequate, run to the more fixed universals, such as ⌒ ① and ⊖*lis*; take the volatile from rainwater; ⌒ ① will give you the *Acidum*, and a ⌒ ⊖*lis* an *Alcalinum*. Consequently, you have here a wide field for exercising in alchemical works.

Every kingdom, however, has its own properties and special qualities, so that they differ among themselves and through these qualities forge both extreme and moderate things, whereby the must allow their heterogeneous nature to be changed into a homogeneous one. For example, animals and minerals are extremely opposed to one another, but the plant kingdom is in between them.[136]

If now the animal kingdom is to become homogeneous with the mineral realm, it cannot easily be done <u>except</u> through its uniting principle or medium, that is, the plant kingdom. Likewise the mineral kingdom can not be made homogeneous with the animal kingdom without plants. Therefore, an alchemist, if he wishes to succeed in his work, must use his intelligence and not mix animals with

---

[136] This our Sons of Wisdom know perfectly well, because when they are asked: Which are the noblest creatures of **all** Nature? They reply: Man, Wine and Gold! If asked further: Of what use is Wine? They answer: As a medium for making gold potable and for transforming it into a medicine!

minerals but first mix them with the medium, plants. Nor should he mix the volatile of the animal kingdom with the plant *Alkali* but mix like with like, the *Volatile animale* and *Volatile vegetabile* should be put together, meaning that he should mix the ☩ *animale* with the ☩ *do vegetabili*. When these are conjoined, he must once again use reason and not immediately pour these united volatile things upon *Alkali* but first the ☩ and only afterwards the *Volatile*. Then he will get a true product, otherwise he will suffer total loss.

From this the artist sees, however, how one thing dovetails into another in the perfect order and not in a confused or confounded way (as many stumble and fall in the Art), but in conformity with the fundamental causes and appropriate means.

For instance, I wish to dissolve gold and will try to dissolve it from the highest to the lowest. Many imagine that they can dissolve gold without a corrosive, and I believe it can be done with water as the dissolvent only, without corrosive. If they torment it first with various mercurial additives and render it crystalline (to a salt) it can be directly dissolved with spring water without any corrosives. They do not understand what gold is, and they understand even less its origin. Likewise they do not understand what a corrosive is and why minerals are usually treated with corrosives.[137]

---

[137] Among most authors there exists a perpetual controversy as to what homogeneous solvent should be added to the creatures of the subterranean kingdom, to make them acceptable to the animal kingdom, so that they mingle internally with human nature , impart to it all their effects, powers and virtues, and can be united with it without harm. Now, however, the mineral kingdom takes its origin in a contractive, vitriolic, aluminous,

Now I will dissolve gold. I divide it into the very finest leaves and pour upon it the volatile ⊡ ⌒ of the animal kingdom. I see that it does not attack the gold. I add ♈ and still it does not attack it. I pour upon it the *Acidum Animale* — it is again too weak. I pour upon it the *Acetum Vegetabile*, that is, the *Acidum*, it still does not attack it. In this way the artist can see that these things are not homogeneous, but distant. For a medium is lacking, grown of and out of the nature of gold. So I go to the mineral kingdom, being its own.

I take ⌒ or ⁰ₒ ⊕, pour it upon the gold, then boil all together. This too does not attack it but only draws out its color, leaving the ☉ white. Now many a man might think, but what is the reason behind all of this? He has indeed used all animal, plant and mineral *Menstrua* and yet he has not achieved anything. The reason is that the ⌒ or

---

sulphurous, arsenical acid. Not only do all writers concur in this, (See: *Des Grosse Hauer,* under the heading *Philosophia Salomonis,* Augsburg 1753, 8, P 27.) (See References -PNW) examination also shows that this kingdom is held together by nothing but small acidic parts, for if you distill ⊕, ☽, ♃, *Arsenicum,* ☿, the *M. Bismuthi,* cobalt water. Therefore it is the will of our great HERMES in his *Emerald Tablet,* "that the superior" should be re-born "into the inferior," that is, into a ⊖ ine body, to make it fixed and △ proof, ana to make this body, soul and spirit remain eternally together as a glorified body. This body can therefore penetrate all bodies and places without harm, and it can act in them as well in △ as in the cold. It is the most beautiful symbol of all the faithful, as after death the soul will again assume its body and will be able to penetrate everywhere with such a glorified body. Everything of a saline nature can again be condensed into a fixed, fireproof body, even if it has become a volatile spirit. For example: Consider the most volatile ⊡ ⊖; how it can be instantaneously be made corporeal by a pleasant and highly volatile acid, such as, among others, ♈, and can increasingly approach fixity. Consider all acids, how much they are in love with the volatile and fixed alkaline salts, yes, and how eagerly they penetrate them, so as to become corporeal in them.

Oleum ♁li is also something distant or extreme in

regard to ☉. For ∽ ♁li or ⚏ris is the most

volatile in the mineral kingdom, while ☉ is,
however, the most fixed. Now an artist can see the
truth of the saying: *Extremum non posse conjungi cum
altero extremo sine medio*, that is, "opposite things
cannot be united without a medium."

Now anyone who has not inspected mountain mines

will say: what kind of a medium is there between ☉

and ♁? Seeing that ♁ is the first matter of
all subterannean red stars. The first and last
matter indeed love each other! This is true, but not
without a medium. Now I will show you clearly what a

great and wide difference there is between ☉ and

♁. Are you well aware that ☉ is smelted out of
the earth, and how a small quantity is extracted
from a hundredweight of ore, likewise how great a
quantity is thrown out? If you know this, I will
honestly reveal to you in brief what intermediates

there are between ♁ and ☉. Namely these: Count

♁ or ⚏ as the first matter of ☉, and for the

most distant of ☉ (I here do not mean ♂ or ♀

♁, but the ☉lar body). ☉, however, consider as
the last matter and also the most extreme. In

between there are these middle things: after ♁ or

⚏ comes Arsenic. Understand, ♁ turns into

239

sulphur; through long digestion ⍀ loses its combustibility and sensitivity (delicacy), nevertheless, it is not yet fixed, but turns into a volatile, mercurial and heavy arsenic. This arsenic turns into a marcasite through further digestion. Marcasite, however, is the next matter of the metal or gold. For marcasite is finally digested into a metal, as all marcasites according to their kind, one more than another, have a fixed grain of a metal. Instead, sulphur and arsenic evanesce and turn into dross. The more fixed, then, these bodies become, or the more alkalized, the stonier the Acidum or vitriol and sulphur becomes, the nobler and more metallic it becomes, as may be seen with ☉ as it is the most fixed alkalized body, and so dense that it cannot be attacked by any ♃. This is because the acidum is more likely to eat itself to death and lose all its virtues before gold would surrender.[138]

From this, the lover of the Art sees that in so far as he should wish to dissolve ☉ with the ⏝ ♄, he would have to turn the ☉ into marcasite, or vitriol. Then the spirit of vitriol would immediately attack it radically and the spirit draws it over into a liquor, but not otherwise. And although ☉ dissolves in sour, alkalized, strong waters, it can yet be separated from them and reduced to its former state as it had been

---

[138] The first constituent of all marcasites and metals is arsenic. In what mineral or metal can we find common quick ☽ except very rarely and accidentally? Instead, you will find *Arsenicum* and ⍀ in each of the aforementioned, be it little or much, but usually in quantity.

previously, namely into a fixed gold. But, if ☉ is

turned back again into its first ♁*lic* matter, as
will be discussed further below, and from there into
a liquor, gold has again reached its origin, that
is, a mineral vapor. This liquor rises over in the
form of a vapor. If now the gold has been processed
thus far and someone wished to have it absorbed into
his body, it would be adverse to him, as it is still
a *mineral corrosive*. To assimilate this into the
animal nature (the body), he must again look for a
medium between the animal and mineral kingdoms. This
is the plant kingdom. Because man cannot consume any
mineral, but largely feeds on substances from the

animal and plant kingdoms, he must change the ☉
into a plant nature and this *vegatabile* must
afterwards be transformed into an *animale*. Thus the
mineral kingdom, by means of appropriate
intermediaries, becomes pleasant for and homogeneous
with the animal, as I have sufficiently shown. One
must always go from one medium to another, up to the
highest, and not at once take the most volatile with
the most fixed. My teachings will repel many and
they will be astonished when they learn that I
prepare the gold with arsenic. But when they prepare

it with ☿ *vivum*, which is very little different

from arsenic, or with sulphur or ☿ ♎, or with

the strongest corrosives of ▽ etc., those are to
them no poisons, they do not harm man at all - so
they think! Perhaps ♎ is weaker to them than
arsenic, although I know that it is sharper. Sulphur
is also totally void of arsenic, although it always
has some arsenic surrounding it[139] and arsenic comes

out of it. The corrosives are likewise so sweet and
mild that they cannot attack the stomach: these are
all simple preparations. I will put yet another plan
before the lover of the Art, to enable him to look
thru the darkness with clear eyes. I have mentioned
that gold is born of *Vitriol*, sulphur, arsenic. If

now, someone wishes to change gold into ♁ in the
proper way, he must drive it back again (degrade it)
with precisely these same constituents by which the
gold has grown, otherwise he will have a very
laborious effort and work. I will not use the gold's
own prime essentials, but others. Each will know how
to find the correct ones. Let someone but consider

---

[139] See *Jo. Agricola* in *Popp.* Nuremburg 1681, 4, P.11, *Tn. de Arsenico,* P. 997, where it is
written: "Without a reason, one should not be surprised that this mineral is so closely
related to △̵, that they are almost sprung from one spring, but in their effects they are
almost opposites. When sulphur is drawn from good and pure pyrites or goldish *minera,* it is
not poisonous. *Arsenicum,* however, is in its whole substance and exceedingly powerful
poison, so strong that almost no antidote can be found for it in the whole world. However, it
is a King of Medicine when it is corrected, one grain or less of it (when properly prepared -
HWN) has the most salutary effect in countless diseases. It is penetrating and tinges the
blood and vital spirit so intensely that it becomes capable of dispersing even the most
pernicious enemy from the body; which other medicines, lacking this power, can never
accomplish. Therefore, you must search in every way to find out how its poisonous nature
can be allayed. Where does ☿̇'s harmful effect and its emetic nature come from, except
from the excess of arsenic it contains? Nevertheless, △̵, so closely related to it in the
sideline, being the true corrector of all poisonous properties in the three Kingdoms, has the
power to take from it all harmfulness and to transform it into a curative medicine. This may
be seen in the *Lapis de Tribus* which can be given to cattle in a rather strong dose, not only
without any harm, but to great advantage.

Not only in the mineral, but also in the animal kingdom does △̵ wield its power by
removing the poison from all poisonous animals and transforming it into curative medicines,
each according to its kind. Spread △̵ upon a very poisonous toad in a pot with a lid. Place a
stone atop the lid, so that the toad, if it is alive, does not escape being driven by pain. Let
thus the toad be burnt by the sulphur, and it will turn into coal. Powder the coal, add to It
again, one fourth of △̵, then let it again gently burn into ashes. Now draw the ⊖ out with
▽, filter and coagulate it. Give four to six grains to a man suffering from dropsy or who has
trouble urinating, and you will see that it is harmless, and, you will at the same time realize
that common △̵ can also ameliorate animal poisons.

242

the *Lapidem Arsenicalem*, as it is called which has been melted into one mass from equal parts of

sulphur, arsenic and ☿*nio*. Of this stone, take one dram to half an ounce of gold. Melt the stone very gently but let the gold heat to glowing rapidly. Pour the glowing gold into the molten mass and they will mix together at once and turn into a brittle substance. If it is often reverberated with sulphur, it will become quite open, like iron, and is afterwards easily dissolved with any sort of acid.

Now let everyone examine these parts, sulphur, arsenic and ☿. For antimonium is a noble marcasite because its ore will show in the test, each time, one grain of gold or silver. Let him give this mass, composed of these three constituents to an animal (now everyone <u>knows</u> that antimony and arsenic are poisons!) and then see just how much it actually harms that animal - even if give a dose of half a dram or a whole dram.[140]

---

[140] This mineral, which is not without good reason held in high esteem in our sacred brotherhood, is called a wonder animal by *Basilius Valentinus,* who says: That It should be considered one of the seven wonders of the world, since not a single man had been found before him who had finally learned all about its virtues, its powers, its operation and its effect, and had fathomed its capability. See his *Triumphal Chariot of Antimony* with *Theodor Kerkringius* annotations, Nuremburg, 1724, 8., P.40 ff. Although this is not quite conformed to truth, as in our school of wisdom the powers of this superb creature have been known for several thousand years, yes, already by the ancient Egyptians shortly after the Flood, it is nevertheless true that in our circle no one had gone so far in its dismemberment as just the Philosophical Benedictine. But just as one single man cannot survey everything, and can easily add something to already discovered things, there have been various scholarly men who have added something to the preparations here and there, improving some of them, taking the praiseworthy trouble to elaborate on them with new explanations. Among these men are, especially, *Alexander von Suchten, Johann Agricola* and *Theodor Kerkringius* who have discovered various not unsuitable processes. Notwithstanding all this, no one surpasses our worthy in-God-reposing Brother Homer. Therefore then, in order to show the world how desirous we are to actively help its poor suffering inhabitants, as much as can be done without breaking our sacred vows, the best of these processes will follow here:

Some solvents which can be used to advantage in dismemberment have already been printed in Chapter XIX, Footnote 84, Chapter XXII, Footnotes 97, and 104. To them may be added the following:

Rc. ♁ P.II. ☿≏ P.IV. ✳ P.VI. °o° ⊕li P. IV. Mix them in a retort as per custom, and distil *per gradus* also as is customary. Then there will be no *Butyrum* but a pure *Liquor.* Keep this, and set the *Caput Mortuum* to dissolve in an alembic. Pour the distillate back upon it, draw it over once again, and this is to be done frequently., always letting the *Caput Mortuum* flow in the cellar, and pouring the distillate back over it in the retort. Thus, the entire substance will go over finally, and it is a direct tincture/essence and *Menstruum.* NB: If it comes into a sweetness with 🜄 through cohobation or circulation, then half of it has to be drawn off, or everything is left together which is still better. Thus one has the ☿*vitae* of *Paracelsus,* for the cure of many diseases, of which something has already been reported above. True, *Kerkring* in his comments on the *Triumphal Chariot,* Page 132, uses sharpened wine vinegar on which he places much hope in the extraction of the redness from the vitro ♁nii. Yet all that does not compare with the Menstruis which our author teaches, among which, is the use of orpiment and ☿≏ as a homogeneous solvent. For ♁ requires a powerful fat and unctuous Menstruum, in view of its sulphurous-oily acid, which should not only dissolve the ♁ but also distil it. Now however, the following points require special consideration in connection with: *(a)* The glass from antimony, *(b)* A S A, *(c)* P H A L A J A, *(d)* The red flowers of antimony, *(e)* The vinegar from it, *(f)* The 🜍 from ♁ and its right fixation, *(g)* The true purification of the *Signetstar, (h)* Philosophic from the regulus of antimony, *(i)* The LAPIS IGNIS.

(a) *Basilius Valentinus* teaches the said *Vitrum* ♁*nii* quite clearly and without any reserve. But one can also make it in another way and *specify* it as one wishes, with whatever additive one likes. Concerning that, the following two methods are especially worthy of consideration:

Take four parts of ♁*nii crud.,* one part of *Caput Mortum* of 🜄, colcothar or red calcined ⊕. Let them flow together, then pour them out into a ♀-basin, warmed and greased with some suet. Thus one obtains a beautiful black black-brown *Vitrum* where the soul of ⊕ is conjoined too. If this *Vitrum* is melted with an equal part of Borax, it will become still more beautiful and fixed, so much so that it melts like wax at the light of a candle.

Rc. ♁ *crud.* Lb.*j.* Let it flow. Gradually add ♁ed corals, crabs eyes, grated or burnt hartshorn, calcem vivam, calamine, steatite, calcined ○, 🜄*sigili.,* bloodstone etc. Of these, take ¼ lb. or four Lot (two ounces); melt it, and you will obtain of each a special *Vitrum.* Only one of these is taken for each kind one wishes to make *(i.e.,* don't mix all of them!)

(b) What the A S A of *Basilius Valentinus* is that it is an excellent remedy for all external diseases, just as his P H A L A J A is for internal Infirmities, has been known for a long time. But since not everyone can succeed in his preparations, the public will be grateful to us for indicating a short, easy and sure way to prepare them, as also their correct usage.

Rc. ☿ *crud.* P.IV., *Caput Mortuum ex* ♈ P.I., or, instead calcem vivam, which is excellent for external injuries. Mix them together and melt them in a crucible until it melts like water. They will turn into a black-red *Vitrum* after being poured into a heated pan or mortar. Powder that *Vitrum* with six times as much *Sal Tartari* or potash. Melt it again like water, pour it out, and it will turn to a yellowish mass. Powder that and mix it with ♆ *viv.* P.IV.

Cement it very hot in a crucible several ♉, then lixiviate it with hot ▽, filter it, and boil it down gently to a ⊖ in an iron pan. Put it in a humid place or in a cellar, together with the pan, and it will become a reddish balsam, Liquor or o o. Keep it in an alembic, and the A S A is ready.

Of such Liquor, one Lot (½ ounce) and 10-12 Lot of flea-bane, or plantain, or *wurzel* (carrots or roots), common celandine, scarlet pimpernel, or heathen anthyllis ▽ or also only lime-▽, that is, only one of these waters must be mixed with the Liquor. With this, one washes wounds, injuries, cancer, tumors, dry and moist gangrene, several times a day. It is injected into fistulas and cavities, and one also puts a double cloth moistened with it over the injury. By repeating this often, beautiful and wonderful effects can be seen. But if this treatment should burn or sting, or hurt too much, it can be softened with more of these waters until it no longer causes a burning sensation or pain when an injury comes into contact with it.

Of the said Liquor, a very excellent ointment can be made with olive and other oils and fats; and of that, a precious curative poultice can be made by adding wax. The ▽ for washing the wound, the ointment for assuaging and removing the pain and heat, the poultice for healing.

(c) It is the same with P H A L A J A which can be improved in the following manner: *Rc.* 4 Lot ☿ *nii pulverisati, 4* Lot ✳, 4 Lot *Cremor* (Cream of Tartar), ♃ri, 4 Lot ⊖ of ♃. Grind and mix everything together into a powder. Put it all into an alembic, pour on it one quart of strong red wine vinegar. Let it stand for one month in a warm place or in the ☉ stoppered with paper, and shaken well every day. Then add to it one quart of strong brandy or of not-too highly rectified ℣, and let it stand again for one month in a warm place but NOT in the sun. Filter it and it is ready. Dose: 6, 8, 10, 20, 30 drops in *conveniente vehiculo*. The P H A L A J A can also be made volatile, thus:

Melt ☿ P.I. with ⊖ ♃ri or ✳ *co fixo* P.VI. and it will turn into a cinnamon-colored mass. Dissolve that in good ⊹, filter it, distill it to an o o; dissolve this again for the third and fourth times with fresh ℣, decanting it each time in BV or M *ad* o o. Then also proceed thus with ℞. Finally, draw off the ℣ to oiliness. Mix this o o with *Terra Sigillata* until it looks quite dry. Then distill it *per gradus* in a retort in an open △. The result will be a yellow or red o o, tincture or Q.E. If desired, this can be further circulated with ℞ to the greatest sweetness, and then the ℣ is drawn off to an oiliness. Now one has the right medicine, flawless, the virtues of which are described by *Baslius Valentinus*. NB: That which has been drawn off is also a kind of radicated *Menstruum* and has *Virtutes radicates*. These two preparations deserve careful consideration, because the Salts being the best *Aperativa* and *Incidentia* are contained within them.

antimonium. If a man were to reflect on the true

(d) The Red Flowers of Antimony - Although *Basilius Valentinus* also teaches how to produce them, the following process is much more vigorous and is done thus: *Rc.* ♁ P.*ij.*, *flores* 🜍*is* P.I. *(limatur* ♂ *P.ij.,* ○ *crud. P.iv,* ✳ *P.vi)* Mix them well together and sublimate them once, twice or four times, and the *flores* will rise dark crimson. Now let these flow into a Liquor in a cellar, on a glass plate. This Liquor is turned into a tincture with the aforementioned 🜹 (that is, the so-called 🜹 *Urinosus,* see Footnote (n) above. It has unbelievable powers for purifying the blood, for abcesses, ulcerations, for internal obstructions of the glands and the nerves, catarrhs, cold feet, infectious spotted fevers (typhus), the plague, etc. When these *Flores* are separated from the ✳ with ▽, they are panacean. They no longer cause vomiting because their arsenic has been ameliorated by the flowers of sulphur, but imperceptibly purify the blood and the whole body by their subtle evaporations, and they have the virtues which *Basilius Valentinus* ascribes to his fixed ♁. Dose: 2, 3, 6, 8, 10 grains in *conveniente vehiculo.*

(e) To make 🜊 from antimony, many have troubled in vain, or else they have been unable to accomplish it according to their desire. But it can easily be made in the following way:

Rc. Take *Min. Hung.* ♁ lb.*iv.*, 🜖 *Hung.*, lb.*viij.*, or also *Sacch.* ♄*,* if you like. Pound (grind) and mix them together, putting the mixture into a lined retort and set it with a receiver into ∴ for twenty four 🝮. Drive it with the first and second degree △, and the ♁ will somewhat open by means of its own sulphuric acid and that of the added 🜖. Let it cool down, and if something has risen into the receiver, pour it back into the retort, and pour on it six (6) quarts of 🜊. Let it digest for one month, or less, in a warm room. Then, distill the 🜊 quite gently in ∴. If now you wish, you can distill the very strong 🝐 🜖 and ♁ into this vinegar, which causes no harm at all, but gives you even more *Menstruum,* and all is ♁*ized.* But if you do NOT wish to do this, then keep the vinegar separate, add another receiver or just the same big one, and collect the 🝐 🜖 and ♁*nii* separately with a low or strong △ as 🝐 and o°o of 🜖 are usually distilled. In 24 🝮 everything has to be finished. Then extract the ⊖ from the ☠ with ▽, filter and coagulate it in a glass dish. Rectify its distilled 🝐 twice or three times over the salts, after digesting it previously each time. Then the vinegar is ready. Now use it as *Basilius* teaches.

(f) In regard to the 🜍 of ♁, *Basilius* makes it with a sharp lye. Yet, it is made more frequently, and faster in the following way:

Rc. ♁ *crud.* P.I., potash P.iv. *vel* vj. Grind them together, then melt it in a crucible to a ▽ consistency, then pour it out into an iron pan, and you will have a cinnamon-red mass. Now grind it to a powder, boil it up in ▽ in an iron pan, and the ♁ in the lye will dissolve. Filter it and precipitate it with 🜊 or any other acid, and the 🜍 ♁ will fall to the bottom. Sweeten it with ▽ until all the saltiness is gone. Now dry it, and you have the 🜍 ♁ in quantity. Filter and coagulate the 🝐*ed* lye and you will get an impregnated ⊖, whose dose: 2-6 grains, is to be mixed with the opening medicines as also among the *Enemata* for the hardest constipation.

Not everyone knows the technique of making this 🜍 fixed, as *Basilius Valentinus* observes. We will, therefore, indicate here the following method which is reliable:

Rc. ⊖ *comm.* and ⦶ $\overline{aa}$ 4 Lots. Mix it well, let it flow in the crucible like ▽. Gradually add the aforementioned 🜍♁ *nii,* well dried, to the flow, 3 Lots. Let it all flow together well during the time of two or three *Paternosters and Aves.* Then pour it off and lixiviate the salts. *(*The German text seems to indicate that a lye is made in the process of washing away the salts. - *HWN*) Edulcorate the sulphur lying at the bottom with ▽ until all its saltiness is washed away. Dry it and put it into a retort, pour on it twice its weight of ⌒ 🜿 *ii,* and draw it off strongly until it is quite dry. Take the dried sulphur out, calcine it well in a crucible and look at its color. In medicine it is *a PANACEA,* better than the common dia-phoretic remedies. Dose: 6, 10, 15 grains with or without a *Specifico,* to be taken in water, wine, soup, etc. But it does not always cause noticeable perspiration if it is used in other diseases.

(g) What the Signatstar is, *Basilius Valentinus* himself explains in his *Triumphal Chariot.* Although its preparation as taught: therein is very beautiful and clear, one or another useful things can, nevertheless, be added to it. For example:

Rc. Take one part of purified *Reguli* ♁♂ *ialis,* brought to its greatest whiteness by ⦶, and three parts of ✳. Sublime them together with the ammoniac, mix it with four parts of quicklime *viva,* make *s.s.s.,* calcine it gently for three ⧗ then let it cool. Boil it down, filter and gently coagulate it into a salt, using an iron pan. Put this ⊖ together with the pan in a cellar and let it flow. In this way you will obtain a vulnary balsam (balm for wounds) for all injuries, and a wonderful remedy for burns; owing to the miraculous curative ♆ -salt. Dry the boiled-down *Massa,* grind it to a powder, it can then be sprinkled on wounds such as: corrosive injuries and open cancers. It is an excellent remedy for improving, subduing and drying all the acidity and corrosiveness of the injury. Combine this with the A S A described in footnote (b) above.

(h) A fine process exists for the ♄ of ♁ thus:

Rc. Regulum ♁ *nii, Sacch.* ♄, *comm. aa* (equal parts of each -*PNW*) q.p. Mix and grind together with half this weight of finely powdered soap. First work it gently in a crucible until the soap is burnt, then more strongly and finally it will be melted clear and can be poured out. In this way one will discover a malleable *Corpus.* If it were not sufficiently malleable, melt it several times with soap until it becomes malleable. *Kerkring* teaches its use in his comments on the *Triumphal Chariot, pg 291.*

Among all the medicines that are made from ♁ there is none to surpass the Stone *IGNIS.* This is why *Basilius Valentinus* has added an entire treatise on this stone to his work, *Triumphal Chariot of Antimony,* in which he not only describes in detail its preparation, but also its effects. However, there exists, nevertheless, some shortcuts and improvements on its manufacture. These follow here:

A SHORTER WAY to turn the *Vitrum* ♁ into the Stone *IGNIS: The Min.* ♁ is detonated according to the method of *Basilis Valentinus.* To lb. *j* of this *Min.* are added 4 Lots of ♁ *crud* grind everything together very well and then melt it so that it flows like ▽ and everything goes faster.

TO EXTRACT and COAGULATE 🜍 and 🜔 BETTER: When all the yellow or red color or 🜍 and tincture have been drawn out of the *Vitro* ♄, a black substance remains behind. Dry this and mix it with half as much *Fl.* 🜍 *ris*. Burn this on a flat Cupel, then again mix the fourth part of *Flor.* 🜍 *ris* with it and let it burn once more. After this is done, extract the 🜔 acid with distilled vinegar or another convenient *Menstruo*. This is now purified by dissolving and coagulating it, and it will finally crystallize. If some 🜄 remains, it can again be treated with common 🜔, as before and thus one can obtain sufficient.

A STILL FASTER WAY: Rc. Pulverized *M.* ♁ *nii or* ♁ *crud.*, one part, and six parts of potash. Grind all together melt it like unto water and pour it off. After it has cooled down, grind it to powder and boil this powder with 🜄 in an iron pan, boil it well. Filter it and 🝞 *irs* it as before with any kind of acid, and you get the sulphur which stays behind in the *fiitro* and in the pan. Dry and melt it with 2 or 3 parts of potash and let it flux again into a watery consistency. Then separate the 🜍 as before and melt what is still in the *Filtro* and in the pan with an equal part of potash, so will you also have the sulphur. Draw its 🜔 out (in the meantime, while you are making the salt, you can make the tincture), and in making 🜔, you will get a good amount of sulphurous gold *rati* thereby. For the smallest child as well as the larger child, it is a true *PANACEA* when their stomachs are filled with or overloaded with mucus. It is equally effective in combating worms. Dose: 1, 3, 10 grains taken with sugar. It induces gentle vomiting, especially in children.

Basilius Valentinus extracts the above mentioned 🜍 from ♁ with **vinegar** (of which there is a fine description given in footnote (e) above). He draws that down and extracts it again with 🜓 , then unites it with his salt , 8 Lots. When the extract weighs lb. *j*, he circulates it for some time, then he abstracts it to a dry powder, out of which he then drives the o o be- longing to the Stone *IGNIS* with a strong 🜂. But this is not as good as when it is only drawn off to oiliness, mixed quite dry with 🜄 *sigill.* , and driven in a retort in an open 🜂. Then you will obtain the above mentioned oil perfectly pure and very beautiful. In addition, it is impregnated with the powers of 🜄 *sigill.*

As far as his ☿ from antimony is concerned, upon which, after it has been 🝞ed with the <u>right</u> o o 🜨 *li*, ♂ tis congeals and thus the famous stone is accomplished. Also, there are many other preparations known from it, but it seems that the one used by the author of *Triumphal Chariot* is the one preferred to all others. With this 🝞ate he unites (as indicated above) the aforesaid o o ♄ and congeals it into a fixed and impalpable (liquid) powder. Although this is absolutely true, one can also proceed in the following way: One takes the 🝞, dissolves it in distilled wine vinegar and then crystallizes it. The crystals promptly unite with the o o. This mixture is put into an alembic with a head, a receiver is luted to it. It is set to digest gently until it coagulates and congeals into a soft, wax-like, fireproof **stone.** This is accomplished through the evaporation of the excess moisture which drips, little-by-little, into the receiver.

Here some stubborn alchemists, who assert that the entire *Humidum* must be dried up without the least wastage, will prick their ears up mightily. However, we need not concern ourselves over this, because *Paracelsus,* clearly indicates this process and *Hollandus,* in addition to *our Annulo Platonico,* teach that the moisture can be drawn off, otherwise much time and expense would be lost coagulating, or, the *Vitrum* vessel might even burst, because only the saltiness can come into the *Coagulum.* For all salt spirits with one body will return to one body or ⊖, namely to the extent that they contain volatile salt. The excess water will be difficult to dry up or will do so very slowly.

Because *Basilius Valentinus* has described the right o°o ⊕ *li* ♂tis for the ≂ tion of the ☿*rii* ♁*nii, it* can be done very well in the following way:

Dissolve, as many as you wish, cleanly washed steel filings in good 🜊 ⊕ ***li*** and distilled wine vinegar mixed together. When all or most of it is dissolved, ≂ gradually with 🜈 ⊡ oso until the solution turns <u>ruby red</u> **and no more**, NB otherwise the corporeal ♂ will be ≂ irted out and the solution will lose its red color and turn green. When the solution is a high red and is somewhat strong, filter and distill it until a thin skin forms. Alternately, it can be left to gently evaporate until this thin skin forms. Following this, set it to crystallize.

Rc. When you have one pound of this ⊕**,** put it into a retort, pour on it strong o°o ⊕ *li,* ¼ lb., let it conjoin (or: cohobate) for 24 ⧗. After this, distill it through the degrees out of the ∴ and you will have a very fine, noble, truly philosophic ⊕ o°o sweet on the tongue, which is used to ≂ the above mentioned ☿, and can also become a most pleasant medicine with 🜈 *Sapienti Sat.*

Whoever wishes to obtain this *LAPIS* △ faster, let him take the above mentioned 1st or 2nd *Butyrum* ♁*nii* with the 🜊 ⦶ sweetened 🜈, (See above notes) add this ⊖ in the manner prescribed by *Basilius Valentinus,* and circulate it until the salt is dissolved in it. Then let him draw off the 🜈 to oiliness. Add the crystals of ☿♁ *nii* according to the above indicated weight, coagulate it through digestion and fixation, in the way described previously, in an alembic fitted with a head and a receiver, until it turns into a beautiful liquid-like-wax stone. In this way one can delight one's mind much faster!

As an extra, I will set down a very excellent work with the above mentioned Butryum of Antimony. If the said *Butyrum,* which has been made with the 🜊 ⦶, united or not united with 🜈, is conjoined in the same weight with o°o ✳ (which has been congealed with quick♆ and dissolved *per deliquiem)* and is mixed until dry with brick and coal dust, and is then distilled in a retort in an open △, an acidic *Liquor* goes over. This is both a medicine and a *Menstruum* which dissolves the ☉ *fulminans.* When it is drawn off to dryness in B.M. and the *Residuum* is sweetened with 🜈, a blood red solution and ☉ *potable* is obtained after a white ✘ left behind. Out of this, the ⊖ is drawn with the just-mentioned *Menstruum,* united with its red 🜍, and thus one has conjoined the red man with his white wife. The same process is used for *Vitro* ♁ *nii.*

By this elaborate, but very useful, annotation, the kind reader can judge how ready we are to serve the entire public, with our talent bestowed upon us by Divine Wisdom. Truly am I convinced that some physicians and apothecaries in love with their old tired methods will have little thanks for our unselfish love of mankind. However, I also know that the conscientious ones among them - of whom there are still many, praise be to GOD ALMIGHTY! - will gratefully recognize our efforts and will benefit from our well-meant back into its prime matter through ⁓⊕li or ○nis. From this the reader learns of the properties of arsenic, how quickly its poison can be removed and thus it is transformed into a better substance. The same can be done with ☿⁓. If it is refined (burnt) with sulphur, its poison is ameliorated to such an extent that it can be used with far greater safety than ever before. It is the same when such poisons are improved with wet spirits, such as: ⁓⊕*li*, °∘∘⊕*li*, 🜍*ris*, etc etc.

We will conclude this long chapter with this, and thus the plant kingdom has been described; which is a true hermaphrodite and *Janus* between the animal and mineral domains. Without the plant kingdom, the mineral can never become homogeneous with the animal and vice versa. An artist is not in his right mind if he tries to make a medicine fit for human beings and animals by utilizing mineral materials without plants or using plant materials without animal things. But, enough of this, more will follow. Let us now turn to the mineral kingdom, where there will be greater surprises given than was given in the previous kingdoms.

---

instructions.

CHAPTER XXIII

WHAT THE BIRTH OF MINERALS IS, AND OF WHAT CONSTITUENTS THEY CONSIST, AND INTO WHAT THEY ARE DISSOLVED.

ARBOR GENERATIONIS MINERALIUM.

1.	Spiritus ☉ & ⊖is cum resolvenda terra, inde
2.	⊕lum sive Gur, inde
3.	♀, inde
4.	Arsenicum, inde
5.	Marcasita alba & rubea ♂ ♁, inde
6.	Metallum, Inde improprie in ultima fixione
7.	Fit Vitrum.

That I here do not discuss the usual prime beginnings should not surprise anyone, but it is here understood that they are: ☿, ♀ and ⊖; the *Volatile*, ☩ and *Alkali*; soul, spirit, and body; heaven, air, water, and earth.

Describing the birth of minerals, I know in advance that many a man will throw in some big objections immediately at the start. However, after he has obtained some knowledge of Nature, her origin, development, and end, he will nevertheless become pensive and think the matter over a bit more. For the reader believes and reasons correctly that I follow Nature and proceed in order, not giving in one inch. Many authors have published their mineral descriptions for the benefit of the world, some clearly, some obscurely, depending on their conception and how they could make the world understand it. I do not hold any in contempt, but praise each. I myself have also derived much benefit from them, and have received many a light when I would otherwise have become stuck. Because the writing of books is not done to haul this or that author over the coal or even to reject him. No, but each and all who have written books did so for the best of the world, although it does not follow that one must necessarily stick stubbornly to this or that author's views. Instead, if one puts together the views of learned men, one can occasionally learn to recognize the intention and basic understanding, which had caused doubts for many years.

Indeed, many senators are elected to the Senate, for what does not occur to one, will occur to another. Likewise, many authors together explain Nature better than one alone. Even if one has done his job well, he still did not know everything, or describe all the facts, or think of everything, as every minute puts other thoughts into his mind. Consequently, what has been omitted by one is described by another, and this one or that one explains it, and based upon his explanation, the reader can correct himself and attain his previously unattained goal. Let the reader also deal thus with my writings. If he does not like one point, another will please him, and there will yet be one in it that is worth the paper it is printed on. If I do not present a good theory in everything, perhaps I am giving good practical directives or teaching some good technique that may become useful to many.

Before continuing, however, I must here say that a great many philosophers usually describe the origin of minerals in the following words:

Vapors rise from the center of the earth, adhere to the cold clefts of mountains and turn into ▽. In dissolving the earth, this turns into a vitriolic-salty or aluminous stone nature, or is afterwards boiled down into sulphur and metals, according to the variety of the subtle earth, etc., etc. They write correctly that vapors rise, but they do not explain what kind of vapors these are or what is their origin and property.

From such descriptions a young student can still not derive much benefit, because such vapors also rise up to us into the air; but how far they differ from those that remain in the earth is known by all those who have practiced looking into the entrails of the great Demiurge. Therefore, if you, kind reader, wish to obtain some knowledge about it, mark and consider the following, as I have mentioned above: The universal seed of all things was ▽ and ⌒ in the beginning. This is not only proven in this but also in the regenerated Chaos, and we have shown how this spirit comes out of invisibility. In the same way everything has sprung out of Nothing.

This Nothing, however, turned into vapor or ▽ in which the ⌒ lay hidden, out of which animals, plants, and minerals grew subsequently, the first two from a volatile seed, the latter, however, from a fixed one.

In the chaotic primordial water, as well as in the regenerated ▽ as also in all waters and kinds of earth, we find two different salts in their innermost center, the visible or tangible world

253

spirit, or male, as well as the female corporeal seed of the macrocosm, that is, saltpeter and salt. We have confirmed that this is the first universal, though not yet specified and not separated, matter of all sublunar things. Together with the regenerated Chaos both are all in all, as it has been proven that they are found in each and all subjects, volatile and fixed, according to the difference in their digestion. This subsequent test will also confirm that they are all in all. A thing that is and should be all in all, must contain in itself the nature and property of all things. It must also be able to unite and combine, copulate and conjoin with all things indiscriminately. In the general view of alchemists these two salts, niter and salt, are mineral, but this is not so. Because the fact that they are dug out, in and under the earth in mountains does not yet prove that they must therefore be mineral, since they are also found in the sea, in lakes and other waters on the earth, in the plant kingdom and also in the animal realm. Otherwise it would also follow that they are animal or plant. No! But, as the test and its result shows, they are found in all three kingdoms as well as in all universal subjects, and they are therefore rightly all in all because they are found in all.

Again, we do not find any salt or subject in the whole mineral kingdom that would be homogeneous with the animal or plant kingdom, except in their transformation, but these two. Saltpeter and salt do not cause any substantial change either in animals, human beings, or most of the plants. Man and all animals can take saltpeter and salt as food. Plants can tolerate saltpeter and salt, likewise minerals, each without harm, except that their weight and measure must be taken into account. Not too little and not too much, is the right measure and aim. These salts, saltpeter and common salt, are an exceedingly great fertilizing and sustaining power; but as soon as they assume an adverse quality, they are the destroyers of all things.[141]

Against them one should put vitriol and alum, which are both counted among the so-called mineral salts. Like the former, man cannot consume the latter without displeasure and a great change in his body, as little as he can consume mercurial or arsenical subjects. Even plants do not desire them, as they die of them. If we give to an animal, a dog or a cat, but one or two scruples of ⊕, we will see that it will throw up or change with great disgust (or: reluctance). In the same way, pour such a solution on a tree or plant and see if it does not perish. Therefore, this once again throws light upon the fact that ⊕ and ◯[142] are homogeneous with all sublunary creatures, and what is homogeneous is the same as their nature; and they consist and grow of that which is of their nature, and are sustained by it; and that of which they grow and by which they are sustained is that of which they have emerged. That out of which they sprang in the beginning is indeed their first matter, and the first matter is the beginning and origin of everything. In dissolution every creature returns to it, so that the first will become the last, and the last once again the first.

---

[141] In the *Compass der Weisen* much useful information is found on this matter. See Part II, Par. 3, Sect. 7 (g). Also Elias Artista's *Das Geheimnis des Salzes*, ("Secret of the Salt") published in 1770 contains much of value in this respect.

[142] The R.A.M.S. version used the word "salt" instead of the 1781 edition's "◯". -PNW

However, both salts, Saltpeter and common salt, are set to work differently in animals and plants, differently again in rocks and mountains. If they had one single disposition, they would be uniform. The natural quality of animals and plants has been indicated in their proper chapters, but here we will also describe the disposition of minerals, that they are no other, or have not originated in anything but an acid or corrosive vapor, or more precisely, in a strongly fermented and acidified ⊕ and ⊖, together with earth finely dissolved by the latter. The subtler the earth is made by these acid juices, and the earthier and more fixed the salts become (that is, such *Acid* become) through the earth, the purer is the metal which they produce. It is known to all natural scientists that not only animals, plants, and minerals are salty in their interior and that the salts are either more volatile or more fixed according to the *generis & specie* of each, but it is also known that the air is nitrous and salty; it is known that the sea and all ▽s are salty; it is known that the earth is salty within and without. Since this is true, a natural scientist will concede that the parts of the macrocosm are salty at their circumference, little or much; that the center of the macrocosm is still saltier, since from the vapors of the center, as proven, salts, mostly the fixed ones, are also often engendered and born at least in part. Now someone will also admit that the center of the earth is no crystal-clear spring, into which nothing but the water of life flows, because one can see the *chasmata terrarum* as well as *aquarum*, through which various kinds of rubbish flow into the center, just as in man's and other animal's stomach. In the same way, various pure and impure, sweet and sour foodstuffs get into the roots of plants. This chaotic mixture or foul cesspool causes a great fermentation with the help of the salts. The stronger the center is fermenting, the stronger the vapors and steam will be. These vapors are carried from the center to the circumference through the interior of the earth. At the surface, the thickest and strongest or most fixed vapors adhere to the walls of rocks, earth and stones and turn into water. That which is more volatile, however, rises to the surface of the earth into the roots of the plants, and that which is still more volatile breaks into the air and reaches the animal kingdom. The most volatile rises high up into the air, forms fog and clouds, and these in turn produce rain, dew, etc.

Such vapors are salty, because the center is salty, and by its fermentation and heating the salts, dissolved by the rain, rise in the form of vapor. The nearer these vapors are to the center, the sharper and more caustic they are. The further removed they are from it, the sweeter or milder they become, because they lose most of their corrosive and most fixed part of it, while permeating the earth and rocks. Since that corrosive is sharp, it attacks any earth and rocks it may meet, no matter which, and always first dissolves the open ones a little, until they are eaten away or dissolved by the vapors that rise up incessantly. When now the corrosive as a volatile acid salt spirit or ⌒ ⊖ attacks the earth, it eats itself to death and becomes corporeal, vitriolic, or aluminous, depending on how the earth is. On the other hand, the earth is dissolved, as the ⌒ is coagulated.

257

The portion of the earth which the corrosive has been unable to dissolve completely, is made subtle, dirty, and greasy in part. Alchemists call it a metallic guhr, or the first matter of metals, but wrongly so, because it is the first and nearest matter to sulphur and arsenic. When arsenic becomes marcasite, that then is the very nearest matter to metals, because metals spring directly from marcasite and not from this guhr which is only the distant matter of metals. This guhr or dirty substance is made ever finer by the rising corrosive vapors and is more and more dissolved. And the more it becomes subtle, the more it congeals the corrosive within itself, and that makes it sulphurous and arsenical. This arsenic is increasingly ripened into marcasite, and the marcasite in turn into metal. Such is the progress of metals, which we intend to show ever more clearly.

When the vapors rise into the cracks and crevices of the rocks, they turn into water because of their condensation (while more and more rise unceasingly). This ∇ contains the intermingled spirit of salt and saltpeter, which spirit is known by all alchemists to be corrosive. Here, however, in the center it is surrounded and diluted by much *Phlegma* and water. Such spirits adhere to rocks and earth by their sharpness, corrode and dissolve them, make them subtle, swollen, sticky, greasy and dirty, and turn them into a moist guhr which lies between the rocks and the earth like meat interlarded with bacon. But often it penetrates outside due to the swelling and adheres to the walls, as may be seen in old galleries and mines. The more dissolved and refined such dissolved earth becomes through vapors and salty spirits, the more it swells, pressing and driving out the remaining moisture by this swelling. In turn, this moisture runs back to the center or elsewhere into other corners and holes of the earth. This swollen earth or guhr now has no peace, because the continually rising and subsequent vapors are attacking it ever more, adhere to it, congeal and coagulate with the earth; and the more such corrosive vapors follow, the more fiery, sulphurous the earth becomes. The more sulphurous it becomes, the higher it swells, and more moisture it drives off, and becomes drier the longer it lasts. The drier it becomes, however, the more the sulphurous component part loses its combustibility and thereby acquires the name of mercury, or rightly, arsenic, which has originated in the sulphurous acidity. It no longer burns, but is still volatile.[143] This volatility, however, is gradually more bound by the central heat between the stones, and thereby congealed and coagulated, so that it is transformed into a marcasite. If the digestion or earthy central heat is strong, the marcasite is congealed into a metal; but if it is weak, it remains what it is or becomes a marcasite or arsenic choke-damp, or pyrite.

---

[143] Here he says something so remarkable and shows what really is the true metal—making Mercurius. The ancient Greeks had already directed us to this matter with the following riddle: "I consist of nine letters and four syllables, understand me correctly. Each of the first three syllables has two letters, the remaining have the others; in such a way that in the whole word five mute letters are found. The number of the whole contains two hundreds or centuries, eight and three tenths or decades, and seven ones or units. When you understand who I am, you will not be inexperienced and ignorant of that wisdom which lies hidden in me." In dismemberment the Greek word "arsenikon" emerges which some scholars have understood common ⚖ed arsenic; but that is not it - but he who is incorporated in our natural and artificial electrum, of which the *Compass der Weisen*, Part II., Par. 1, 2, 3, and its commentator, (who thanks to the grade he obtained in the sacred brotherhood, is bound to know better than all those pseudo alchemists who dare to criticize him) deals in detail.

Now one must know that when Nature has got so far that she had made ⚏ or arsenic, she has already filled the crevices and chasms and dissolved and caused so much earth to swell that the crevices are full to the brim. Then the earth does not admit any more vapors or moisture, and is no longer in dissolution. Now begins the desiccation, fixation and coagulation, and passes from there to a metallic nature or fixity. This filling of earth caverns, crevices and fissures appears to me like bees filling their cells with a little honey until they are full of it. Then they close the cells. Likewise, Nature sends one vapor after another, whereby she causes more and more earth to swell. This earth is full of acidity and is increasingly filled with it, so that the *Acidum* and the earth are so much intermingled that one cannot really see what they had been before, in regard to their primal matter. The acidity and the earth have formed into a third substance which differs altogether from the first. It is indeed the same with this birth as with the animal and plant, except that here it is planned that more fixed, harder and stonier subjects are to be made. But Nature works in the same order as those: for she first forms soft parts from soft and moist vapors, until she advances ever further and even hardens them into a stone, just as the subtle, soft, juicy young oak shoot or tree hardens more and more into wood as hard as stone. The difference in mineral creatures, however, is this: First, it depends on whether they get much or little corrosive. If they get much corrosive, they become ever more fluid. If they get little, they become and stay solid. If they get pure subtle earth or rock, and the subtler they can make these, the more noble

the metal is that they make. And again, the grosser
the earth the corrosive spirits etc. and the less
they work it, that is, render it ever subtler and
nobler by dissolving it, the baser is the metal they
make. Again, the more inadequate the digestion and
central heat is, the less they are dried, coagulated
and congealed. Therefore they remain stuck and turn
into marcasites, pyrites, arsenical, antimonial,
bismuthic and suchlike ores. Instead, if digestion
is strong, it produces stable and metallic ore.
Again, if the digestion and central heat are uneven
and too strong at the outset, the earth is indeed
dissolved but not volatile, and is immediately fixed
and coagulated and advances further to a metallic
nature. This is called *Embryonata sulphura*, such as
bloodstone, emery, magnetite, Tutia, calamine, etc.
If the digestion is weak from the start, however, so
that the earth and her caverns are soon filled and
the vapors can no longer get inside, they go
elsewhere. Because the earth does not get a stronger
degree of the central fire, it stays as it started,

volatile and open, such as ⊕, iron pyrite, etc.,
as has been reported above.

But if Nature provides the same grade of heat
throughout the four seasons or divisions, of the
year, she makes nobler metals with the help of the

moderate corrosives, such as, ☉, ☽, ♃, and ♀.
From this, an artist can "judge the difference
within the human race." as they say: Many heads,
many minds. Although we are all human beings, we are
yet not one like another, especially in our minds.
Each can see God's greatest miracle in this human
race, how he had created so many millions of human
beings, and yet there are hardly two among millions

and millions who are totally like another in one, two or three ways even, let alone in regard to their whole body organism. Just as Nature makes great differences in this race, she also does in the mineral "race." It would be impossible to describe everything; let everyone contemplate it himself. One

♁, one ♃, one arsenic, one marcasite and one metal is not like another. Consider only the difference in ☉; how it is found in various nuances of its color, depending on whether it is created pure or impure by Nature. Thus also, one

☽ is finer than another; the same for ♀, ♂, ♃,

♄ and ☿. One is finer than another, and likewise among the baser minerals.

Just as the earth and rock are the mother and foundation, or the vessel, of minerals in which fossil ore bodies are made, so vapor or ⌒ is

⊖ lis is their food. ♁ or vitriolic guhr is the

root, ♃ and arsenic the stem, marcasite the flower or blossom, metals, the seed, the completed birth and offspring.

That ♁ is first born of ♃ and arsenic can be proven by guhr, if it is lixiviated, filtered and coagulated. Then one finds a vitriolic-salt after the kind of earth that had been dissolved. I do not call it ♁, not that it is common green ♁ as can be bought from shopkeepers, but because it has a vitriolic or aluminous taste.

That ⚥ or arsenic originate in this way (because in the white metals, such as ♄, ♃, and ☽, yellow burning ⚥ is seldom found or, if so, very little of it, but more often white arsenic and aluminous ♁ or ○), can be seen during dismemberment. When the *Acidum* or the sour vapor is driven by △, it distills first. It is followed by the flowers of sulphur, then the arsenic, then the volatile marcasite. The fixed marcasite flows into a regulus and slag. That marcasite is made from arsenic, can again be seen in dismemberment, because bismuth and antimony driven into flowers are quite arsenical and volatile, and also have the total nature and quality of arsenic.

That metal is made from marcasite by a tedious fixation, can be seen by the fact that almost every marcasite gives off a fixed grain of perfect or imperfect metal in the assay.

Thus the lover of the Art sees once more that Nature advances gloriously and beautifully through intermediate stages and not from one extreme to another, but from the volatile vapor of its kind to an ever more and more fixed nature. In comparison with the plant and animal vapors, this vapor is fixed. True, many authors have taught that is the

♁,[144] the guhr or prime matter of metal. Some have also added that is the root and mother of metals;

---

[144] This symbol from 1781 is missing in the R.A.M.S. edition. -PNW

but because they did not make any distinction, error and confusion have arisen from those statements, of which the lover of the Art knows no way out. Not every alchemist has the opportunity to go into the mines, and even if he did go into one of them, not one out of a hundred would have any knowledge of them. He may well see the mountain walls, the ore and the rocks, that the ore is black and white and is this or that metal. Further he cannot penetrate, and it is also impossible to do by mere looking. But if he breaks off a specimen of ore, puts it in the

△ and sublimes one thing after another, he can then further examine those parts and judge what they are and of what they comprise. In general, if there is some liquid, it is acid, vitriolic, sulphuric; if there are *flores*, it is generally ⚲ and arsenic.

⚲ is recognized by its combustibility and acrid smell. Give a little bit of arsenic to a dog: If it throws up, it is arsenical. Thereupon quickly give him a lump of butter and mithridate mixed together.[145]

Marcasite is known by the fact that it does not rise as high, but that a volatile part such as cinnabar or ☿ ♎ *tat* has risen above the more fixed, over the feces. Melt these sublimed portions and the feces together, and you will obtain a brittle, reguline mass. This is the marcasite. The more fixed, however, will turn in part into slag with which the stone-mother is mixed. The latter causes a good part of the dead head and the metal to be likewise turned into slag and glass. The dead

---

[145] *Save the dog!* Easy-to-use arsenic test kits are available for about US$20. -PNW

head, however, which settles among the slag, is partly (fixed) marcasite, partly metallic. In refining it, the marcasite is driven off the metal, and the metal remains.

That many writers have called ♂ the root or prime matter of metal is not totally wrong,

especially if they understood under ♂ the marcasites or the marcasitical kinds. Otherwise,

however, ♂ is a marcasite which got stuck for lack of more ripening. In this way a lover of alchemy can sooner reach his goal: because the volatile always goes ahead in the fire, and the more fixed part follows at higher heat.

Above we said that metals are born of vapor, of a salty, spiritual vapor or spiritual salt. We added that such is a corrosive. Now I will take for granted, as I did above, and again remind the reader that all and everything is born of salt and saltpeter, and that all and everything will be reduced to niter and salt in the ultimate dissolution. Because this is known, I say that these salts also reside mixed in the center of the earth. They are fermented and driven into a volatile vapor by the central fire. Because this vapor consists of saltpeter and salt, I should very nearly call it *Aqua regis macrocosmi*, the royal mineral primordial water, but I will leave the naming to every alchemist. Let him call it what he likes: Some call it the mercurial and sulphuric vapor: the salt they call ☿, and ☉ sulphur, etc.

But here there is a hitch, namely, that I said that Nature ferments and produces dissolved corrosive salts. This is a point against which the whole world screams the contrary. In general alchemists do not wish to know of any corrosive but want to have everything sweet and nice; and yet there are so very few who possess this treasure of the sweetening and the *Modum dulcificandi*. It is this that the whole world contradicts. But how can I prove this against general opinion, since the whole world is against it, and they will never find a corrosive in and on earth in a natural form, that is according to their way of understanding and that of the common laboratory workers! Answer: I have proven above that the salty water vapors rise again from the center to the bowels of the mountains, adhere to the earth and eat themselves to death with it, coagulate and congeal themselves, whereby they turn into a greasy turbid guhr. The *Acidum* dissolves the earth, and the earth coagulates the *Acidum*.

When an artist admits that salty vapors rise, he must confess that such salt is a dissolved salt. All alchemists call such a dissolved salt ⌒ ⊖ *lis* or ⊕ *tri*. If then it is a spirit of salt or saltpeter, whichever, they say and admit themselves that ⌒ ⊖ and ⊕ are corrosives, especially if those spirits are separated and rectified from all *Phlegma* and excessive earth. The more those spirits are separated from wateriness, the more corrosive and caustic they are; the more wateriness there is, the less the corrosive is noticed. Let someone pour one lb. of ⌒ ⊖ and ⊕ into a pail of water and then see how much of the corrosive he can sense.

Even if one swallows a few drops of the said spirits in a spoonful of water or any other *Vehiculum*, one can already see that the corrosive is milder. Why then should the great mass of water in the earth not dampen and hide the sharpness of those corrosives to such an extent that it cannot be noticed? Instead, if the pail of water which contains the corrosive is concentrated through evaporation of the water, the corrosive will become ever greater and sharper.

Consequently, the quantity of the water renders the corrosive insensible. This blinds Messrs. the doubting Thomas-brothers so much and detracts them from the real reason and main foundation, that they will never get behind the true prime beginnings of Nature.

Therefore, as soon as these corrosive vapors reach the earth or rocks, they adhere to them and attack the earth by dissolving it. In so doing, they turn it into a vitriolic or aluminous salt, which can be proven by manual work. Take a corrosive, whichever you like, ⌒ ① or ⊖, or ⅁; throw into it some earth which the corrosive can attack. The sharpness will enter the earth and adhere to it; the earth will be dissolved and the corrosive coagulated. One learns of this if the moisture is evaporated to one-third and the rest put in a cellar. Then it will crystalize into a vitriolic salt or crystal like the earth. By this ⊕ one can see that the corrosive has eaten itself to death with the earth by dissolving it. The artist will also see that the corrosive still contains excessive wateriness after it has been dephlegmatized in the best possible way. When the corrosive has dissolved the earth, distil the moisture into the receiver through the head, and you will get sweet insipid water. Or, if there had been too little earth and the corrosive could have been dissolved more, a corrosive may well also go over but so much weaker than the first that it is almost all pure water.

That this ⊕ or guhr is made by the universal corrosive of ⊕ and ⊖, and not by the corporeal salts but by spiritual and dissolved ones, is shown by the ⊕ itself. Let an artist treat earth with salt that is not spirit but body, whatever he does, he will not turn it into such a vitriolic substance in all eternity; but he could do so by using any kind of acid or dissolved salt, or the salty spirit of saltpeter and common salt, vitriol, sulphur, or alum, yes, any sharp rectified plant vinegar. That this is true, and that the mineral guhr is made of corrosives must be proven by its restoration into the prime matter. Distil that guhr or ⊕ lixiviated from it, and look if it does not give you some corrosive water. For in what anything originated, into that it must again be dissolved and restores: *Ex quo aliquid fit, in aliud rursus resolvitur*. Minerals are engendered of corrosives, and are again restored into corrosives. Let someone distil marcasite, iron pyrites, alum earth or other ore - he will always obtain a corrosive *Liquor*, be it much or little. If an artist wishes to know the component parts or prime beginnings of metals, he must not inspect them when they are in a molten state, since most of their primordial substance has been separated from them by the fire.

If he considers the minerals and ores when they come out of the mountains and have not yet passed through fire, he is sure to learn the difference. Let him just take such guhr or marcasite, iron pyrites, ores of arsenic, orpiment, alum, antimony, lead, tin, iron, gold, copper, silver, or mercury, and let him distil it with the strongest fire. He will everywhere find little or much corrosive water. But the more open and nearer a metal is to the guhr, the more water it gives, because fixation drives nearly all excessive moisture out of those ores, so that they become fire resistant and almost indestructible. The less moisture a metal has, the more durable it is, because the excessive moisture, which has to be separated from it, is a tool of the World Spirit. As long as that moisture is in the metal with the spirit, it keeps on rousing the spirit to action, since it cannot operate in dry matters as it can in moisture. This is the reason why animals and plants are forever in a state of alteration and instability: because of their excessive moisture. As they grow, they simultaneously move toward their destruction. The same applies to those ores which still contain moisture. Yet they are much more durable than plants

and animals, for instance: ♂, ♀, ♄, ⊕, ○.

This is the reason why the ancient philosophers, seeing that all species of animals and plants die so quickly, some quicker than others, felt compelled to look for a balsarnic vital and universal spirit, and they did indeed find it in minerals where it is concentrated in quantity. Each and all, stone, skin and bone, and whatever there is

in every ore, is found coagulated and fixed, the
world spirit or spirit of life.[146]

But because they saw that also among minerals
some are instable and not durable, they chose those
which they considered the most durable, such as ☉

---

[146] It is an age-old truth founded in Nature, that the PHILOSOPHER'S STONE, the age-old
magical UNIVERSAL STONE, originated in the universal primal substance or prime matter of
all metals. That is why the COSMOPOLIT says somewhere: In cavernous metailorum occultus
eat, qui est lapis venerabilis, colore splendidus, mens eublirnis & mare patens, that is, "In the
caverns of the metals is hidden he who is the Venerable Stone, of a splendid color, a sublime
spirit, and an open sea." Here some who are so eager to impute contradictions to the philosophers will object and say that it is explicitly written in the Book of Saturn (Theatrum
chem. Vol. I, P. 244ff), that our Stone, which is the MERCURIUS OF THE WISE, is not found in
the caverns of the earth, being no other than body and spirit. But there is not the slightest
contradiction here; for in the first instance the Cosmopolit speaks of the true substance from
which it is drawn, while in the latter he is considered when he has already become a double
☿ (Mercurius duplex) by processing, which occurs only after the union of man and woman
has been accomplished (See "Grosse Bauer," P. 31). Since a large number of scholarly natural
scientists indiscriminately reject all metals, no matter what natures they have, and only keep
to nonspecified products of Nature, this does indeed cause confusion in the inexperienced.
But all one has to do is make a clear differentiation and not mix the Lapidem universal magici
opens maximi with other tinctures, and the difficulty will soon be resolved.

The Universal Tincture is made from an astral subject, in which the ♒ mundi as ☿,
the heavenly △ as ♁, and the humidum astrale as ⊖ are concentrated together, by means
of the pure primordial perfection, through a second separation by the artist. The matter is
called Astral or Universal.

Instead, while the matter for the universal tincture has indeed a homogeneous
origin or beginning with the universal astral material, it obtained a determined nature after
exhalation of the purest light by the △ of Nature, that is, a specified ♁. It can be cleansed
from its curse by an experienced artist instructed in our schools, and be brought to super
perfection in the kingdom specified by the Creator. And this matter is still all greening, still in
possession of its ☿ and multiplying seed, and can still generate many living beings of all
species and kinds through its ♒ ☿ ii.

Finally, the matter for the Particular tincture has indeed also its beginning in the
great SHAMASIN, but it has already received the middle and end of the destination assigned
to it by the Creator through the Nature - △, so that it has really already become a specified
body, whose substance has been abandoned by the ♁ ♒. Therefore it is impossible to
animate its multiplying seed without renewed assistance from a life-giving spirit, and such
matter is everything that has been brought to its perfection destined by Nature, specified by
♁ through ☿ and congealed by △.

270

and ☽ and almost all precious stones since precious stones are only available in small quantities, they stayed with ☉ and ☽ and prepared from them medicines for lengthening life.

But every lover of the Art should mark that just as minerals can discard their excessive moisture of themselves in a natural way, he has the power to separate it even more by the Art, and not only in minerals but also in animals, plants, including universals. When the artist dismembers, divides and purifies such subjects, he can see for himself how all excessive *Phlegma recolaceum* is easily separated by rectification; the spirit, however, thickens, concentrates and gets sharper. It can also be pressed so hard that it cannot be consumed with harm, except in smallest dose, as I will describe in my other book in *Anatomia & corruptione rerum*. If therefore, a reader has to prepare a remedy for sustaining, preserving and also lengthening his life, he has no reason, if he does not wish to do so, to run to mine galleries or earth growths and animals, but he should run to himself. His own urine and excrements are powerful enough to prepare with them the most glorious medicine for himself and his neighbor, for they contain the world spirit as truly as gold and silver and the very carbuncle itself. But you must separate the excessive moisture from them and put the pure constituents together. When it is still moist, draw it off in B.M., and you will find at the bottom a treasure above all treasures for your health.

That against all objections, an effective corrosive is found in the earth, can be seen by everybody in sulphur. It corrodes, drives and chases imperfect metals, especially ♂ and ♀ back into their prime matter by its sharpness. Is not its smell as sharp in the nose, and does it not bite as sharply as a corrosive? Does it not violently attack the lungs, so that man can hardly sweeten it and drive it out by long coughing and much saliva and slobbering?

Is it not a much subtler corrosive in a dry state than while liquid? By this may be seen the difference between its own acid character and that oil of vitriol? It has such a delicate and penetrating acidity that one cannot believe it unless one pays special attention to it.

What else is arsenic but a corrosive? Does it not eat through and corrode all metals and does it not even spare ⊙ and ☽ ?

Can one not see clearly that the subterranean vapors are so caustic in some places that they drive the miners from the mines, unless they are ready to suffocate? If there is no corrosive in the earth, why then are many a miner's clothes so eaten away as if he had marinated them in ▽̶ , as soon as he but leans against a certain spot?

If one considers the nature of the subterranean waters and examines them, and concentrates a good amount of them through distillation, such as the warm baths of sulphur, alum, vitriol, and saltpeter, likewise the warm acidic springs, are they not extended corrosives? And if someone were to concentrate and heat them just a little, then put into them a dead chicken, would the solution not injure and peel off the feathers together with the skin and flesh? Sweet, and also salt water as they are, on and above the earth, would never do this.

One can indeed see it in the raw: When someone bathes too much in such waters, they attack his nature and skin violently, so that he often looks quite horrible and his skin peels. Acidic waters cause a similar reaction if one drinks too much of them, since, after the death of a patient who used the acidic waters, it has often been found that his muscles were quite loose and so soft as if he had been marinated like poultry, and one could have detached them from the body without a scalpel.

By concentrating several quarts or tubs of those waters, one can see how little of such a powerful active substance they contain, while even such an amount of water nevertheless is still so strongly effective.

Chemists cannot understand, far less the squared gas-bag Aristotle of the Peripatatics, that this corrosive can hide (especially as no miner speaks about it and no historian has until now written anything about it, or if he did, very little). The reason is, as I have said, (1) the great quantity of water, (2) the earth, which absorbs and coagulates the corrosive, (3) because no corrosive can ever be felt as vapor but, instead, as water. This is proven by the following experiment:

Take $\overset{\circ}{\circ}\,$ ⊕ *li* or $\overset{\circ}{\circ}\,$ ⚷, or ⌒ or $\overset{\circ}{\circ}\,$ ⊖ *lis*, or ⊕ *tri*, etc. Of that, pour lbj. into a bucket of water, stir it well. A man can drink of this without harm, and thereby one can see that the corrosive is not noticeable. Likewise, it is contained in the earth (rocks etc.) insensibly also.

Now take this water, pour into it j. or ij. lb. powdered chalk; let them boil together, then pour the water off from the chalk and taste the chalk. You will find that it has become salty. This salt comes from the corrosive which the chalk had absorbed and congealed, but part of it has stayed in the water. Let it evaporate and crystallize, and you will find vitriol produced by the dissolved chalk and congealed by it. Consequently, it is contained in the earth. If the corrosive is diluted by water and evaporated, it cannot be noticed by the nose, except (common burning) sulphur, which is a pure concentrated corrosive. Take ⋎̶ or ⋎ℛ, ∿ ⊕ℛ *li* or its oil etc. Put it into a small open dish, set it over △, let it evaporate in a room, and it will cause tremendous vapor to rise, so much so that one dram will fill a whole room with vapor and steam. This vapor can be inhaled by anyone without his noticing the least sharpness. But if one were to give to a person just j. or ij. *quintlein* (substantive) *in liquido* on his tongue, one would see how he would jump about, worrying, that the fire might penetrate through him.

The higher such a corrosive is driven up into the air as vapor, the more it gets intermingled with the air and sweetened and chaotized by the air's volatile salt. From this the lover of the Art again sees - and should take careful note of the fact - that through circulation not only the air but each and all dissolved things return into the prime

matter, or into chaotic ▽ . This is not only true
of animal and plant evaporations, but also of

mineral ones, and those which are dissolved in ▽
or vapor, for example, animals and plants by fire,
water and earth. How many animals and plants are
everyday dissolved in water, partly by cooking them
for human food, at which time also vapors emanate
from them, and partly by consuming them by fire
while cooking, when thick smoke escapes through the
chimney like clouds. In that fire, cooks often burn
animal bones and drive the smoke through the flue.
Just as they use butter and lard if the fire does
not burn well. From what has been said above, we can
see clearly that Nature cannot make a metal without
a corrosive. If Nature had to make metals from a raw
and corporeal salt water or saltpeter water - which
might be feasible, since all lightly dissolved
earths can easily be altered by warm saltwater - but
as Nature must previously work at them for a hundred
years, she would probably here be busy for at least
a thousand years. If salt is spiritual and
dissolved, it attacks twenty times more than not
dissolved, even if it has only been dissolved in
water.

All one has to do is take a corrosive, or salt
made spiritual, dissolve it in some earth through
digestion in a sand cupel, in a small alembic. Now
take the corporeal salt out of which the aforementioned corrosive was made. Put it into water to
dissolve it, pour it over an equal amount of earth,
set them both immediately to dissolve, and you will
find the difference.

When now both have been dissolved, one will obtain a bitter, astringent ⊕. But with corporeal salt you will get nothing of this quality.

Quid? Dissolve a metal as quickly with dry or wet corporeal salt into such a truly mineral ⊕, either by melting it, or in water, and use a corrosive. There will be a great difference. A corrosive will immediately begin to attack it and turn it into ⊕. With a fixed ⊖, you will never get a ⊕ of the same strength and taste as the one derived from the corrosive.

But if someone were still to doubt and say that it is not true that Nature makes minerals through corrosives, then I say to him: first he should meanwhile not believe it until he is forced to do so through many mistakes and errors; second I refer him to the more volatile and much weaker species, i.e. animals. There enough sharpness can be found thereby to give testimony to the mineral kingdom by means of comparison from the lesser to the greater.

If man had no sharpness in his stomach to attack the food, how could he achieve such a highly-amazing, fast putrefaction? Supposing a person is given some easily soluble metal, such as ♂ and ♀. Is it not true that the *Menstruum* immediately attacks it in the stomach in order to dissolve it? But because it is adverse to him, it causes a convulsion and the stomach discards it again by vomiting.

Consequently, everyone can see clearly that sharpness is necessary for every dissolution. If the sharpness in man is great, it is still greater in plants, and greatest in minerals, which must have the strongest digestion, because they must boil the raw fixed earth. Plants, on the contrary, require for their constitution a tender earth, putrefied already long before (i.e. soil). Men and animals, however, have to digest and boil down the very tenderest and juiciest plants by means of the acidity present in their stomach.

Such sharpness is called a dissolving, caustic, acid substance, etc., *menstruum corrosivum a corrodendo*, because it attacks the subject, crushes, powders, dissolves and renders it subtle and finely divided. Do not alchemists call the strongest rectified ⟨symbol⟩ and ⟨symbol⟩ a corrosive, which is yet in its expanded form an excellent tonic and medicine and an invigoration of all natural forces. Thus we can see that although these animal and plant spirits are consumed by everybody when diluted and dilated, they are nevertheless extremely sharp in their concentration or contraction and rectification, so that they must be taken in the smallest dose. If then there is within ourselves and the other animals and plants such a great sharpness, who would then doubt that minerals must necessarily have three times more sharpness in order to dissolve the raw earth? I have said that the *Acidum* or mineral *Menstruum corrosivum* that is, the subterranean acid salty vapors dissolve the earth, thereby turning into a greasy guhr together with the earth, which guhr is now vitriolic or aluminous, acidic and styptic. The more now this guhr is dissolved, fermented, coagulated and congealed by the ever following caustic and acid or ∿ual-salty vapors, or by the volatilized vaporous salt, the more and

more it becomes sulphuric and alkali (simple ▽ or *Phlegma* is unable to dissolve that earth and to turn it into a metal without spiritual salt).

Since the *Acidum* accumulates in the guhr, and the guhr in the *Acidum*, the more the guhr gets Acidum, the more it becomes sulphuric and liquid. Thus this is gradually more and more digested and congealed from stage to stage. It becomes arsenic, and this turns into marcasite which is the closest

matter to metal, just as 🜍 and volatile arsenic are the closest matter to marcasite; the guhr, however, or its vitriolic constitution is the

closest matter to sulphur and ☿ *vivo*, i.e., arsenic.

When ⊕ or guhr is overwhelmed and dried up by the acidity, it turns into 🜍. This may be seen by the fact that when ▽ or ∿ ⊕ is often distilled over common ⊕ at the third degree of △, over the still head, it becomes drier each

time. If it is then put on heat, the 🜍 will immediately be perceptible to the sense of smell.

When now ⛢ and arsenic are digested into
marcasite, either together or alone (⛢ is sooner
born due to the accumulation of ☽ric or ⊖ty
acid), the marcasite becomes a metal by the length
of the digestion and maturation, coagulation and
fixation, and this in accordance with the digestion
and accidental (geological) environment.

This, then, is the beginning and end of the
mineral and metallic birth, described after
theoretical principles. Now we will write
anatomically about the praxis, as much as possible,
and confirm our theory, although it has already been
sufficiently shown elsewhere, so that it could not
be revealed better to a well-instructed disciple.
Since many honest men dare enter our Art like
simpletons, I will say the following:

Thus, Rc. some ore, as it comes out of the
mountains, that is, before it has undergone any

smelting, such as ☿, ♄, ♃, ♂, ♀, ☉ or ☽,
etc. Wash these clean of all mountain earth, or
leave the mountain earth with them - it does not
matter. Powder them small like millet-seed, and not
into dust, because it would lie together too
strongly and thickly and would congeal more than
dissolve, as the emerging or rising vapors would
become stifled. Put this powder into a strong retort
in the open fire, add the recipient and apply heat
by degrees. Then you will see rise over first a very
little bit of water, which is the excessive
moisture; after this comes a strong vapor, which is
the mineral corrosive. It settles in the receivers

and condenses into a corrosive water. Thereupon vapors rise once again, but not as volatile as the first, because the volatile always comes first and always followed by the more fixed parts. Such vapors rise but little into the recipient; instead, if the neck of the retort is long, they settle first in the neck. The subsequent ones are more confined to the stomach of the retort, while the following can stand the fire ever better. After this, the more fire resistant parts are left in the bottom of the retort, many or few, depending on whether the subject had been very volatile or fixed.

Now examine everything that has gone over and risen, also that which has remained at the bottom, and you will find in the recipient:

(1) The excessive moisture mixed with the corrosive, which is a sulphurous or vitriolic, sharp, salty *Liquor*, much or little, as the ore had been dried, coagulated or congealed strongly or only a little, in one word, a ⌐⌐

    ♅ or 🜍. (♀)

(2) You will find flowers at the entrance or beginning of the neck (i.e. where the retort joins the receiver). The first are quite volatile. Test them on coal: If they burn like sulphur and smell like sulphur, they are called sulphur; but if they do not burn but flow and have an arsenical smell they are called a volatile arsenic.

(3) Above half the neck (i.e. that half of the arm of the retort nearest the body of the retort) you will again find flowers that are somewhat more fixed than the first. They are a fixed arsenic.

(4) At the beginning of the stomach of the retort
or its firmament (i.e. the upper part of the
body of the retort) other flowers have become
distilled. They are still more fixed than the
previous. These are called a volatile marcasite
or a marcasitical fixed arsenic; for the more
fixed sulphur and arsenic are, the more they

lose their first name and acquire another. ℞
changing form and then called sulphur, arsenic
is then called marcasite, and this is then
called metal. They acquire this difference by
becoming ever more fixed. At the bottom of the
retort there remains a threefold mass.
(5) First, the more fixed marcasite, which is
closest to the metallic nature and which turns
directly into metal.
(6) Second, the metal grain, which has sprung from
the marcasite (regulus).
(7) Third, the stone-mother, matrix (gangue) in
which the metal had grown and was congealed, as
in the great philosophical vessel or glass.
This matrix turns into slag or glass in a big
smelting furnace.

In this part there is still a fixed salt, which
is lixiviated with water. It can be considered the
mineral *Alkali*, which had been concentrated and
congealed by the fire of the risen vitriolic spirit,
although only a little, but according to whether the
aforementioned subject had been more or less moist.

Now remove this last part from the bottom of
the retort. First, wash the salt out with water and
dry to a powder. Melt this powder with a strong fire
and it will leave a regulus, and (a second) slag

281

will be on top. Now boil the dead head with ♄ and refine it as metals are generally refined, and you will find the metallic grain. But the dead head, although it has greater resistance in the refinery than its preceding parts, must nevertheless finally

flee away because of the force of the fire. This ☠ is the more fixed marcasite and mercurial alkalized part, or the congealed and alkalized sulphuric acid, and this is the first, direct matter of the metal, out of which metals are born directly by long fixation.

Yet this practice does not at all apply to all metals but mostly to ☉, ☽ and ♀ ore. These can be refined in the greatest heat, according to the grade of their fixity and strong alkalinization, because sulphuric acid becomes the more alkalized the more it is congealed and mixed so thoroughly with the earth that not the least bit of *Acidum* can be noticed about it. It is the same with gold refined to the highest degree, which I call alkalized because it cannot be overcome by any acid, unless the alkalized gold is aroused with its homogeneous marcasitical or salty *Alkali*, which can then again turn into an *Acidum* through the sulphuric, vitriolic or nitrous acidity and not otherwise.

From what I have said above, the reader can see the structure and dismemberment of metals and ores. In the same way, he should carry out his investigations and not immediately use mere fire, since he would then drive the vitriolic, sulphuric and arsenical parts away, which are the vital and life giving spirits of the ores.

The lover of the Art will also see how Nature proceeds so beautifully through intermediate stages, both in the animal and in the plant kingdoms, from the watery volatile parts ever more to the more fixed, up to the most fixed and hardest dry parts.

Now let him also consider the metal which has, been drawn out forcibly in so many ways, how little it is in comparison with the separated remainder. Metal is only the ninth part, and this is of such a small quantity that the things which had been associated with it previously, exceed it a hundredfold. One can indeed see how little gold and silver a hundredweight of gold and silver ore furnishes: hardly a few ounces. Instead, much of the excess goes up in smoke and remains as slag. Now let him reflect that if Nature must produce some inferior metal from the aforementioned kinds, how long it must take her with her slow digesting, let alone if she is at work with perfect ones. From this the lover of the Art can see how many degrees of fixation from the volatile vapors or the vitriolic guhr are required, to get to the more fixed marcasitical nature, and only after that for the metallic, let alone for the best birth. Each and all are nevertheless born only of salt, or actually a double salt, ⦶ and ⊖, that is, ⦵ [147] of their spirits and rocks. Through such stages Nature goes from the remotest end to the other in lawful order, in everything as well as in these creatures and Individuis. First, she makes the softest, then she

---

[147] This sign indicates the universal ⦵ of the philosophers, or the prime matter of all created things, from which may be seen from what constituents these things took their beginnings.

hardens this more and more up to its perfection in bones, wood, metals and rocks, as it has been sufficiently explained.

I should probably relate here the constitution of each metal in particular, but from what I have said everyone can examine them himself and muster up his brains. If he knows the general constitution, he will probably also determine the particular. Yet I will here teach him how he can immediately recognize any metal or mineral, whether it has excessive moisture or not, that is, if it is highly fixed or has a middle or acidic nature, or which metal is acid and which alkaline, and which has both natures simultaneously.

So, let him take some ore, a universal or a piece of metallic rock etc., whichever he wants, and add an acid corrosive and an alkaline corrosive. We recommend most and in general ⌒⊕, or ⋁̄ and ⌒⊖ *lis*. If then something is dissolved by the ⌒⊕ or ⋁̄, you may infer that its *Acidum* is still open and not congealed and alkalized, as may be seen with ☽ and ♄, for like dissolves like. But consider a metal to be alkaline and fixed if it cannot be dissolved by such ✝ but with an *Alkali*, mixed together with the acid. Judge by this that the *Acidum* in such a metal or mineral is totally alkalized or congealed, so that it must be awakened by its homogeneous spiritual *Alkali* and acquire the acid nature, so that, after it is opened, the *Acidum* can also transform it into its own nature and return it into its first vitriolic nature and matter,[148] such as ☉ and ♃.

As against this, ores and metals that can be dissolves by acids as well as alkalis are called hermaphrodites. These had begun to become alkalized or fixed but remained stuck, so that they are fixed and not fixed, alkali and acid. Therefore they can be joined to and especially attacked and dissolved by both solvents. They are ♂, ♀, ☿, ♃ might also be counted among these, but *Alkali* dissolves it more easily. ♄ also dissolves in ⌒ ① or ♈, but in ▽ it even dissolves into water.

But now someone will say: So I am to recognize ♄ and ☽ as sulphuric, because ♀ is considered acid, while they are yet mercurial; and ☉ and ♃ are to be considered mercurial, although they are mostly sulphuric? To him I reply: Be satisfied with what you can see with your eyes, which your brain will sooner believe. But leave that which only depends on speculation to others to worry about and to tire their reasoning until they change their mind. Remember this at all times, all ores originate in *Acido universalio corrosivo*, which becomes increasingly alkalized through fixation and exsiccation. Try to understand such *Acidum* and *Alkali*, and you will in a short time learn more than all mercurialists, sulphurists, salinists, or

---

[148] Whoever wishes to alter bodies out of their nature, must first bring them back into a salty and alum nature, thereafter dissolve them, or else he will not accomplish anything.

vitriolists will ever get to know. Follow in the footsteps of Nature and reflect upon her ways. After that, name them, and let go of the other names which cause the confusion of all things. But if you do not like my opinion, I do not wish to deter you from the other.

Some will say, however: If metals do not consist of ☿, 🜍, and 🜔, and have not grown from them, how then can we change them again into them and drive them back into those primordial constituents, as all old philosophers admitted that they consist of.

Answer: That metals can be returned into those prime beginnings, I readily confess; but that they consist of them according to the rule and the direct universal law of Nature, I do not find true. For I do not find common ☿ in any mine except in its own ore, which it would penetrate and impregnate with its *astro*. Salt and sulphur I find in most ores but no common salt, nor common sulphur, as their symbols show, but vitriolic-sulphuric-aluminous salt and pyrite intermingled with arsenic and marcasite etc.

Few artists understand the secret meaning of the prime beginnings ☿, 🜍 and 🜔: because the patriarchs understood things differently from the calculations of the modern smart alecks. Do they not say: All things are made and composed of mercury, sulphur, and salt? And that of which they are composed, into that they are also dissolved? If you now wish to conclude: The symbol of ☿*rii* is supposed to be common mercury, and sulphur the burning sulphur, and salt the common salt, you will go wrong far above heaven and earth. Who can find common ☿ in the animal or plant kingdom, since their volatile part is precisely designated by the symbol ☿? Who will find in those kingdoms burning yellow sulphur, or common kitchen salt, which are precisely designated by the symbols 🜍 and ⊖ ?

Can you now see from this how the world is mistaken and how the world's reasoning sometimes interprets a thing wrongly, which is then immediately followed by many thousands? The old sages did not understand it thus: For even if they designated common mercury by the mercury symbol, it is by far not the general mercury, and the same applies to common sulphur and salt. And although common mercury can be made of metals, it happens by chance and not by virtue of the natural composition of metals. If that were true, I could turn ☿ into ▽, or ☍, ☌ or ♁, salt, oil, and ⌒. Consequently, must I consider ▽, ☍, ☌, ♁, salt, oil, or ⌒ the prime beginnings of Nature, out of which Nature forged the metals? No, far from true, and such a man will in no way upset the natural action of Nature, the order of Nature, with his accidental practice, but indeed bring confusion to the whole of Nature:

Inde mundus errorum plenus.[149] They are forever running around in circles, never turning their eyes to the center. And as they do and describe it, they also teach others, and thus the blind are leading the blind, and afterwards all fall together into the pit. One follows the other, and among a thousand, not one learns the truth. That is why so many expenses are incurred on account of potable metals and mineral medicines, let alone alchemy, because they treat minerals and metals with ineffective solvents from the animal, plant, and mineral kingdoms. And while they use a right and homogeneous Menstruum to do it, they believe the mineral or metal should immediately change into a sweet sugary oil, in which they are terribly mistaken and then proclaim that the Art is no good.[150] For no one has ever reflected upon the fact that the mineral kingdom is by nature totally acid and must necessarily be corrosive if it is to dissolve raw rocks and earth, digest them and boil them into metals. None of them has thought about this, for they wrongly imagined that they had to add the mineral-corrosive medicine directly to the animal nature. They did not consider that Nature herself has indeed put up and hung up a self-evident barrier and curtain between animals and minerals. This barrier, that is, the plant kingdom, as an intermediate nature, has ever been evaded by them, or if they did use it, they dealt with it so contrariwise that they were blind with seeing eyes. They did not notice when they tried combining heterogeneous things, and they did not unite. An intermediate was missing, which they cannot find due to their peculiar blindness. That is why they have

---

[149] from there men wander -PNW

[150] In another passage of his unpublished writings, our author says: The insipida menstrua, which some scrupulous ones demand, are the true idols in Ebronand and the secondary gods who entice you away from Silo until you will again be forced to return to the right source of Nature.

prepared nothing but caustic medicines, or empty precipitates and powders, with miserable results. But what they achieve thereby, they know themselves.

Now back to our purpose. In the chapter on the plant kingdom I mentioned that acid is astringent, styptic, and blocking: but here I have said that it liquefies. To avoid being proven guilty of contradiction, however, I will also elaborate on this point. Where Nature has added Acidum to a large amount of earth, we can see that it produces nothing but contractive and styptic things, as is evident in ♂, ♀, bloodstone, red earth (hydrated iron oxides), magnetite, emery, bole, etc. Because the small amount of acid adheres so strongly to that earth and dissolves it, it congeals and dries it into hard-to-dissolve ores when the central heat is too strong and sudden. But where the heat is not all to strong and the *Acidum* or corrosive is a little stronger, it produces somewhat liquid ores and metals; and where there is an abundance of acidity and absence of a longer or stronger exsiccation and fixation, the ores and minerals remain open, so that they are not very durable, such as ♂ and ♀.

Where, on the contrary, there is an abundance of acidity, liquid minerals are produced, as may be proven by manual work. Rc. Chalk or quicklime: Pour upon it some ⌒ ⊕, ⊕, ♀, ⊖, or their acid caustic "oils." Distil it again strongly to complete dryness, then calcine it on a cupel, under the muffel, or in a crucible, and it will turn into a styptic and become an infusible solid. Since it was styptic before, it is now far more so, so that the Acidum can be heated and congealed with the chalk.

Now pour some more *Acidum* on this earth and distil strongly. Calcine it as previously, and you will see that it will become a little more liquid than before. If you pour yet more acid over it and congeal it, with it, it will become increasingly liquid and will finally flow like salt.

To emphasize this point and to remind the artist of it often, I say that he should know that the stronger the *Acidum* works in the earth, and the oftener and stronger it is exsiccated and congealed by the heat, the more that *Acidum* is alkalized, concentrated, and made heavy. First it is called

⊖, then ⧋, and finally fixed ☿. For just as ☿ had previously been a volatile and very light vapor, so it becomes fixed and very heavy afterwards. The more this *Acidum* is alkalized, or the more it goes down into fixity, the more it changes its name:

First, it is called vapor, then guhr, then

⊛, then ⧋, then arsenic, after that marcasite, and finally metal. The fireproof ☩ turns into ☿, as may be seen with all marcasites. It can now be shown in the form of a fixed powder, and it is

rightly called ☿ ⎯⎯ *tus*. If common mercury is precipitated with an acid, it also turns into such a

spongy powder, and as it has not enough ⧋, acidity and fusible fixed arsenic, it becomes glass by reduction. That glass then contains the most fixed metal, the best tinging ⟁, whose fusible *Acidum* has gone and escaped. An alchemist should take careful note of this, and Metallurgist ponder over it day and night: How can he join these two, (1) an arsenical subject, i.e., a ☿ous liquid one, (2) a *Praecipitans*, such as they are easily found. One has only to consider ♄. Is it not saturated with ⚴, and is it not the noblest subject to reduce and metallize the fixed powders which are otherwise hard to reduce? But its ☿ must be strengthened still more, otherwise it is all too "influxible," as may be seen by its glass. Lead (☿)[151] must be turned into such a liquid Vitro by a homogeneous additive that it also melts like wax at candlelight. Nearly everybody knows the *Praecipitans*. For ♂ ⌣ is ♀ and ♀ ♂, as also ☽ ⌣ₛ ♄. ☽ ⌣ₛ ☉ and ♃, as also ♀ ⌣ₛ ☽, and ☽ ⌣ₛ ☿. However one must not take the fine bodies but their refuse. The fine bodies do not ⌣ in the flux but intermingle with the other bodies to which they have been added. But what kind of refuse this is? Let everyone ponder over it. It can easily be found, and it is not only sold or made in all shops but is even thrown away as useless in mines. Pick those things out and acquire them.[152]

---

[151] Note: the RAMS edition appears this way, but note that the symbol shown is "mercury" not lead (♄). –PNW

[152] So that the reader should not let his thoughts digress and lose himself too much over

But someone will say: You may well write about metallurgy and the genealogical register of metals, how Nature proceeds in stages in order to produce metals and minerals, but if you were to take these component parts and produce a metal or mineral in just these stages, you would probably hesitate. Answer: If it is true that Nature forges metals solely from corrosive vapors and earth or rocks, she nevertheless does not use one kind of earth or stone, nor one kind of heat and suchlike obstacles whereby she produces different minerals and metals; and although Nature has as her final aim to make

⊙, the intervening obstacles produce other forms. And just as Nature cannot make a specific form that she plans, as it depends on the earth or rocks, I will describe such an experiment as a favor to you, and teach you to make a metal from earth or a rock - which is metal - no matter what kind of metal is the result, because I do not promise you a specific one.

---

this page but concentrate on specific objects, keep them narrowly together and use them to advantage, we would explain hereby that the Author understands that excretion or discharge from the body out of which, with which, and through which they are produced in the bowels of the ▽ by Nature. But what kind of a discharge this must be, we learn to infer the mineral kingdom of Nature from the others according to the Author's principles. Everybody knows the seeds of the animal and plant kingdoms, and we know that they are discharges from their own bodies. The chaotic seed of the universal kingdom is a discharge of all four elements of which it consists, and we have also already seen where we can find it unspecified and in particular. Consequently, the seed of the bodies of the mineral kingdom is likewise a discharge from them. It is precisely this and no other discharge that our sincere Homerus wants us to understand here. But we must well consider and keep it deeply embedded in our memories that it is not the body which we call seed in common parlance but the active ∿ that begins, continues and accomplishes the natural business of generation in its own way, which is the seed in the real sense. Just as a mercurial, arsenical, or sulphurous spirit accomplishes this business in the entire mineral kingdom, engendering all the bodies and kinds of mountains that pertain to it, so the author orders us here to take just that discharge in which this ∿ has collected most abundantly and is still to be found most open. But what kind of natural products this discharge is, is explained thoroughly and at length in precisely this chapter.

Therefore, Rc. A pure water pebble. Heat it often and slake it always in ⌒ ⊖ or ⊕, so that it turns completely into powder. Put this powder into a glass alembic, put upon it its weight in ⩗, made of one part of ▽̵ and three parts of ⌒ or °° ⊕ or ⏚ris. Digest it in sand at a low degree of △, then draw it off to its oiliness. Let this crystallize in the cellar, and you will obtain partly ⊕ or crystals, partly some subtle, spongy earth. Thus you have the guhr and the ⊕. Set this again in sand in the fire, and frequently draw off the abovementioned ⩗; or better, take that out of the ▽̵ and ⌒ ⊕li, draw it off to dryness, so that it melts together into a liquid stone which will be very brittle like sulphur when cold. If it is thrown upon burning coal, it will burn and leave a sulphuric stench. Take this, ♄ it, set it in the sand in an alembic, again pour ⩗ upon it, but not much, only enough to dissolve it. Otherwise you will make it volatile and rise above the head in the form of a *Liquor*. Draw the ⩗ off as before, so that it will melt in the third and fourth degree into stone. Remove this stone, powder it, put it into an alembic, pour distilled rainwater upon it, set it into moderate digestion, and let it stand for one month. Then a glistening metallic ▽ will settle at bottom, and it will increase more and more and be of a marcasitical, fine granular kind to which arsenic is admixed. Put this ▽ into a crucible between calcined and pulverized pebbles, cement it *per Gradus rotationis* (in rotating steps -PNW), so that the crucible will finally glow strongly. Then break it open, remove the *Macasam* and boil it in a cupel

with ♄. Thereupon refine it, and you will find a grain of gold or silver, but it will be of little use if you are looking for wealth through this process. I warrant that you will lose house and home in a short time. One can only try it *curiositatis gratia* (to satisfy your curiosity -PNW). You say: Nature has no furnaces, sand, cupels, alembics,

crucibles, etc. Answer: Give me the central △, and I will bring you a vessel made of stone for it. Bring me the central vapors in large quantity, and I will also arrange such a performance and generation. But you, wait for a hundred years, and you will finally also hatch something. An intelligent artist does not reproach another with such an impossibility, because the Art can never imitate Nature to a hair, for either he will do it faster or much more slowly, so that a thousand artists will not obtain their object according to the likeness of Nature, but something similar in *similibus Principiis homogeneis*.

But someone might say: Why do you take pebbles and not some other earth? Does Nature really make metals from pebbles? I would have thought that stone would be the vessel for making metals and not the matter. To him I reply: There are very few alchemists indeed who know flint. If they did, they might sooner learn the Art. Flint is nearest to lead, also to gold, because it is a sticky, mercurial-alkalized substance, a *gluten mineral* which stands every △.

It could rightly be called the ☿ of metals, which lacks nothing but Acidum to become metallic: It is the *figens fixissimum*. Just give the flint a metallic color or, as they call it, ♁ in fusion, and you will see how very eagerly it accepts it and totally colors its body with it. If you give it more, and still more, it will finally make a *Regulum Antimonii*. Driving this one off, you find the grain which it has produced during congealing.

If one wished to use it in the Art, however, one would have to increase its fusibility by its like, for otherwise it would need all too strong △. Therefore increase its fusibility with slightly more liquid and suchlike things, so that it can fuse with them to quite a liquid condition, like liquid salt. Then it will have taken a great step forward in its ability to congeal volatile things and to change dry powders into a metallic nature and kind. Only, it is said in connection with the flint: *In metallis, cum metallic, per metalla & earum genera fiunt metalla* - "in metals, with metals, by metals and their species metals are made."

Let someone just pick up some mineral or ore and examine it in the above-mentioned manner, and consider the first, middle and last. He will indeed find different subjects, wet and dry, volatile and fixed things, likewise fusible and nonfusible or refractory ones, and easily fusible ones, according to the mineral taken. For instance, ♄ and ♃ ores are more fusible than ♂ and ♀ ore; silver and gold ores are in the middle, they are neither too fusible nor too refractory. The artist must well consider the degrees of fusion in flint. If it is too refractory, he must add to it a subject that is more fusible by one degree. If it is still too refractory, for his work, he must give one that is ever more easily fusible, until the fusion is suitable for his work. Then he will find the *Sigilium Hermetic* which prevents the volatile heaven from breaking Out into the abyss, because that *Sigillum* is not only the lock to shut in, but also the band to bind the volatile.

Flint is a noble subject which Nature elevates above ☉ in its fixity. In addition, it is the basis and beginning of the durability of all precious stones. It is a pure water, a water of durability as it melts in the strongest fire like non-combustible oil. Nature has elevated it in the highest degree, because Nature does not go beyond stoniness and glassiness but rather goes backward. Just as the Art cannot go further than to the making of glass; after that, it must again go back to the first.

Let flint be highly recommended to the man who wishes to achieve something in a hurry. In it and its adhering ability, just as in the crystal, which is a transparent flint, there lies the main agent of all durability. We can see this in the growth of all minerals of which flint is the mother, but this is not in its raw state but only after various preparations. It acts differently when it is raw than when it has been changed into water and oil, differently again when it is a salt, and again differently when it is a refractory or easily fusible glass.

Whoever understands this stage of Nature, will be able to work independently. He can make the fixed volatile, and the volatile fixed, like Nature herself but much faster. What Nature achieves in a thousand years, the artist can do in a thousand days, yes, even faster. Whoever understands the origin correctly, can return the metal into marcasite, the marcasite into arsenic and sulphur, the latter into ⊕, the ⊕ into a corrosive vapor or into the prime matter; or else he can change such a vapor back into ⊕, the ⊕ into sulphur, sulphur into arsenic, this into marcasite, and finally marcasite into a metal, and that metal into the extremity and final aim of Nature, namely, into glass or stone.

As food for thought, I will explain in somewhat greater detail. Thus: If I wish to turn metal that is already refined, molten and separated from its brittle parts by frequent smelting, into marcasite, I must again add to it what it has lost, that is, in the very order in which it had grown and in which it has lost those parts. Then the refined metal turns again into that which it was, and how it came out of the mountain. During smelting the metal lost marcasite, arsenic, sulphur, and vitriol or ⌒

⊕ *li*. If I now wish to change the metal into marcasite, I must give it its own marcasite out of its mountain or a similar one. And just as marcasite exceeded the metal in weight and quantity, I must here also add more marcasite and thus one has to take everything into account. Therefore, Rc. Metal - Add to it marcasite or *Regulum marcasitum*, and melt them together. When they are joined, give it ⌒ or

°₀° ⊕*li* or ○*nis*, as the metal is red or white.

⌒ ⊕ or ○*nis* with its excessive quantity will again return it into what it had been in the

beginning, namely, into ⊕. And when it has been

processed thus far, the ⊕ can be totally changed into pure vapor or corrosive water, as it had been in the beginning. Thus, the last is the first, and the lowest has become the uppermost, *inferius factum est eupenius*.

And thus it is done: Out of ⊕*ic* ✝ he can

make ⊕, out of this — arsenic, out of that - marcasite, out of that metal, and out of metal - glass. In brief: You must alloy the metal with its sulphur, arsenic and Marcasite; then add its stony matrix in an overabundant or equal quantity, melt them together, and it will turn into glass.

But now let everyone take note of this main point: Just as the artist always takes a larger

quantity of volatile parts to turn metal into ⊕ and ⌒, because it has to become volatile, so he must here take a larger quantity of the fixed and a smaller quantity of the volatile, otherwise he will not succeed. If I wish to make the species fixed, I must not overload them with abundant quantities of the volatile. Instead, if I wish to make them volatile, I must not take so much of the fixed but much of the volatile, otherwise more is congealed than made volatile.

In this way one must make things volatile or fixed, or else little is achieved. Just see the peculiar alchemical additives taken by those who intend to congeal common mercury with perfect metals, when without reason or reflection they take seven, eight, nine and up to twelve parts of volatile ☿ to one part fixed or perfect metal. Do they then not see that it is against Nature and all her rules? If a man wishes to congeal, he should rather do the contrary and take twelve parts of the fixed and one part of ☿*rii* or the volatile. When this is fixed, it will in time increase in quantity, so that he may add more of the volatile, which will then be of benefit. But first he must take patience. But those laboratory workers have still some fog in front of their eyes: Because they do not see that although the ☿ adheres to the metal, it does not conjoin in time & *per minima*. By this they should notice that a medium is missing. That, they should look for, because ☿ is a volatile and more concentrated metal. ☉ and ☽ are also metals. However, no metal will form a true alloy with another without their intermediates, which are taken from them in the foundries. Go there and take it, or something similar.

This is the reason why the world is so full of such mistakes, since they put together the most volatile as *extremum*, and the most fixed as the *alterum extremum*, and they wish to make a conjunction immediately, but then they see in their work that they have gone wrong everywhere. Let a man just take the volatile sulphur and put it together with ☉. Then set them together into the fire, and he will see the sulphur fly away without harm to the gold. But if he took the intermediates like arsenic and marcasite, and let them melt, he would immediately turn the ☉ into dust. From this they should learn to make like with its like.

Indeed, there are plenty of such intermediates at hand. Does not ♁ exist for the red *Astris*, also the yellow and red arsenic, ♄ and gold marcasite? For the white *Astris* alum, white arsenic, bismuth? From the above, everyone can become wise.

We have now explained somewhat the mineral kingdom and endeavored to indicate a few main points concerning its origin and completion. But the main point in this kingdom is the following:

Whoever undertakes to congeal and make fixed, to coagulate and thicken something, finds the best shortcuts in this kingdom but, as I everywhere admonish, through the intermediate stages and not directly from the highest to the lowest, nor from the lowest to the highest level, although one need not pay too much attention to this, because every kingdom has its adequate congealing agent, as will be shown in the dismemberment. If someone wishes to

congeal something, he must not take the most
volatile and the most fixed together, but the
volatile, the fixed, and most fixed, that is, the
intermediate grades. Only in that way and in no
other can the desired end be obtained, though in
everything sooner and better through homogeneous
things than through alien, heterogeneous ones. Thus
and in no other way does the desired harmony of a
concentrated Quintessence appear.

We will therefore conclude this first part and
begin and deal with the subsequent other part about
the destruction and dismemberment of natural things,
which follows their generation, in order to set it
against the preceding one for greater clarification,
because, destruction and finally regeneration
follows in the wake of generation.

End of Part I: The Golden Chain of Homer
On the Generation of Things.
(DE GENERATIONE RERUM)

PART II: THE GOLDEN CHAIN OF HOMER
INTRODUCTION

Just as the first part deals on the whole and in general with the generation of things, so in the following part the destruction of things is described on the whole and in general, from which each may draw his conclusions in particular. As I did with my own hands, so I transmit. If anyone is to gain from it, let him be grateful to the giver of all gifts and not to me. Let him at the same time eternally endeavor to practice, aside from the preceding, the highest commandment, love of his neighbor, without giving offense to friend or foe. Because I did not inherit this from myself and out of myself but received it from the Supreme and his ever-present holy guardians, I am transmitting it as my inherited talent to the sincerely and seriously hoping individual, to try his luck with it. But should he incur damage, or should he not attain his goal, let him excuse me because I cannot manipulate with him personally; however, let him not lose heart because no fruit ripens before its time and no child can represent a man: likewise, a beginning alchemist cannot be a perfect philosopher. Therefore it is said: we learn by making mistakes, and in time turn from imperfect pupils into perfect masters.

CHAPTER I

IN WHAT WAY NATURE RETURNS THE ALTERED CHAOTIC PRIME BEGINNINGS TO THEIR FIRST CONSTITUTION, WHICH IS ⊕ AND ⊖ AND HOW SHE TURNS THEM AGAIN INTO VAPOUR

Just as we have proven above, Nature generates everything out of the age old and then reborn Chaos - rain or dew and snow water - and everything comes into existence out of it, be it still volatile as it comes to us from the air to the earth, or already somewhat fixed and corporeal which can be seen in the form of saltpeter and salt; so, in reverse, everything be it volatile or fixed is again destroyed, dissolved and corrupted by just that chaotic water, and returned into its first nature from which it had originally sprung, namely, into saltpeter and salt, these into ▽ [153] and ▽ into vapors. Nature achieves her births by ascending from the said prime beginnings to-her specific perfection. Then, however, she reverses and again destroys everything in descending to the prime beginning.

But how Nature dissolves again those fixed seeds of saltpeter and salt into water and then transforms that water into steam, we have sufficiently shown in the first part, in the chapter about the effluence of the earth, and in connection with other species, especially the birth of minerals, so that it is not necessary to repeat here how they break out of the center of the earth into

---

[153] I suspected that the R.A.M.S. edition might have inadvertently substituted ▽ for △ here. Therefore I checked printed copies of the 1757 and 1781 German editions, and the 1762 Latin edition, and found that the R.A.M.S. edition is consistent. -PNW

the air in the form of steam, etc. Having completed
our instruction on universal things, we therefore
begin with the next animal sphere, how its various
products return to their first nature and again into
destruction.

CHAPTER II
IN WHAT WAY NATURE DESTROYS ANIMALS

Animals rot, turn into worms and maggots, these into flies, and these dissolve, in accordance with the goal set to them, into the first nature of the universal limbus (lower plate of transit), that is, into a salty-nitrous or chaotic nature, after that into water and steam, from which spring dew and rain, in which again a highly volatile Nitrum and salt are generated.

Animals are of a very moist and juicy nature and kind, full of volatile salt. Consequently, as soon as their balsamic vital spirit has volatilized, they go into putrefaction, swell up and begin to smell bad. For the volatile ⊖ breathes out and spoils the air with a foul stench. Everything becomes slippery, moist and wet. For the sake of brevity and so as not to instill disgust into an honest man, I will stop with this kingdom and instead explain better that of the vegetables, because it is almost one and the same. But whoever desires to learn more, let him just go whenever he wishes to a place where a dead animal lies and look everyday at the changes in it - and he will see more than he cares to. Maggots are creeping in the mass, which stinks horribly; and these maggots, after they are well-fattened, are changed into gnats and flies. For if someone takes well-fattened maggots, locks them into a glass and feeds them some stinking meat, sets the glass in lukewarm air, in a weak sun with glass closed only with a piece of perforated paper, he will see in a few days and hours how a race of gnats or flies grows out of the maggots and how this race is changed.

This is partly due to the volatile part; the more fixed part, however, which is not so volatile, turns into water and earth from which saltpeter and salt, can be extracted. These left-over parts, saltpeter, are in every subject. When it starts in its utmost reduction, the volatile smokes, breathes and disappears in the air as vapor and thus into its chaotic origin (being coagulated there). The fixed-part, however, enters earth and water, to be there also changed into its first seed nature again, i.e. ⊕ and ⊖. Out of these reduced animal constituent parts, vegetables grow again, so that the animal kingdom becomes vegetable in descending, as was taught in the first part of this work.

But because the legs in animals are very hard and coagulated, Nature is occupied with them for a much longer period until she can turn them also into mold and earth, as will be noted in the section dealing with ligneous growths.

CHAPTER III
IN WHAT WAY NATURE DESTROYS VEGETABLES

Vegetables can be observed with somewhat more pleasure than animals, because they fall off after they are withered and are wetted by rain or dew, which arouses the inherent ✝, that is, makes it volatile. It becomes lukewarm, warm and hot, partly because of the inherent vinegar spirit, partly because of the supervening solar and central heat which exhales and radiates constantly from below upwards, like the sun does from above downwards, as one can well feel such terrestrial warmth in the cellars in winter. This roused ✝ passes through the *poros* (porous parts) of the plants and stirs up or heats the volatile, so that it will evaporate into the air and be there transformed into its chaotic nature. The more fixed and hard part is softened by it, turned into slime and juice which creep into the earth and mix with it, in order to await there a rebirth and thereby be transformed into the more fixed chaotic seed-nature, saltpeter and salt.

If the heat does not dry it out too much, the volatile part of plants also turns into worms and maggots, and then into gnats and flies, which is an indication that the vegetable kingdom aims at becoming animal or volatile. This can also be seen in the still green trees and plants, when the excess exuding juice begins to putrefy, out of which now whole nests of worms arise, and from these beetles and various other flying insects. This typifies the destruction of the smaller vegetables.

With the bigger plants, such as trees, however, Nature has a harder struggle to return a withered tree into the primary matter or a chaotic water. For a tree lasts for many long years before it decays into mold and dust. But how does Nature deal with them? In the following manner:

First, when the tree's growth-causing spirit has died, it withers. Its root no longer draws juice as food for-the tree, but it loses its absorbing power, gives no more food to the tree and can no more separate the fine from the coarse. Therefore the leaves fall off, because, through his *poros* (porous parts) the tree is filled inside with foul vapors which begin to decompose and gradually soften its parts. As said before, they change its root moisture into its *contrarium* (opposite). For as Scion (descendent) as the balsamic spirit has left the tree, its primordial components act in an opposite way, that is, toward dissolution through their reverse destination. Because the tree has lost its nourishment and the power to drive out what is injurious and opposed to its nature, they attack the tree altogether, affect it with dry rot and mold, and make it inside quite spongy and soft from the core to the outermost bark. Outside, however, it is attacked by heat and cold, sun and rain. The sun heats the tree through and through, so that it often bursts open with the heat because its sustaining moisture has left it and has turned into a *contrarium*.

When the rain comes and wets it - and because the tree is heated and dried up by the sun - it absorbs the moisture avidly, and this is to its own doom, since the moisture enters through the *poros*, putrefies inside, as no resistance is offered by the now departed power of preservation. Afterwards the sun returns and again heats the tree thoroughly, thus opening the poros of the tree and making room for decomposition to penetrate it completely through the opened pores, thereby causing it to rot totally. When it now rains again, the way had already been cleared for penetration the first time, so that the next time it can take a big leap inside and infect the whole tree with putrefaction. This is done by heat and humidity.

Cold pushes it still further into destruction, for it quickly attacks it through and through as its inherent warmth has gone. For when the sun shines and warms the tree, the cold melts into water in the penetrated poros, and this water sits in the tree's heart and core, begins to putrefy, makes the tree turgid, brittle, rotten and moldy; and this is continued by Nature until finally the tree decomposes altogether and at last collapses into mere mold and dust, and this is then the vegetable calcination. In the animal kingdom it may be observed in bones and legs, where the same thing happens. But this does not occur in one go, so that we might perceive it with our eyes in a short time. No! It takes place very slowly, so that such a calcination often lasts through the lives of three men and even longer when it is hard wood, because there is always a little and again a little that disappears of the tree.

We see a faster example, however, in water willows and elms which frequently produce such mold on account of their excessive moisture; but when the tree is calcinated and has turned into mold, it

decomposes all the faster and returns into its prime
nature, namely $\oplus$ and $\ominus$, and indeed often in a few
months and weeks. Gardeners, therefore, like to use
mold as they would use dung.

But if the tree is turned into sawdust by man's
art and hands, or is otherwise chopped up and wetted
with foul rainwater, it rots just as fast and what
is more interesting, even faster than the soft parts
of the vegetable plants. If one takes sawdust of a
tree, moistens it with foul rainwater and lets it
stand in lukewarm air, it gets quickly putrid, bad
smelling, slimy and finally turns into a thick
water. Unless it is prevented, it will become full
of worms and maggots, finally flies, and when these
have flown away, a small earthy moisture remains, as
I have experienced with some plants and woods. If
that is prevented, however, the gardens can be
manured with the decomposed growths or the component
parts can be separated chemically. And such is the
natural separation or calcination and transposition
of vegetables into their prime nature.

But someone might ask why I use rotten
rainwater for this and what it is in the rainwater
that assists putrefaction, or what component part in
rainwater causes this putrefaction. To him I will
reply: First, just as I have proven that everything
is born and generated out of and with chaotic water,
so everything must also be destroyed again with and
by it, because it is indeed the identical fermenting
ferment of all things, although many chemists not
unjustly mix it with leaven or brewer's or wine
yeast.

Secondly: What the putrefying component is,

however, the reader may know from the following: *Alkali* is balsamic, *ergo* what destroys is the volatile and the sour. Since rainwater apparently is more volatile than fixed matter, there is consequently also more *Acidum* than *Alkali* in it. It follows that the volatile and sour is the putrefying component.

Just as the juicy parts of animals putrefy quickly and the hard and dry parts more slowly, the juicy vegetables also decay faster and the hard ones more slowly; likewise, minerals decay still more slowly and are more durable than all the preceding. For everything that is of a juicy and moist nature, putrefies faster after the death of its balsamic vital spirit; but whatever is of a firm, thick, hard and dry nature, causes Nature to labor harder and longer, and that owing to the direct command of the Creator, because water and moisture are an instrument of the all—acting Spirit for putrefaction, and putrefaction is the principal key for opening and locking all and everything in Nature.

CHAPTER IV
HOW NATURE DESTROYS,
CORRUPTS AND TRANSFORMS MINERALS

Each and all in heaven and earth is made of water and spirit, and this water has two principles ⊕ and ⊖, and it is these two that forge without hammer or tongs in their proper mother vessel - in the whole world, each and all, whatever there is natural, visible and invisible, all animals, vegetables, minerals and universals. For if these are spiritual in the air, they are attracted by man's breath and transformed into his nature and seed, and thus they become animal. If, however, they fall upon the surface of the earth as dew and rain, they turn into vegetables. If, however, they reach the depth by means of water, through the crevices, cracks and air-holes of the earth, they generate minerals, and the difference among all these is, as often said: The more volatile these two, saltpeter and salt, are, the more they produce animals; but if they are between fixed and volatile, they produce vegetables; the more fixed they become, however, the more they produce minerals.

The ores have a stony nature due to the effect of both seeds, of ⊕ and of ⊖, through the intermediary of the water. How then will Nature break a stone per ee, and crush it and turn it into dust, ashes and water without a hammer or iron? In the following way:

Nature has two main tools with which she makes and breaks everything. One is fire and air, the other water and earth; one is the ☉, the other the ☽. One is the inner central heat, the other the inner central water. Fiery, hot is the ⊕ for it is the pure concentrated sunbeam and the essence of the sun, its child and offspring, or a coagulated sun, because it is totally fiery in its action, although it may look ice-cold. Cold and watery, on the contrary, is salt, the true mother of attraction, an offspring and child of the ☽ which intensely desires the ⊕ as husband for begetting. Without him she dare not give birth to a perfect body because of her earthly fixed-watery property. Through these two, therefore, spiritual and corporeal generation and destruction, according to the difference in all things, is to be expected and hoped for.

This said, we will now see what kind of quarrymen Nature has. According to the above understanding, Nature has a △, the ☉ or inner central heat. This fire warms and heats the rocks, stones and earth through and through, so that they often almost glow. One has only to touch a stone or iron upon which the sun can directly shine in the dog days, and I believe he will soon take his hand back. Water or cold follows such heat and wets the heated stones, where heat and cold now meet, causing a *contrarium*, because the heat has heated the stones through and through. When then cold water runs over the heated stones, the heat tries to escape immediately; but because the heat cannot escape so fast through the stone, it is driven toward the cold, and the cold toward the heat, because the cold

313

resists and drives the heat inside the stone. Thus a force is generated by the struggle of these two, causing the stone to break into pieces. By that, water and fire become one, because the fiery Stone falls into the water, and the water becomes warm through that fiery stone, so that fire and water are within each other. Through the opening of the stone, the water becomes warm and the fire cold, in which action the *pori* (pores) of the stone are opened, so as to prepare and allow even better and more admittance to fire and water in the future.

For example: In the summer the sun shines upon a rock and heats it strongly. Then follows some rain that wets it. Thus the stone bursts and is cleft asunder, it peels off in scales and breaks into small pieces, although not all at once but by and by, as in the destruction of trees. Therefore understand, Nature moves very slowly. These small pieces are again heated, and again wetted by moisture and rain, so that the slates burst into still smaller pieces and are finally exploded, scaled, crushed and powdered into sand and dust by this continued occurrence. This dust or former stone is again heated, wetted with rain, and finally begins to decay because of its absorbing much rainwater. It becomes salty or nitrous and like sal ammoniac through its own innate ⌒. For its own coagulated salty ⌒ is aroused by the rainwater and moisture to act against its own subject. And the stone goes to its own ruin just as vegetables and minerals go to their death.

Later, the salt of the earth is added and then the *sperma volatile duplicatum* (duplicate volatile seed) from the rain and dew. When it has come to

pass that the stone has become dust and is salty, it
has already begun to acquire another nature, namely,
vegetable. For now plants and trees grow out of it;
these plants and trees decay in turn, and then they
change into worms and maggots, and these into flies,
gnats and caterpillars. Or animals feed on such
growths, and in this way the stone is transformed
another time, that is, into a vegetable, and from
there into an animal. This animal putrefies and
dissolves into a chaotic, salty-nitrous, watery,
vaporous, hylealic nature. Now the stone is again
the prime chaotic matter.

Thus you see how Nature goes back step-by-step,
albeit slowly. In the same way she goes forward
without using much force, but gracefully, without
much noise, not with hammers and axes, or beating
and pushing, but she accomplishes everything through
and with fire and water. If Nature could have the
tools with her as we have in our Art, she would act
as fast as we. One can see this on the highest
mountains, where Nature slowly, daily and
continually breaks off small stones, also big
pieces, sometimes also sand and dust. The farmer
notices this sooner than a doctor behind his stove,
since such an action and production must be ascribed
to Nature and to no one else. For no human being and
no four-footed animal, no matter how heavy, can get
so high up, and the weak birds are sure not to
continually break off stones weighing 10, 20, and 30
pounds and powder them with their weak feet. But if
Nature could pour plenty of salt or salt water upon
such heated stones, she would turn the biggest
mountains into small hills.

For if we calcine a stone in our Art and quench

(chill) it in saltwater, the stone crumbles into pieces, even if it were as big as a house and it were possible to calcine and chill such a one (one like that). If now we calcine and chill the crumbled stone several times, it will become ever smaller the more often we do it; yes, finally it will turn into slime and water, which is clearly shown in the infallible praxis.

If then we distill these salts into spirits and dissolve stones with them, they suddenly become all water. Thus the reader sees that the calcined stone can be changed into water by the Art in a few hours, while Nature has to work for long years before she achieves her end. In this way, the stone reaches its prime matter much sooner, namely, a salty nitrous watery nature, which water can again become steam by distillation, and this steam can again turn into water. From this we can see the difference between Nature and the Art, and from these different levels of Nature and the Art everyone can learn both the generation and the destruction of all things. If now such stones turned into salt are mixed with earth and are commended to heaven or the air, plants grow out of them, as said above.

Nature does likewise with the subterranean creatures or metals. She heats them and bursts them open with water in which, as in all waters, a salty seed or essence lies hidden, be it much or little. It attacks the mineral or metal as its offspring and turns it into slate or rust, so that it gradually becomes rust and crocus, which she dissolves in time into the nature of a salt, and finally into $\triangledown$.

But someone will say: That may be true for stones;

316

but as to metals, which are such tough, compact and finely made bodies, Nature cannot do much,

especially since ☉ and ☽ are hard to destroy by the Art, let alone by Nature. I say, speak and teach with discrimination. When I speak of natural things, I do not speak of artificial ones. In regard to minerals and metals that are locked in Nature's workshop, that is, in their mother, in mountains and rocks, and have never yet been in the fire nor been separated, where mountain, stones and metals are still together, Nature goes one way with stones, for she now works backwards where previously she had worked forwards. She destroys those minerals much sooner than stones, whenever she can get around to doing it, because they have an open salt which Nature need only arouse with water and salt in order to work backwards.

But as far as the metals are concerned that are artificially prepared, separated in the fire and finely made, I myself say that Nature has to do more work with them. For their excessive moisture, as much as they may have had of it, has gone away and

escaped due to the tremendous △; however, more from one than from another. It is therefore necessary for her to work for a long time to return them into their prime matter, because the latter has been almost totally taken away - and partly locked

within and concentrated- from the ☉ and ☽, as well as sulphur, arsenic and marcasite. On the other

hand, Nature can sooner destroy ♂ and ♀, turn them into rust and crocus, because they still have excessive moisture in them, are open and are very easily changed into verdigris and rust. Likewise ♄,

and ♃ which turns into white lead.

Isn't it by experience that we know that gold
and silver buried under the earth are aroused when
the salty moisture of the earth awakens the sour
spirits of ☉ and ☽ into action; that is why one
has found only their *electra* or even only some dust
instead of ☉ and ☽. When therefore gold and
silver have been put in places where many arsenical
or marcasitical vapors evaporate, they are sooner
destroyed by Nature. We can see this in the Art,
which must necessarily follow Nature in just these
steps, when we melt sulphur, arsenic and marcasite
together and let them flow, then put glowing ☉
into it, so that the gold turns into all powder
which is then immediately dissolved by the salts or
salty vapors or spirits and reduced to its first
nature. Thus it is with all things.

When they meet naturally or artificially that
which is their own, they acquire it for their
preservation or destruction; without it, however,
they retain their nature for a long time until the
like of it comes along - as it does not remain
outside - be it sooner or later. For Nature never
stands still but works without intermission. She
makes one, breaks another, and that till the end set
by God himself is reached.

If this field did not become too wide, I would
impress upon everybody how Nature herself carries
out the transformation of things, so that no doubt

318

would remain as to how this transformation can change one into another, just as it has been inserted now and then in this and the first tractate, and as it can also be inferred from it in no vague way. For if the Elements can be transformed into each other, Heaven or Fire into Air, Air into Water, and Water into Earth, and vice versa, it is certain that their offspring can also be transformed into one another, because they spring from precisely those Elements, and their difference exists only in the way of fixation and volatilization.

Only, let no one believe that I am here teaching that one can also transform a devil into an angel, or an angel into a devil, or that we wish to make something out of nothing. No, but we only take the things produced by Nature, divide them into certain parts, and we put those parts together again as Nature herself serves as our model and shows us examples. I could well write here about such transformations, but not about metallic ones, how to make gold from ♀ or ♄ only such as Nature does by changing minerals into vegetables, and these into animals. I am saying that, however, if the world is worth it, for future publication in another tractate, so that everybody can see it and grasp it with his hands.

Here a peripatetic would like to see me prove that a cow can turn into a donkey, or an ox into a man, a man into an ox, and a donkey into a cow. To help him and for the love and honor of the scholarly world and our Art, I must solve this knot. If then a donkey were to become a cow, it could easily be done if the cow were to take a donkey for food, or the donkey the cow. But since the cow's and donkey's nourishment comes from the vegetable kingdom, we must first transform the donkey or the cow into a *vegetabile* and then give one to the other for food. Let the donkey or the cow rot dead under fresh earth, so that the earth be manured by them. Out of it vegetables will grow. Have one or the other eat these. In that way the donkey, after being made into a vegetable will be changed by the cow's *archaeus* into a cow, and likewise the cow into a donkey, and likewise also even the peripatetic.

How an ox turns into a human being: We do not eat beef everyday and transform it into our human essence, so that the whole ox discards within us its specification of a head of cattle and is totally transformed into human nature, without retaining the slightest trace of an ox. For if the ox were to remain an ox and did not discard its form, we human beings would all have to become oxen, owing to our continual eating of beef. And such is also the case with other transformations. If man is also to become an ox, he must rot in the earth and become a plant which the ox must eat. In that way man is transformed into an ox.

But the peripatetic will say: "Alchemists do not understand transformation thus by means of intermediates, but by means of their tincture they turn all unlike metals direct into gold and silver." This one does not understand the nature of ores, else he would judge otherwise. For the tincture is a medicine that does not cure the whole metal as it grows in the mines, but only its purest mercurial part which has been cleansed of its dross by much and strong fire. For all scholars know that the philosophers do not take the ore from the mountains and throw their tincture upon it, but that they first separate from the ore its excessive corrosive moisture by fire, then sulphur and arsenic, afterwards marcasite. Only then do they take the malleable metal which must be separated out of so many parts. For in the big blast furnace the excessive moisture, ⍙̱, arsenic and the volatile marcasite vanish in smoke and into the air, up into the universal Chaos. The rest, however, turns partly into slag and partly into a regulus.

In turn, they purify the regulus from the more fixed ☺ which they also call slag, until they have extracted the pure metal grain. That is what the philosophers take and transform into something better with their overripe tincture, namely, into gold and silver. This transformation can rightly be called a healing of the metallic sickness: ♄ has melancholia, ♃ contraction, ♂ gall and bitterness, ♀ an irritable liver, ☿ epilepsy, ☽ dropsy. Their medicine cures such diseases into a moderate sun nature.[154]

For metals seem to me to be no different from the marrow in legs. When a man has melancholia, his marrow is also infected with it, and a doctor applies his medicine to the marrow and not to the legs or the flesh. For if he can heal marrow, he can certainly also heal other sicknesses, because marrow is the most remote of the whole body. It must be

---

[154] This is also the reason why warmth changes these metals into the said nature and not only frees them from their inherent invalidities but also raises them to their super perfection which then heals all these diseases at the root.

penetrating medicine that circulates through so many digestions and places of digestion to the marrow, since the majority, especially the vegetable medicines, mostly get stuck in the third or the fourth digestion and their power is changed and dissipated in the veins and again escapes through the *emunstoria*. Consequently, none of it can either penetrate or reach the marrow.

And just as all men are descended from a perfect pure man's own seed, who nevertheless assume different shapes, complexions, properties, diseases, etc. by chance, so, although all human beings have sprung from a single seed, one can nonetheless see that one man by no means resembles another perfectly in his constitution and mentality. In the same way metals are all born from a single seed, the universal ☨, but hatched into different complexions and shapes or forms by different mother vessels (wombs), and therefore do not acquire their differences because of their substance but by chance. They are all metals and born from a metallic seed, but that which is accidental distinguishes one from another, as human beings differ one from another in regard to what is accidental and not in consideration of their substance. For one is melancholic, another sanguine, a third phlegmatic, a fourth choleric, and a fifth has too much of this or that. Thus ♄ is melancholic, ☽ phlegmatic, ♂ choleric, ♀ sanguinic. Therefore they need a tempering medicine, so that they may be brought by it into a solar substance, that is, elevated by the Art into a perfect condition of their nature. This the philosophers do by their medicine, and they cure the purified metals and not the ores of ☉ and ☽.

Just as there are different kinds of marrow in the legs, the best marrow being in the very center of the medullar cavity. The other marrow, which is not quite as good, is at the outer layer of the cavity adjacent to the bones, or on the porous leg bone. That, however, is on its way to becoming the perfect marrow. Such a spongy, porous or sieve like leg bone is covered by hard cartilage. This cartilage covers the synovial capsule into which the synovial fluid is secreted. This fluid is, in a certain way, the prime matter of the cartilage and the marrow. Therefore, the physician does not try to cure water on the knee of the synovial capsule, or the hard cartilage, or the sieve like leg bone and its marrow, but the <u>best</u> marrow. For he knows that if his medicine penetrates the best marrow, it will also heal the weaker parts as much as their nature requires it. Yet, it still does not turn them into marrow, but only changes the malign character into a benefic one.

Likewise with metals or minerals: The medicine or tincture does not intend to heal the sulphur, arsenic or marcasite, but the metal, even if it were thrown upon sulphur, arsenic or marcasite, it would not turn them into pure gold and silver, but into a pure solar or lunar nature, just as the base marrow is transformed from its deficiency into a better and healthier nature and finally into the best marrow by digestion and maturation. Therefore, such a

solarized 🜍, arsenic and marcasite can also be made into gold or silver by digestion and maturation, but

323

not into pure gold and silver as metals are
transmuted by casting and melting, etc.

We will now descend from the destruction or
putrefaction to the division, conjunction (fusion,
union) and regeneration, first of the universal
chaotic water and subsequently to that of all things
in general.

AMEN!

CHAPTER V

THE DISMEMBERMENT, SEPARATION, REUNION AND REGENERATION OF THE CHAOTIC WATER INTO A FIFTH NATURE.

In the first part we have explained the beginning and origin of Nature, how everything was born of Water and Spirit, or of the universal Vapor or Chaotic Water, and was divided into the four universal prime beginnings or Elements, and also how these four regenerate this divided Chaos hourly, without intermission, by command of the Supreme, and fashion it into the universal seed of all mundane things for the birth of all animals, vegetables and minerals.

Now we will here consider dismemberment in general, and to be in order, we will begin with the regeneration of the universal Chaotic Water or rain, as a mirror and model of what follows. We will divine and separate it into its parts, examine it by the art of Vulcan and (because it is impossible to fathom it completely) analyze, dissolve, melt and separate it into its volatile, fixed and half-fixed parts, then join, coagulate and fix these separated parts again, so that everybody can see how the most volatile can be made stone-fixed. The fixed can be made volatile, Heaven turned into Earth, Earth into Heaven, the volatile into the sour and Alkali, and likewise in reverse order, whereby a *Harmonia concentrate, Quinta Essentia or Magisterium Universii* will result. According to this example all subsequent must follow as children follow their mother, namely, as animals, plants and minerals follow the universal.

DISMEMBERMENT (SEPARATION) OF THE
REGENERATED CHAOS OR RAINWATER

Take, therefore, rainwater or snow water, whichever you wish. This water is, then, the Universal Seed or *sperma* of all things; and it is nothing else but Water and Spirit. Collect it in a new wooden vessel and filter it into various containers so that no coarse dirt or matter gets into It. Set it in a place that is not too warm nor too cold, but, rather, lukewarm. Cover it with a cloth so no impurity can fall into it. Allow It to stand thus, for about a month and it will putrefy and become foul-smelling. At this point it is ready for the separation.[155]

FIRST SEPARATION (No.1)

Now stir this ▽ thoroughly with a stick, put it into a copper alembic, affix a head and a receiver. Slowly distill one subtle water *Ceiementi* off after another until half of the liquid remains.

Now you have separated Heaven (△) and Air (⊟) with their subtle waters from their casing) or shell. This then is the volatile. The acid and alkali (the Water and Earth) stay behind.

---

[155] Some let it stand until the putrefaction is over. One knows this when the water no longer smells bad.

SECOND SEPARATION (No.2)

Now take that which remains in the copper alembic and distill it further into a different receiver until what is left is thick as honey. That which comes over is the element ▽, the plentiful, coarse *Phlegma* which goes over before the Acid and the Alkali, and follows after the volatile. That is, after the volatile comes the element water. After the water comes the acid and alkali.

THIRD SEPARATION (No.3)

Further, remove the honey thick material from the alembic, put it into a retort, give gradually fire in the sand, and first will come a *Phlegma*, then a sharp spirit like vinegar. That is the Acid. This is followed by a thick O O which belongs to the Acid. The acid is an extended oil, but this one is a concentrated Acid. These may partly be called the essential and partly the elemental waters, and partly also the volatile Earth. The reason for this is that Water and Earth are always together and neither is without the other. They are also only one matter and only differ in regard to their volatility and fixity, or their liquid and dry consistency. Consequently, these parts may also be called the more fixed Heaven and Air, as I have sufficiently differentia-ted above in the first part. Let the reader go back there and get good advice about this whole idea.

When now all the liquid parts have been drawn off *per gradus*, (in steps) there remains a black ☠ in the retort, a real coal which can be lit like all other coal, and it is a virgin macrocosmic Earth or *Alkali*.

Now we have the Chaos divided into four parts; Heaven, Air, Water and Earth, or into the volatile, the *Acid* and the *Alkali*, or into a very volatile Water, a coarse water, a sour spirit or vinegar, a thick bad-smelling oil, and the coal in which the alkaline salt is hidden.

Collect these parts and keep each part separate, as a special Element. Thus everybody can see what the seed of the whole is, into what prime beginnings it can be divided, and what is the origin of all natural things.

Just as the one and simple Chaos is divided and separated into four parts, those four parts can again be divided into several parts or degrees, namely, each part can again be divided into three parts by the subtle, the subtler and the subtlest rectification, as will be explained next.

RECTIFICATION OF THE PARTS OF HEAVEN

FIRST RECTIFICATION

To do this, take the first distillate (from separation No. 1), put it into a long non-cut-off retort, add a head, set it in B.M., put a receiver on, distil through the first and second to the third degree $\triangle$, and a clear, bright, volatile Water will go over. That is Heaven mixed with the finest Air. What remains in the retort is the coarser Air. Keep these two separate again; then, the first rectification is done.

SECOND RECTIFICATION

Take that Heaven and rectify it a second time in B.M., as before, and distil half of it. Then the Water is still subtler than before, and you have now made Heaven still subtler and more volatile.

THIRD RECTIFICATION

Take again that subtilized Heaven and distil it anew to half. Now you have rendered Heaven most subtle and it has a great diamond-like sparkle.

Regarding the half that has remained in the distillation, distil it also once over, and keep each distillate separate with its label or name. Call the most rectified Heaven *Coelum seu Volatile subtilissimum*; the next one, which had remained as half of the first, call *Coelum seu Volatile eubtilius*; the third, which remained after the latter, call *Coelum seu Volatile subtile*. Thus you have divided Heaven into three parts.

RECTIFICATION OF AIR

Now take the coarser Air which was left over from Heaven during the rectification and pour it into the distilled Element of Water which went over in the separation of the Chaos (i.e. separation No.2). Put these two into a retort, set them in B.M. and distil *per Gradus* four, and. the Air will rise over. The coarse Water does not easily rise in B.M., especially in such a high retort, but it will do so in ashes and in a low retort.

Now you have distilled the Air out of the

Water. That must also be separated into three parts like Heaven, that is, three times, each time distilled to half in B.M. Then mark these also with labels, like Heaven. Call the most rectified *Air Aërem seu Volatile subtilissimum*; the next *Aërem subtiliorem*; the third after the first *Aërem subtilem*, and put them in order in their proper place.

RECTIFICATION OF WATER

Now take the Water left from the Air, put it into a cut-off but not too low retort with a head and a receiver, set it in ashes, distil it by the first or the second degree of heat and the most subtle Water will rise. Collect this distillate as the first part. Then distil again the other part of the water from the second to the third degree of heat and put it also separate. Distil the third part of the coarsest Water from the third to the fourth degree of heat, and then you have also rectified the Water. Now name the subtlest first Water *Aquam subtiliseimam*, the other *Aquam subtilorem*, the third *Aquam subtilem*, and put it in order next to the separated and rectified Air. Although I should ascribe the remaining liquid parts to the Element Water because they are moist and watery, nobody will blame me if I count them with the Earth, since they are easily made earthy or coagulated.

RECTIFICATION OF EARTH

Therefore, after you have separated and rectified the three Elements, Heaven, Air and Water, take now also the Earth and divide it also into three parts by rectification, thus: Take the products of No. 3 separation in the initial separation of the Chaos, namely ☩ or Acidum with its *Phlegma*, the oil, and the mass burnt into coal. Pulverize the coal and stir the oil into it. Put it into a retort, pour the Sour upon it, add a receiver, and distil the ☩ or *Acidum* in the sand by the first degree of heat, until you see oily drops. Now put the ☩ into a separate glass container. After that, collect also the oil specially and put it into its own glass container. Finally, give the fourth degree of heat for two hours, then let the fire go out and the furnace cool down. Remove the retort from the furnace and take out the coal or ▽ , and you have also rectified the ▽ into its parts. Call the acid *Acidum Terram subtilissimam*, the oil *Terram subtiliorem*, and the coal *Terram subtilem* and put it in order next to the Water.

Now the Chaos is completely separated and rectified, and has passed through putrefaction, separation and rectification, i.e. dissolution. It must now pass through coagulation and fixation, and thus through rebirth into a fifth nature or *Magisterium* and *Arcanum*.

Here someone might perhaps ask me what I intend to do with the coal which is usually calcined and reverberated or burnt to ashes, and the soluble salt extracted - otherwise the coal would be of no use (in their way of thinking). To him I reply: Let him be patient until he reads the following pages, when I will tell him why I was moved to do it this way.

COAGULATION, FIXATION AND REGENERATION OF THE CHAOTIC WATER INTO A FIFTH NATURE, MAGESTERIUM OR ARCANUM.

Now you have first separated our parts out of the Chaotic Water by separation, and out of these four parts you have extracted twelve parts by rectification, namely, three parts of each in order. Take then the coal (the subtle ▽) mix it with its subtler ▽ (i.e. oil) in a retort, add its very subtlest ▽ (i.e. acid), and you have united the earthy parts. Put them in B.M. through three degrees of heat for four days and nights giving stronger fire one day after another (the degrees of heat are explained below). Then from the third to the fourth grade of heat, add a still head and a receiver so that if something rises, it goes over into the receiver. Meanwhile the earthy body or *Masea* will intermingle and a conjunction will take place: the sign indicating that crystals will sprout when the glass is lifted from the B.M. and put in a cellar; or, when there is no more smell of acidity, it is a sign that the mass coagulates and becomes fixed. When this has happened, set the retort in ashes (the retort should be cut off and not be too high) and draw the moisture off gently, so that it becomes quite dry, but that its sour vapors or the oil do not rise. Therefore, keep the degree of heat very gentle.

Many alchemists are mistaken in the grades of the fire and now do too much, now too little. But so that a lover of the Art may not risk any mishaps or have any doubts in this matter, I will also disclose this to him. Set all your furnaces up with four or

332

six registers, and when you begin distilling, open first two or three valves, so that the matter to be distilled gets going. When it does, close two valves but leave one open as the first grade. Now let the matter go in this grade as long as it will go, and if it does not go any more, open the other valve so that it gets going again. Let it also go until it stops of its own. Do the same with the fourth, fifth and sixth, and when you open a valve and if the process does not resume within one or one and a half hours, open another, and when it is in motion, block the first one again until it becomes once more necessary to open it. In this way one cannot go wrong.

Therefore, as I have said before, draw off the moisture from the Earth, and if any of the Sour or the oil were to rise, pour it back again. But take good care, because if you give a strong fire so that the oil rises, it will coat the whole retort and you will lose a noble liquid part of your Earth. Watch carefully, therefore, the degree of the fire. Take note: the noblest part of the Chaotic Water will coagulate and congeal, and it will let go over in distilling what is too much for it or what is in excess, which artists should remember as a very necessary point. Nature does not take more in one go than she requires, as here during coagulation and fixation, and if the matter has once been coagulated, fixed and dried completely, she needs moisture again, and when that is given to her, she again takes of it as much as he requires and relinquishes the rest. Let everybody take careful note of this, and he save a great deal of effort, time, work and expense.

When now the *Acidum* and oil coagulate on the coal and nothing rises but some tasteless water without acidity and strength, discard this water, because Nature herself has let go of it as an excess. When this has been done, give a somewhat stronger fire, so that the matter can well dry out in the glass and become quite dry. This is the philosophical calcination and reverberation which must be done frequently in this way, whereby the Earth coagulates, congeals and becomes thirsty. The dryer and thirstier it is, the more it desires its own moisture: for Heaven must moisten the dry Earth, or otherwise it cannot viably bear fruit.

Therefore, take the three rectified parts of Heaven, Air and Water, for they are to wet the Earth. Pour them together in the right weight, sprinkle the Earth into them, and thus the Earth will be dissolved. Coagulate the ▽ in the following way: Rc. *Coeli subtilissimi* 3 parts, *subtiliores* 2 parts, *subtilis* 1 part. Pour all three together into a glass vessel, and once Heaven has descended into the other, just as it was said in the first part that the subtlest Heaven is always caught and congealed in the thicker. Likewise in the Air, Water and Earth in the course of its descent, we see that it finally becomes quite earthy, as will be explained here. When this is done, take *Aeris subtilissimi* 3 parts, *Aeres subtilioris* 2 parts, *Aeris subtilis* 1 part. Pour these also together. Take now aqua subtilissimi 3 parts, *Aqua subtilioris* 2 parts, Aqua subtilis 1 part, and pour them together. When now each part is again united, take the Water and add to it the Air and then Heaven. When Heaven, Air, and Water are together, it is the Ambrosial Nectar, or the Drink of the Gods, which must rejuvenate and vivify or regenerate our aging process.

Therefore, pour enough of this water on your dry Earth, so that it first becomes thick like honey, stir it well with a wooden stick, then pour more water on it, so that it becomes like thinly melted honey, and it will have enough moisture for its growth for the time being. Set the retort in B.M. in the first degree of heat, let it digest in it for two days and two nights, so that the Earth becomes well softened and dissolved. Then distil the moisture again in B.M., and when nothing will go any longer by these degrees, set it in ashes and do as before, so that the Earth may once more become quite arid, dry and thirsty by slowly increasing the degrees of heat, yes, that it might burst open or cleave with dryness. However, do not over-do it at first, because it is still quite volatile.

When it is dry again, give it once more fresh $\triangledown$ as before and proceed in everything as before with watering, digesting, distilling, drying and gentle reverberating in ashes. Continue with this imbibing and coagulating until the $\triangledown$ is well impregnated by Heaven, Air and Water, which may be seen by the following:

When you believe that the Earth has absorbed much of Heaven, Air and Water, pour one hand's width of the separated water upon it, set it in B.M. day and night, let it dissolve, and distil it to one-third; then let it cool down and put it in the cellar to crystallize. If it has grown many crystals, and has taken as much as it can coagulate

of the volatile Heaven, Air and Water, and has also
made the Earth quite subtle if it shows this sign,
as it will soon do, it is time to fix it.

After this, take the retort, distil all the
moisture in B.M. and finally in ashes, dry the Earth
well, give it a somewhat stronger fire, and it will
reverberate at the bottom of the retort and will
become brown or red with mixed colors. This drying
up and reverberating will be finished in one day (12
hours). At night, remove the retort, carve the
matter out with a piece of wood on a grindstone
(mortar), grind it well together, very gently, and
put it back into the retort. Pour its separated
water upon it, or fresh water, enough to make it
thick like honey, set it in B.M, draw off the
moisture, then coagulate and exsiccate the ash, give
a somewhat stronger fire so that they reverberate
and acquire a color as before. Then let it cool
down, take the Earth out, and grind it again
together. Put it once more into the retort, wet it
with its separated moisture as before into a thick
honey, set it in B.M., then in ashes, coagulate,
exsiccate and reverberate it.

Continue this work until the Earth is
altogether of one color during the gentle
reverberation, because then it can already tolerate
a stronger fire. When this happens, remove the Earth
again from the retort, grind it fine, put it back
into the retort, moisten it with its drawn off water
and set it in ashes. First draw the moisture off
gently, then coagulate gradually - always gently -
and finally reverberate somewhat more strongly than
before. Thus the Earth acquires a more fixed color
at the bottom, as you may see when you take the

retort out. When it is cold, remove the retort, grind the Earth fine again and continue in everything as before. It is now important that the Earth be reverberated more strongly and once more acquire one single color, thereby becoming more resistant to fire. This imbibing, coagulating and reverberating must be repeated until the Earth gradually becomes fiery-red and fixed in the ash due to strong calcination, for then it can be even more strongly reverberated in the sand *per Gradus*, until it is finally so fixed that it can stand the open fire. Then the *Magisterium* is finished.

You must remember, however, not to rush immediately into the open fire from the sand-grade, but to set it before in hammer scale (dull red heat) in the fourth and fifth grades, and when it has stood this, lock it into two crucibles and let it go in the reverberating fire per gradus for four hours. Then take it out, and Heaven and the very subtlest Water have turned into a corporeal and fixed stone, and it can now be said after Hermes: *Vie ejus integra est, si versa fuerit in terram*, its power is total when it has been transformed into Earth.

This is now the universal chief medicine of which 1, 2, 3 up to 6 grains (English pharmaceutical measure) will heal all diseases at the root, being the radical moisture, and the natural, animal and vital spirit, which produce the whole animal balm of life.[156]

---

[156] Those who do not admit any universal medicine will here laugh and scoff at it as being an impossible thing and an absurdity. But true hermetic physicians do not pay the least attention to these jeerings, because they know from our well-founded philosophy, that all diseases spring from one single primary cause, that is, the weakened or interrupted efficacy of the *Archaeus* or Vital Spirit. Consequently, they can also be cured by remedies that are in harmony with the said Vital Spirit which are able to restore this efficacy.

By this general example the lover of the Art can see how the most volatile water vapor has turned into a most fixed stony body, and how the invisible, intangible has become visible and tangible.

Let now the reader take good note of this example, for all animals vegetables and minerals follow it: They must first go through putrefaction, then must be separated, rectified and again coagulated, fixed and be reborn as a transparent glorified body, each by equivalent components in each of these kingdoms, as will be shown later.

Perhaps, however, many a man will say that this work appears to be very venturesome because 1) It is very long and tiresome; 2) It goes directly against the basic rules of all philosophers. He may well speak of putrefaction, separation, distillation, rectification, conjunction, coagulation, fixation and regeneration; but after the separation, the philosophers united the first Beginnings in a certain measure, locked them in a phial and luted it thoroughly to prevent any air, let alone water, from escaping, and they boiled it to perfection in a stove, in a glass, and by a regimen of the fire, also in a vessel, without touching it further. This one (Homerus), however, commands us to join the parts and always to distil them, again to imbibe, to dry up, to coagulate, reverberate, to remove the mass from the glass, to powder it, again to imbibe, dry up, coagulate and reverberate, from B.M. in ashes, from this in sand, then in hammer scale, afterwards to put it in the Open fire, which methods not a single one of the philosophers has taught. At that, he does not indicate anything about the separation of the feces but leaves fat and dirt and everything together, which the philosophers most emphatically and strictly order us to get rid of, or else the work would sooner turn into poison than into medicine. The philosophers also say that one should never let the heat go out, as otherwise the work would be spoiled - and this one (Homerus) interrupts the heating incessantly.

Answer: That this work is long and tedious, I admit myself, and I have not described it here to make the reader necessarily proceed in this way but so that he may see how the Chaotic Water can very conveniently be divided into its grades of fineness and volatility as well as into those of corporeality and fixity. Nor do I desire to direct anyone to follow this way, except if he wishes to undertake it *curiositatis gratia* (for the sake of curiosity). Aside from this, there exist shorter and more amusing ways, some of which will be shown later.

Now the reader will remember that I said in the first part, that the great primordial Chaos had been divided into four parts, into Heaven, Air, Water and Earth. These four parts were again subdivided and separated within themselves into their grades of fineness and coarseness, as may be seen in Chapter 9 of *Liber* I about the "effluences of the earth." By the previous process, we wished the lover of the Art to understand how to refine and condense these grades, so as to make him see that the finer always rises and can be separated before the coarser, which is immediately followed by the coarse, and then by the coarsest. I have here described this way of proceeding solely to let everyone see how Nature works in lawful order by means of her levels and grades, namely, by incessant ascending and descending, and so that the lover of the Art might better cognize and grasp with his hands the work of Nature.

Just as young students have their levels of comparison, Nature also has Hers; that is, the most subtle and volatile, the subtler and more volatile, the subtle and volatile. Again, the thick and fixed, the thicker and more fixed, the thickest and most fixed. Nature ascends in volatilization, (in order to transform the most fixed into the more fixed and the fixed, from there to transform the fixed into the volatile, the volatile into the more volatile, and this into the most volatile) in order to fix something volatile, and she also descends in precisely the same order, changing the most volatile into the more volatile, this into a volatile, the volatile into a fixed, this into a most fixed. She does not desire to turn the most volatile into the most fixed directly but through the aforementioned middle stages.

If then someone wishes to work this process curiositatis gratis it is up to him, and he can divide it into such parts; but he is not forced to do so.

If the reader were to think that I act against the philosophical rules, I would tell him this, as above: I am not particularly aiming at the secret of the sages, but I am a natural philosopher or lover of the fire-art, who follows in the footsteps of Nature, and as Nature works, so will I also work. And I do not deviate one inch from Her, neither to the right nor to the left, no matter what the sages have written, and I know their ways full well. But because I do not heed theirs, nor despise them, but follow my own, and am also sure that they are in accordance with the laws of Nature, I do not want it said that I have led anyone away from the philosophers way, but only leading them to ponder somewhat over my own path. Whoever does not like it may withdraw again at the threshold, so that he does not go astray because of my way.

I do not enclose the moist and the dry in a phial according to the usual way of the philosophers, and coagulate and fix such in constant digestion and circulation until it dries up and all of it turns into earth in a steady fire. This has been taught me by Nature, whom no philosopher despises or holds in low estimation; but whoever reaches the natural way and purpose has won and has shortened his work.

Nature, in order to cause the Elements (Water and Earth) to bring forth their fruits, provides the seed in the form of water from above, of which the earth takes and retains as much as it requires for growth, driving the remaining excessive water back into the air as steam and vapor by means of the lower and the upper warmth, that is, the subterranean central and the overhead heat of the sun. There it turns into water again, falling and dripping back upon the earth. Again, the earth absorbs as much as it requires for growth, the rest being again driven up into the air in the form of steam, vapor and fog; and it carries on this perpetual circular course until the Creator's Will coagulates and fixes everything together in a fixed stone. With this impregnation and distillation of the macrocosm, that is, the great world, all the fruits of the earth now grow, each according to its characteristics. For when the earth is dried up and reverberated by the sun, heaven provides moisture again and wets it with rain and dew. After that, the sun comes back and once more dries, coagulates and reverberates the earth, so that it becomes thirsty and again attracts moisture.

This action of Nature should serve as the best model for coagulating and fixing for every artist, just as he learned the best model for dissolving and volatilizing in Part I, Chapter 9. Every thing takes as much fire and water as it requires, and no more. It lets go of the rest because it was too much for it this time, and therefore not needed.[157]

Such laboratory wok has emptied the purse of many a man who dared coagulate and fix all the moisture of his intended product, and he burnt many cartloads of coal in so doing; in addition, he often made the subject bubble by too much heat, so that the glass cracked and his treasure flowed into the ash. Thereafter he was overcome by fear and grief and gave up the ghost from melancholy.

O miserable life and time! If the poor human beings but observed and learned from Nature's course, who indeed works daily under their noses and labors incessantly and presents Herself to all natural philosophers as an exemplar and forerunner! To be sure, I cannot hold it against anyone, because at first I also believed I could achieve everything by relying on my brains; but the result taught me

---

[157] From this we can see what models are given in the alchemical works by way of cohobation, of which the most superb examples can be found in our sacred schools.

the contrary, until I hit upon this way and grasped the method of solely copying Nature. After that, my eyes were somewhat more illuminated, and as I have received it and as I have proceeded, so I also communicate it. Whoever then wishes to go the way of Nature, let him follow this tractate. He will yet get some satisfaction, that is, as much as he hopes to find in others. And if he were to find one or another point too difficult, let him run to Nature and reflect. There he has a wide field to reflect on.

Of course, many will call: Back to Nature! Back to Nature! Yes, show me someone who ever truly examines Nature! There are many, yes, a thousand Nature-writers, who want to describe all things. Yes, it is true, they have done their share; but very many, that is, most of them, have described only the skins and hulls and not the kernel, only the exterior and not the innermost, and through those writings they have - although innocently and unknowingly - ruined and led astray many thousand individuals who followed their tenets, because they explained the author's meaning now this way, now that way, and arranged and understood it according to their own way of thinking.

If someone wished to describe everything with all the intricate details, it would be impossible to summarize it in a brief concept, for it would turn into a folio volume. I admit that I am unable to do so. For who would presume to take upon himself such an indescribable work of all things? But in this book I am showing lovers of wisdom as much as Nature permits, and I have arranged my exposition according to the works of Nature, so that a man on the wrong path may immediately return, not only to me alone but also to Nature and her works. When he has understood a point, he may then direct his reflection further and confirm it by practice. Then he will find the right path and reach the desired goal.

One can indeed see how long many have been busy getting their subject to the stage of solution; how much time, expense and power they consume and burn, let alone how long they have worked until they coagulate and fix their conjoined *Liquidum* into a powder, since some wait whole months and years for a single subject to oblige them (by coagulating etc.), and when the time comes, it is *lari fari* and nothing (stuff and nonsense).

If then such a man is to be helped and his subject is to be coagulated faster, he should himself carefully consider his subject, and of what constituents and parts it contains, that is, water and spirit. Whether the spirit is concealed in the water in the form of a salt or an oil or a delicate powder or of whatever form it may be, it does not take more water than it requires to form a body or become coagulated and fixed. It lets go of the rest through the power of the fire. This excessive water must also be taken from the spirit by distillation, as Nature shows us in giving water to the thirsty, parched and dried up earth. Of that, the earth absorbs as much as it requires. The rest, however, is drawn off again by the sun and the heat. This an artist should carefully note, but he should not draw the moisture off with a strong fire, only with a gentle fire, in B.M., and he should cohobate this until the earth can stand greater heat. Now it no longer requires moisture, for it must increasingly accept dryness and proceed to coagulation and fixation. Then the excessive useless part rises up and out, and the seed or spirit thereafter coagulates ever faster. This process is always hindered by the Aqua recolacae, which can only very slowly be transformed into earth.

Some will say however: "How do I recognize that the spirit in the water attaches itself to the fixed body, coagulates and congeals, since as much water goes over as I have poured on?" I admit that I myself have found it hard to acquire that knowledge. But take note of the following:

Water being a vehicle and a visible tangible body in which the spirit or seed lies hidden invisibly, is the sole means of uniting all things by joining them with itself, because all moist, liquid things can more easily be conjoined in their innermost than the dry ones. This water has contained within itself, in a hidden and invisible way, the spirit and seed and its power, and water is a *Vehiculum* of the spirit. Those waters are either subtle or coarse, depending on whether they have been extended, refined or thickened, and the seed or spirit is volatile or fixed. According to those differences, water takes its character from the seed, and the seed acts differently in different waters.

For example: ⧖ is a water, vinegar is also a water, oil is also a water, everything volatile is a water, but of like quality as the coagulated or

dissolved spirit. The way the spirit works in ⧖,

it works differently in vinegar, differently in oil, differently in salt, and differently in the acid corrosives.

It is of course obvious in such waters that they are dissolved and in the state of liquidity and still have a rather great amount of moisture. If they were coagulated, they would be dry, and alchemists call the *coagulata*, "dry things." Therefore their useless and excessive moisture must be taken from them by distillation, that is, in such a way that the spirit or the sharpness contained in such a *humido recolaceo* does not go over with the rest but remains and coagulates. The *Humidum*, however, must go over quite empty, insipid or without any taste, like an empty tasteless *Phlegma*, without any sharpness. In that way the seed coagulates instantaneously and so fast that the artist is overjoyed by it a thousand times and also becomes a thousand times more eager to take up and practice the alchemical Art, because he perceives truth and is himself guided by it through further contemplation.

Therefore, learn here and ponder this point very carefully, and prefer it to those which would derive advantage from this Art in another way: Water, or the useless part, is by no means the main part in coagulation, but the spirit or the seed contained in the water is that which alone coagulates, concentrates and gets fixed through its own intermediaries. This means that the Volatile coagulates and gets fixed through its own acid or alkaline part. The essential components discard the useless excessive water off themselves and retain for their constitution nothing but what they require to form or preserve a body of an incorruptible Constituent moisture.

They retain such an attracted *Humidum* so firmly within themselves and together that they flow with it steadily in all fires like wax, without smoke. It may be seen in silica and glass that, when the excessive moisture has been driven out by them to the utmost, they retain no more than they require, that they flow with it in the very highest and strongest degree of heat like oil, without any loss of stability or fixity, as long as they are not pushed back again by Nature or the Art.

Let a lover of the Art consider this: it would be an insurmountable task for an artist, as well as for Nature herself to coagulate all water or all excessive moisture, as much as every *Individuum* contains, into earth, dry powder or a stone. It can be done, but so slowly that it would be a waste of time for the lover of wisdom and that the greatest age would be short in doing so. Yes, let someone just try and shut some rain or spring water in a phial and set it to coagulate. He may well find some earth, but in half a year or a whole year he will notice little or no decrease in the quantity of the water or its coagulation.

Therefore, we rightly follow Nature which in the animal kingdom does not turn all moisture into animals or animal parts; otherwise animals would not give off any *Escrementa urinosa, sudorosa, mucilaginosa and stercorosa*. Nor does all moisture stay with plants, or else they would not have any *Excrementa resinosa, picea, aquosa*, etc. This may be seen in the larger growths, such as trees, especially in the spring, when their barks open because of the excess, and the excessive *Humidum* drips out in different forms. Likewise, not all moisture remains with minerals and stones during their growth, otherwise not so many rivers, fountains and springs of various compositions would flow out of the mountains. If they all remained with the growth of the subterranean creatures, all the water in the mountains would turn into rock and ore, and none would reach us. Similarly, not all rain, dew, snow etc., is for the growth of creatures, or else the central terrestrial heat and the sun could not sublimate and attract any *Vapores*, vapors or steams, while they are everyday producing such abundant vapors, and form such abundant dews, rain and snow, and again precipitate them upon the earth. With that, how-ever, Nature wants to show us by macrocosmic impregnation and cohobation that she does not give in one go so much moisture that creatures have enough of it until their perfection. No, but she is constantly cohobating a little, by constant impregnation and desiccation. Nature observes this rule, and we should also observe it and not undertake to coagulate our to-be-dried earth inundated with water. Instead, we should only gently distil the excess from it after impregnation as Nature herself does, and cohobate thus until the earth can take stronger heat. Then it no longer requires any more moisture, for it must increasingly assume a dry state and move toward coagulation and greatest fixation.

By this everybody sees clearly that water is only a cover or a casing of the universal seed or spirit (as has been sufficiently stated above in the first part), but that water itself is not the seed

or spirit. Water can therefore not be all
coagulated, but the earth demands only that part of
the water which is the spirit itself. Nature herself
does not desire more than she requires, and if a
hundred buckets of water were to be poured upon half
an ounce of earth, all the volatile water and
Humidum would indeed be removed by distillation and
the earth would alone remain. Yet the earth would
not coagulate in itself more than it needs to
retain, and it would let go of the rest. But if the
water had also contained earth or fixed parts, it
would remain with the earth as its equal.

Thus it is with the seed or universal spirit,
as well in *universalibus* as in *specifices* and
individuis. If this spirit is made fixed, it takes a
volatile spirit of its kind and draws it to itself,
so as to make it its like and congeal it too. It
will discard, however, almost the same amount of
excessive water in which the volatile spirit had
been hidden. Thus like joins like, and like attracts
like, as the saying goes. *Natura naturam ambit and
amplecitur, natura natura gaudet* - Nature embraces
Nature and surrounds her, Nature rejoices in Nature.
In the same way one disagreeable thing repels
another if an enduring unity is to be made thereof.

As long as the tasteless $\triangledown$ is present, the
seed or spirit cannot be rightly or permanently
united in a body, and there will be no immortal
union, permanence and fixity. This may be seen in
the easily corruptible and dissoluble animals and
plants, which have a great excess of *Aqua recolacea*,
and even minerals are likewise not rid of it to the
highest degree. As long as the *recolacea* or the
excessive tasteless moisture is not separated from

them, they are always subject to mortality or decay, dissolution and change. Animals and plants decay and easily putrefy due to their accumulating excess water, which is a curse, especially if they get more of it from outside, such as rain, snow, water, etc. In the same way, minerals decay, because such moisture is everywhere more or less admixed already in the mineral and also added in other ways.

Let the reader recognize that the ▽ *recolacea* is the hammer or anvil of the implanted spirit or seed, by means of which it is roused to act, because it can never rest in the water but causes various changes, one after another. But if the spirit coagulates and becomes fixed, and its excessive moisture is thereby taken from it and dried up, as may be seen in minerals, metals, stones and precious stones, glasses, etc., it is lulled, contracted and brought to its highest potency, in which it stays stable and incorruptible until it is again aroused by the same moisture. After this, it endeavors to resolve its coagulated body back into its first nature. Then it returns to its workshop and its tool, by which it changes the *generato* into a *corruptum*, until it once more generates something else out of it.

Someone might here reproach me that the excrements expelled from the bodies of animals, plants and minerals, which Nature herself expels and discharges by means of her appropriate secretory way, are not *Aqua recolacea* or a thing without power or substance, but that those waters are still full of the seed or the spirit. Such are the urine of human beings, the Gummata and resins of trees and the various mineral waters from the minerals.

So I reply: (1) Because Nature found them superfluous for maintaining the generated body, she wished to expel them. (2) Because Nature, in accordance with the will of the Creator, does not yet intend to undertake the transformation into the Fifth Essence, as man is able to do through the Art. And (3) Because Nature directs man to the out flowing discharges (urine etc.) without damage to his body, and to diligently seek therein the necessaries of like for his body, so that he does not need to attack the body itself but only its discharge, which is just as full of power and virtue as the body itself.

In the animal realm Nature has given the body the discharges which are especially urine and feces, also perspiration and mucus, stomach and lung, saliva the tears of the eyes, and earwax. In the plant realm, the trickling out *Gummata* and *Liquores*, the flowers, seeds, leaves and stalks. And it is not necessary to take an animal's entire body or to dig up a plant's root, since the above-mentioned discharges contain just the same powers as the roots.

In the same way, Nature has given man different metals for different purposes, and out of the less expensive metals, poor people can still derive great benefits. Instead of gold, the laboratory workers have goldlike marcasites, goldlike vitriols; further

the goldlike iron pyrites, as well as fixed ⚇ and an unripe volatile which are found in antimony. In addition, bloodstone (hematite), emery, lodestone (Magnetite) - all of which share in the heart, and courage strengthening solar essence.

351

Thus it is with all red *Astris* (the word Astris means literally "star"), ♂ and ♀: instead of their metallic bodies, one should take their offspring and hybrids; and likewise with ☽ and all white *Astris*. Just as ☿ contains ♃ *embryonatum volatile solis* (the volatile embryo of the sun), bismuth contains *embryonatum lunar volatile* (the volatile embryo of the moon). Galmey (calamine), and tutia also contain the fixed lunar ♃. Is not ○ a ☽ nor ♁ ?

Therefore, the lover of the Art sees that Nature has not provided us with only one subject for the preservation of human nature but with many different ones, and more than we require. Consequently, it is not necessary for us - unless we wish to - to take the subjects and their bodies together with their roots and everything, but their discharges offer us more than enough help, if only we know how to use them properly. Where Nature stops, however, the artist should begin and drive off the excessive *Humidum*, as Nature shows us in the mountains and presents us with examples of how we can attain the Quintessence and incorruptible permanence. There She herself forges the most durable bodies, which cannot be consumed, or if so then very slowly, either by water and air, or even by fire. This is what the artist should take to heart, and learn from his grandmother (Nature) Herself, whom most people have up to now neglected, to achieve a long and healthy life.

Now someone might ask, because the *Aquas recolaceas* are to be distilled off, whether there is

352

no purpose in Nature's having them, or if they are so devoid of the spirit's power and virtue that they cannot be used for anything. Further, whether the spirit or seed does not also turn into a *recolaceo*, or the *recolaceum* into seed. To answer these questions briefly - because it is not really necessary for the main point and is more a speculative amusement than a useful discourse - I say that the *Aqua recolacea* can never be totally separated from the spirit or seed and in such a way that it would no longer contain at least some hidden powers or rays of the spirit. On the other hand, it is impossible for the spirit or seed to be separated so totally and perfectly from the *Aqua recolacea* that no matter how stone-fixed and dry-coagulated it be, it (the spirit or seed) will at all times retain a trace of the water.

Water and spirit are one, as I said above, so that the minutest droplet of ▽ and the infinitesimal mote of dust are altogether filled with spirit, just as the spirit is altogether filled with water. Now someone will say: But that is a contradiction. If ▽ is altogether spirit, there is indeed no superfluous discharge, and if water is altogether spirit, or the spirit is totally water, water is indeed nothing but all seed. That is so, and that is as it must be.

Look now, as I have said enough above, you must understand the difference and the *Distinctionem termini*. Water and spirit are one single primary matter issued from God. In its essence spirit is not at all different from water. Consequently, water cannot be separated from spirit, but they are one

353

and remain one always and ever; be they in a liquid or a dry form, they are altogether one.

This is then the difference and the confusion which man has made for himself: that we have divided this subject or matter into two, according to name and not according to power, and these two into four, and these four into countless separate things, all of which are nevertheless nothing but one single thing and, as said above, the whole difference consisting only in regard to the degree of fixation or volatility. The more fixed and coagulated spirit or water becomes, the greater power to act it acquires (i.e. after it is again given an aqueous vehicle to act in). If it possessed such power in its extreme dissolution, in the form of dew and rain, as it has in its extreme coagulation in gold or in the philosopher's stone, rainwater would be a universal medicine, that is, a raw one, and men would no longer toil to resolve the *individua* or coagulated seed and transform it into the Fifth Essence or *Magisterium*. But since it does not possess that power while in the form of dew or rain but only when Hermes' saying is fulfilled: *Vie ejus integra est, si versa fuerit in terram*; Its power is total when it is transformed into earth, therefore the diluted spirit and water must be concentrated, coagulated and fixed. Then it has *vim integram & fortitudinem fortissimam,* that is, its total power and the strongest strength of all strengths.

That I call this ▽ a *recolaceum* is not to be understood according to its inner, but according to its outer part, that is, according to the name given to it and not according to its indwelling power. When names have the value of words, many say: This thing is of no use to me. Should it therefore be altogether useless in everything? No, but if it is of no use to you, it is of use to someone else. If it is not suitable for this, it is suitable for something else, just as the cornerstone that was rejected finally became the most useful stone and the foundation for the building. Likewise, the ▽ recolacea which, although it is called a useless *Phlegma*, yet contains within itself this power to act and is the concentrated spirit's own *Vehiculum*, by means of which, if it is injected into a sick body, the concentrated spirit or Quintessence is again awakened and joins with and strengthens the sick *Archeus* so that the latter can drive out its enemy. But the reader must understand my view and not immediately try to charge me with contradictions.

This, however, is the true reason why we separate this Aquam recolaceam from the rest, because it is a spirit or seed still embedded deeply in the *hyle* of the original matter, which has not yet become specified enough or become salty by putrefaction and fermentation.

Saltiness is the beginning and cause of all coagulation and the first in the earth to be transformed into precious stones. Therefore, because this water is devoid of saltiness, it cannot be coagulated and made earthy, or only very slowly. Spirit, however, has a salty and spermatic, coagulating nature. No matter how volatile it is, it can much sooner be coagulated than the useless, powerless water. But when the latter also becomes salty by fermentation, it will behave just like the seed and spirit. Therefore, because it cannot be coagulated or only incredibly slowly, we separate it in order to hasten and shorten our work by distilling it off, not that we reject it as useless but because it hinders, retards and delays our work.

The Creator has also created the smallest speck of dust and the smallest droplet of water to his honor and glory and the benefit of all his creatures. But that we call them useless is to be understood in the sense that they are superfluous and therefore unsuitable for our work. The reader must well grasp this discourse, because it does not contain a single useless word. What he does not understand immediately, he must reflect upon until he understands it.

Now someone might also ask whether the *Aqua recolacea* will turn into seed, or the seed into *Aqua recolacea*. The seed is dissolved in the aqua recolacea, for the seed and the *recolaceum* are one, but we human beings separate them *cum termini compositions multiplicata*.

But to allow a lover of the Art to see with his own eyes that this is so, and that only the sharp and salty seed can be quickly coagulated and not the ▽ *recolaceum*, let him take note of the following example which will make him grasp with his hands what he cannot see with his eyes in the hylealic or chaotic subject (i.e. water), as its separated volatile watery parts are nearly all totally alike in taste and smell. In wine however, there is a considerable difference, by which he can well notice how the powerless tasteless water differs from the palatable or sharply perceptible spirit.

Therefore, take ripe grapes, make a juice of them, and let it ferment - which is its putrefaction - and it will then turn into wine. Or take some wine that has already been made, the older the better. Put some of it, as much as you like, into an alembic and draw the burning spirit of wine over. Rectify it, so that it can ignite gunpowder, and you have separated the Volatile.

After this, continue distilling to the thickness of honey. Mix this with brick powder, from which the fine dust must well be washed off, and which falls to the bottom in the water. This mixture must be dry, so that it can be formed into a ball between one's hands. Then put it into a retort in ⸪, add a receiver and distil through the grades. You will first obtain a coarse *Phlegma*, than a sour ∽, like wine vinegar, and that is the *Acidum* or the vinegar.

This is followed by a thick, stinking oil in the open degree of fire. In the retort there remains a ☠ burnt into coal, which is the alkaline part. Take it out and rub it to powder between your hands. Now put water into a deep dish and throw the powder into it. The brick powder sinks to the bottom, but the coal will swim above upon the water. Remove the coal with a feather and keep it; but filter and coagulate the water and you will find the alkaline salt of tartar.

Take that salt and coal powder, both well dried, and stir the bad-smelling oil into them. Then put it into a retort, pour the Acidum or vinegar upon it, set it in B.M. for one day and one night.

357

After adding a head and a receiver, draw the moisture or the Recolaceum over in B.M. - gradually, everything that will go. Discard that, then open the head and pour the ⍦ upon it, or the Volatile, put the head and receiver on again and distil slowly in B.M. A pure *Phlegma* will go over, or an *Aqua recolacea*, and the sharpness of the ⍦ remains with the seed or the *Acido* and *Alkali*; or, if some sharpness should still go over with the water, the ⍦ will nevertheless be so weak that it will never again ignite gunpowder as before, and the reason is that the earth has absorbed as much of the ⍦ as it needed, and it lets go of the rest.

By this work the artist can perceive how the seed or salty ⌒ is coagulated and congealed and how it relinquishes the superfluous useless matter. The universal water or rain and its volatile components are almost identical in smell, taste and color and have no noticeable specific quality or sharpness like the specified realms, i.e., the animal, vegetable and mineral kingdoms. The ⌒ ⊡, the ⍦ and the ⌒ ⊕ have a noticeable sharpness which, when it coagulates upon its *Alkali* by means of the Sour, leaves the rest empty, sweet, without taste or smell, like common well water. From this it follows that there must be something special about its sharpness, and that is the spirit or seed, which has taken a salty binding nature through putrefaction and fermentation.

When the artist has coagulated the ✝ and ☿ the upon the coal and the salt, and has drawn off the *Recolaceum*, let him only think of how much *Recolacei* and how much sharpness or seed he has obtained from his distilled wine; and he will find that the *Recolaceum* by far exceeds the seed in

quantity. Let him previously weigh the ☿ which he rectifies so as to ignite gunpowder, and when he has poured it upon his fixed parts and has drawn off the *Recolaceum*, let him weigh the *Recolaceum* again and he will then see how much sharpness or seed had been contained in this *Recolaceum*, although the artist

believes that the ☿ is now rid of all *Phlegma* because it ignited gunpowder. Then he will

nevertheless find that the ♎ has almost as heavy a *Phlegma* as it weighs, and that its sharpness is a very small part that allows itself to be coagulated and congealed.

But to give some praise to the *Recolaceum*, I say that it is an excellently purified radical moisture which consumptives should take as their beverage; and those who suffer from compulsive thirst. It replaces the lost Humidum; but take only

that *Recolaceum*, out of which the ☿ which ignites gunpowder has been coagulated and congealed. This is a very pure *Mercurius vegetabilis hyleosus seu insipidus volatilis incoagulabilis*, etc. This then is proof to the reader that only the seed, the spirit, the sharpness, or the salty seed-essence in all things can be coagulated and not the Recolaceum. When an artist separates the Recolaceum from the seed, coagulation is at hand in one instant, which is immediately followed by fixation.

I have said that when the earth was saturated with ⟡, it still let some sharpness go over with the ▽ recolacea, and many like to coagulate and congeal this residual sharpness that goes over. That is easy to do, as I said before, if one dries the Coagulum or the earth in ash by gentle reverberation, and makes the earth thirsty; that is, if one dries it very gently, the Alkali or Coagulum seu terra desires eagerly to attract once more the seed that has gone over, in order to coagulate and congeal it, and it now lets go over what is useless and without any taste. If even now a little sharpness were still to go over, make the earth dry and thirsty again by reverberation. Repeat this until it has attracted all the seed from the Recolaceo, and the Aqua recolacea is without taste and smell, like clear and tasteless spring water. Then you have the Magistrium vini, the Quintessence and the Arcanum vegetabile. It is the same with all plants and also all animals, as will be taught later. When coagulation has taken place, fixation follows, which means that it is to be reverberated in ashes ever more strongly until it can stand the fourth grade. Then one sets it in the sand until it has withstood that grade, and so forth, as was said before.

Now the lover of the Art sees that coagulating so much water is folly in the Art and against the order of Nature herself, who serves everywhere as a model for us. She takes to herself what is agreeable for every subject; fast, eagerly and impetuously not slowly, although she seems to reach perfection slowly. In the process of perfecting she rushes quite eagerly, as you will discover in your practice.

By this one may see *quod natura gaudeat natura propria, natura recipiat naturam, natura amplectatur naturam, & contrarium seu non necessarium ipsomet repellat*, meaning, "that Nature rejoices in her own Nature, that Nature accepts her own Nature, that Nature surrounds her own Nature and repels what is contrary or unnecessary to Her."

That I do not have a stove or a constant regimen of the fire and jump from the B.M. into ashes, from there into sand, hammer scale and the open fire, and in this way interrupt the heat in that I again imitate Nature, who teaches me that if I wish to congeal or coagulate something, I should not always boil it in water, since that procedure softens everything and does not congeal. Because I intend to make my medicine ever more fixed, I also give an ever stronger degree of fire, like Nature herself, since a moderate heat does not produce a durable fixed body. Because I see that ash is stronger and better than the B.M., sand hotter than ash, and the hammer scale and open fire still stronger, I also notice that the stronger the heat, the more it congeals the ⌒, and the more the *Recolaceum* goes out of it, allowing the spirit and seed to advance to its extreme stone and glass coagulation.

In the warmth of water Nature makes watery, easily corruptible fish and toads, etc.; in half-dry and moist heat she makes more durable animals; in

361

the air, in dry heat she makes growths and plants;
and in yet stronger heat, minerals. By this we may
see that Nature uses the water vessel for fish and
their species; for the perfect animals that live on
and above the earth she uses a moderate, dry heat;
for plants, however, a stronger and dryer heat,
because they have dryer and harder bodies than
animals. Then we can see that the sun constantly
strongly irradiates them and the earth upon which
they are growing, and because they cannot get out of
the way, they are strongly heated and dried up,
while the animals being mobile, avoid such heat
partly or altogether, and seek a cool breeze or take
shelter in the shade.

Minerals, however, require a yet greater heat,
yes, the strongest inner central heat, whereby they
are coagulated into rock in various ways. The nearer
the ores are to the center of the earth, the
stronger the heat that they must endure. In an
animal's stomach, a great natural heat is
concentrated, especially in winter. The greatest
natural heat is concentrated in the stomach of the
great Demorgorgon (Demiurge - the fires of hell) or
the center of the earth. Otherwise, Nature could not
distil or drive the excessive vapors and steam up to
the uppermost surface of the earth. The nearer then
a subject lies to the center, the more strongly it
is congealed, provided it is not prevented from so
doing by the frequently rising moistures, as one may
indeed find the best and strongest metallic veins
toward the center, and subsidiary veins, however,
toward the circumference. This is so because the
higher the central heat rises the more it is cooled
off, so that it cannot congeal everything perfectly.
This is the reason why in most places many ores are

extracted that have not matured perfectly as metals, but as vitriol, alium, iron pyrites, marcasite, blende, composite ores, etc.

But someone may object and say: If Nature works at minerals in such strong heat and dryness, why then is there so much water on earth? Reply: That there is much water beneath the earth and even more in its center, is true; but that water should rise into the crevices and clefts of the earth where Nature intends to form metals, as frequently as it breaks out above the earth, that cannot be. If water often reached such places where Nature wants to make metallic guhr and flowed together in great quantities, it would dissolve the guhr and its vitriolic salt, and would sweep it off the surface. Then empty vaults and caverns or holes would remain, because the water would prevent all metallic growth. Because the water does not fill all places in large quantity, nor is very abundant where Nature works at metals, Nature fills such places with her corrosive vapors, which adhere to stones and rocks, attack, dissolve and turn them into guhr. The vapors keep on producing such guhr until the caverns of the earth are totally filled, like the waxen bee-cells. When then Nature completely fills these places, no humid vapor can again penetrate into them. Therefore, such guhr becomes ever more concentrated and coagulated, exsiccated and congealed, until it turns into a rock which is rich in ore, which now braves and resists all water and fire.

Regarding the vapors, however, in such a place where there is accumulated water, such as subterranean bogs, pools and lakes, the mineral seed does indeed get into them and is also accumulated

there, but it drowns. When the water flows out above
ground, it is precipitated and coagulated by the
cold air, and it turns into a hydrometal (literally
"wash metal," probably a silicious precipitate
bearing some metal sulfides). What does not flow out
above ground, however, but stays in these waters
will nevertheless be precipitated in its time and
will turn into various *Electra*, *Gummata*, mineral
juices and mineral wax. But whatever returns to the
center causes further sublimation to the surface and
it is carried to the place destined for it by
Nature, there to assume its ore type and form.

Therefore, if Nature were thus impeded by the
subterranean water, why then does the miner dig up
nothing but dry, stony, hard and rocky ores and no
soft, greasy things? Nor is any water found with the
ores except the subterranean vapors which resolve
and accumulate in various subterranean crevices and
sometimes flow out between the metallic veins as
tiny springs.

Because I interrupt the heat and thus cool the
work, the artist should not think that I am here
trying to produce animals. I am aiming at stone-
fixed things which do not corrupt so easily, since I
am shown the way by Nature who is boiling during the
day, warming the creatures by the sun. By night,
however, she moistens and cools them and interrupts
the heat without any harm. But the reader should
chiefly take note that the Art does not demand to
produce exactly like Nature, for that would be
useless, superfluous and meddlesome, because Nature
herself relieves us of such pains. The artist aims
at another production, which is a rebirth into a
fifth, lasting, immortal and transfigured essence,

into a spiritual body that penetrates through everything, and a spirit made corporeal. Because man is always in an active state and has a short life, the Quintessence separates the corrupting moisture from creatures or the useless excess. The artist turns it into a stony, salty, easily soluble medicine which dissolves conveniently in any moisture. When that is taken into the body, it penetrates it like smoke in the air. In the same way, the medicine is to penetrate the whole body, from the stomach to the most distant leg and marrow. That will then result in a complete healing of diseases. Then the *spiritus animales, vitales, naturales* and every part of the body is invigorated, so that Nature herself becomes strong enough thereby to drive her enemy out through such strengthening. For sick Nature, or the invalid *Archeus*, does not need anything except some aid, some invigoration, when the diseases have overcome and vanquished him, and he must therefore succumb, until he gets some powerful enough help to resist the enemy. Then the *Archeus* deals with the illness together with all remaining natural forces and the additional help. They begin to fight with each other until the illness is overcome. He, however, being the victor, resumes his rule anew until he reaches its previous standing. A scholarly physician knows full well that Nature does not need more than a tonic to make Her powerful enough to help Herself. Such tonics, however, cannot be obtained in a better way than through such a rebirth into the Fifth Essence, since everything is pure and clean and a fixed, yet spiritual medicine.

Furthermore, the fact that I have not taught the separation of dregs will cause many a scholar to run me down, because the separation of the impurities has gained ground so much that everybody wants to do nothing but separate feces, although he neither examines them nor knows that they had contained the kernel while he has been left with the hulls. This, I say, will startle a scholar and even more so a beginner. I do admit that I do not separate the feces, nor do I wish to know anything about dregs in Nature. Although all philosophers for the last two or three thousand years, say that they have separated the feces, I nevertheless say today that Nature has no such impurities, but that everything she has made is pure, good and healthy, and must stay together and cannot be dispensed with.[158]

In order to define what I mean by impurities, I say the following: Something adverse, originating in an outside source, is added to a subject. For instance, if I offer a man a rock, a mineral or a corrosive as food, everybody will immediately see that Nature has not meant those to be food for human beings. Accordingly, they are adverse and heterogeneous and harmful dregs for man. Nature has not destined the mineral subjects or poison to be man's food but plants, such as bread and wine, animals, such as beef, veal and also mutton, etc. Those then are homogeneous and propitious to man. That is why each accepts its like and rejects what is adverse to him in the form of a discharge, feces, *haeterogeneum*. Such an excrement, how-ever, is not negative and altogether an impurity, or *res sive terra damnata* (a thing damned by the earth), so that it would not be useful to anything; but if it is fit

---

[158] Whoever will read what has been said in *Compass der Weisen* (Compass of the Wise), Part II, Par. 3 (d), P. 236 ff., about this reproach, will not regret reading this exposition of Homerus.

for nothing here, Nature has destined it for
something else.

Therefore, whatever does not directly belong
together, such as minerals and animals, are against
each other and consequently a mutual impurity,
adverse and heterogeneous. Now, however, although
minerals and animals are directly heterogeneous and
resist each other *in sua specie & individuutate*,
they are nevertheless one when viewed in the
universal sense or in regard to their inner essence,
and identical in their innermost, because they all
originated in a single first matter or primary
element and can easily be made identical by specific
means, for instance, via the plant kingdom.

To prove both theoretically and practically,
however, that no feces can be found in the nature of
things but that everything is composed both of an
individuum and *universale* which are absolutely
necessary for its existence, I say first that each
and all things have sprung from the purest God and
have been made by Him, out of Him and through Him,
the *Puro purissimo* (the purest pure). If God then is
pure, all creatures sprung from Him must necessarily
also be clean and pure. For out of Himself God
aroused his Holy Word "FIAT" which became a very
pure and clear vapor, and this vapor condensed into
a pure, clear, crystalline water in which no
impurity could be perceived, and out of this pure
clear water God made all pure and good creatures.
Had they been unclean, he would not have said
Himself: "And it was good." That it had been good
after Creation and before the Fall of Adam,
everybody must admit, as practice also clearly
shows, that no *terra damnata* (damned earth) or *feces*

were present, and yet the same creatures existed out of the primary matter before the Fall as exist today and existed after the Fall. And from where should the feces have come after the Fall? They were not in the world, for everything was good. From God, the pure God, they cannot have come either. From where then do the feces in Nature stem?

There are some who say that God, upon Adam's sin and after his Fall, cursed the world because of Adam, and that this curse brought the feces into the world. Those who understand it in this way do not understand it correctly. A curse is indeed the opposite of a blessing or good and prosperity. The more God blesses the world, the more the world prospers; but if He curses, blessings and God's power decrease, to punish man's sin, but not that it becomes impure and that he throws feces into Nature. No![159]

---

[159] It would be superfluous to make the reader remember and comprehend by an elaborate elucidation that the author wishes his discourses against feces to be understood only as concerning the inside of all visible creatures but not that which adheres separably to their exterior. Within, they are totally identical in their smallest parts, or after their greatest dissolution, in their highest volatility. They are an identical water (2 Petr. Ch. 3, V. 5), from which they have all taken or received the substance of their bodies; to form the variety of their so innumerable manifold degrees of coagulation, in so countless different shapes and properties. In this regard, our Homerus is therefore totally right in stating that there are no feces at all in the whole of Nature and creatures, that is, that there does not exist an excrement that would be of no use to anything but would in all respects be a *terra damnata* with no other determination than the curse. Also, this rebuke only applies to those sophists who, while intending to separate the feces, throw away an essential part of their work by mistake and thereby must necessarily miss their ultimate goal. True masters of the Art do indeed speak of feces and accordingly teach their separation from the pure. But since they aim at shortening their work in that way, they therefore mean us to understand by the term feces, only that excess which would hinder the shortening of the work, if the essence stayed with the feces. Homerus, on the contrary, with the same intention, teaches us to separate the same excess under the name of his *Recolacei*. So everybody must recognize and understand of his own that Homerus deviates from the other philosophers only in regard to terminology, but that he in no way deviates, let alone contradicts them in regard to substance.

For example, a plant, animal or mineral possessed hundredfold virtue and potency before the Fall of Adam. Except for some of it, God has taken this power and virtue from plants, etc., and what he has taken, he has locked and hidden within the boundaries of his treasures in the greatest *Mysterium*, so that man should not again make a god of himself through the knowledge of good and evil and its wrong application (wrong motive) by knowing how to apply it for "good," but even more so for evil, as Adam let himself foolishly be led astray in trespassing near the tree.

And you should know that God has not taken any power away from any creature, but from man, whom he had set into the world as the Lord and ruler, and before the Fall had bestowed upon him the knowledge to recognize everything good at the first sight of any object. From this man, I say, God has totally taken away and confused his understanding of these natural and supernatural things, so that he knows and recognizes of Nature as little as a dumb animal, but that he must first learn it by long experience, a good education and teachers, since by nature and from birth to cognizes and knows nothing at all. That feces were mixed with animals, plants or minerals - No! Man can now never again recognize it as before in his first and innocent nature. He can never again look into the core of creatures as before, because his mind and all his senses have become blunted as a consequence of the first Fall. Therefore, he also does not know with any certainty how to name creatures, but he forms nothing but conjectures and has no certainty in anything, unless the Creator himself privileges him-with a special Grace.

From this darkness of his mind he concludes that feces have been thrown into Nature. Supposing I admit that an artist should separate all imagined impurities from a thing. Does he then immediately believe that he has fathomed all its powers and virtues? By no means. Ask yourself what use you could make of such a creature and how you would apply it. What will you answer? "I have pondered over it," or "I have read it," or "I have heard that it would be good for this or for that. It is to be applied in such and such a way." By this you do indeed not recognize its true power, that is, not of yourself but because it has been told you by others, otherwise you would not have known it. Consequently, it is only your own and other persons' conjectures, and not your own long (direct) experience.

By this you see that you have hardly experienced a power after all. But tell me also the other. That you don't know. Why? The curse it is that darkens your understanding, that has tagged the dregs or ignorance to your understanding on account of Adam's sin, but not to any other creatures of themselves. The understanding has been taken from you and the wisdom to recognize it and use it for your best. If you had not heard it from your forefathers or read it, you would know less than nothing of it, like all other dumb animals. That is the curse; that is the dregs that we can nevermore separate without God's own Will, and it is because of this that God said to Adam: "At the sweat of thy brow shalt thou eat thy bread." That is as much as saying to every descendant of Adam. See! You have been born quite unreasonable and irrational. Learn through hard work, sweat and toil, seek through long experience, try everything, and whatever you find of good and evil, size it up. Discriminate! In so doing, you will recognize and learn from long work if it is suitable for you or not. What is useful and suitable, use it for your own best and the benefit of all your Adam-brothers. But take also note of the evil, but not to the detriment of your neighbor, only so that others by recognizing it can also avoid it.

If you had previously (before Adam fell - if it had been God's will) been able to recognize everything immediately at first sight without any trouble and with complete power, and had been able to use everything with innate wisdom and without any uncertain conjectures and errors, you must now experience and seek it at the sweat of your brow. This is the impurity that has driven a pile into the eyes of the understanding of all children of Adam. It is this that Adam forfeited by his bite into the apple. This was the freedom and *aurea libertas, privilegium humanae naturae* - "the golden liberty, the privilege of human nature."

I have proven theoretically that I cannot find any feces in Nature. Now, however, I will demonstrate it practically and prove it, for instance, by the creatures of the plant kingdom.

Gently, through the alembic, distil the ⏝ and the excessive *Phlegma* from a previously decayed herb or plant to a still rather moist Liquor or mass. After this, drive the settled Liquor through a retort and gradually distil everything you can, that is, the coarser Phlegma, the Acidum and the thick oil, and a ☠ resembling coal will be found in the retort. The artists take this and burn it to ash. They leach the ash and throw away the rest of it as feces, and this ash is Faex to them. As it is in the plant kingdom, so it is in the animal realm. In the mineral realm they take the ⊖ out of the ☠ after distillation, and throw the rest away as garbage.

O you alchemists! What are you doing? Indeed, you burn the hop and malt. What are you doing? You reject the coal which contains the best and embryonic 🜍, the more fixed plant 🜍, the best and most fixed tincture of the plant and animal kingdoms. Likewise with minerals. They are not feces but a very strong medicine, capable of clearing up fixed chronic diseases. And you throw those away! That is precisely the cause of, and main reason for all your mistakes, why you cannot cure the more fixed diseases. You say yourselves that fixed diseases must be removed by fixed medicines, yet you throw away the best tincture, the fireproof 🜍.

But you will reply: Of what use could the coal and ☠ be? Moreover, there exists no solvent that attacks and dissolves coal. What is one to do with it? Since you know no way to use them to advantage, I will show you one, which is in accordance with Nature.

When you distil all parts of a subject of the animal, plant or mineral kingdom, you will obtain, after the volatile ⌒ and the *Phlegma* which rises with it through the alembic, a sharp, sour *Liquor* from the *residuo*, like an *Acetum distillatum per retortam*. I here call it also the ☨, the ✝ or the plant *Azoth*; likewise, in animal things I call it the ✝ *animale*; in mineral things, the ✝ *minerale*. In the distillation of an animal or plant this *Azoth* or ✝ is followed by a thick, bad-smelling ⚬ ⚬.

Then the coal remains as a ☠. In minerals, however - because they are strongly fermented and coagulated bodies and do not have such a highly volatile spirit as the two preceding realms but a subtler *Phlegma* - there follows a strong, corrosive, sour spirit. This is followed by a yet stronger corrosive *Liquor* which the alchemists call a corrosive oil. The ☠ remains at the bottom.

So that an artist may understand, however, what coal is, and not become mistaken, let him take note of the following. Then he will easily be able to understand all things and differentiate between them. It is the following: Coal is pure ☦ or coagulated ⚬ ⚬; ⚬ ⚬, on the other hand is coal dissolved into a liquid state, which can also immediately be turned back into coal. When its moisture is removed through a high retort in ash, with a gentle △, there remains at the bottom, after a distillation done by degrees, a coal-black matter which had been oil before. The *Humidum* which had been removed from it, is a sour vinegar. Thus the lover of the Art sees that the sour is indeed also a dissolved oil made delicate and thin.

What else is the volatile spirit but a subtilized *Acidum*? Thus an artist sees that the component parts do not differ in regard to origin and matter but only in regard to solution or coagulation, a fineness or coarseness. Therefore coal is a coagulated ⊙, ⊙ a concentrated or *Azoth*, *Azoth* a coagulated or concentrated volatile spirit. Reversely, a volatile spirit is a rarefied and subtilized vinegar, vinegar a thinned oil, this latter, however, is a dissolved coal. But when coal is burnt to salt and ash, it acquires the highest degree of fixity because of its resistance to fire, but salt and ash differ from coal by the degree of their fixity. When then the ash and coal are melted into glass, the subject is transformed into the highest degree of perpetual and indestructible fire resistance.

In order to deal with coal anatomically, the artist must take care to change every subject back into what it had been before, and this must be done by precisely the same out of that which it had originated. Thus, for example, coal had previously been oil, the oil vinegar or *Azoth*. Therefore, coal must again become oil through oil, and the oil must again become vinegar through vinegar, because it had been vinegar before. That this is so has previously been proven, since all thin parts are more and more thickened, coagulated and congealed by digestion and, *e contra*, all thickened things are likewise dilated by digestion with dilated components that harmonize with them, provided those are added to them in a preponderant measure.

Therefore, whoever wishes to transform coal back into oil according to the laws of Nature, must take 2, 3, 4, yes 6 parts of its own oil to one part of coal, etc. Rc. One part of coal, powdered fine, then mix it with 3 or 4 parts of its own bad-smelling oil. Pour upon this 6 parts of its own ☩, set it in B.M. to boil in a high retort with a head and recipient. Then the oil will open up the coal, while the ☩ dissolves and thins the O O, so that they thus turn all together into a *Liquor* and afterwards go over together through the retort. If now you wish to make this still more volatile, pour some of its own volatile spirit upon it and digest it in B.M. Then put it back into the retort and it will rise more quickly. It will go over through the head more and more after you have added to it a great deal of the volatile. Thus you see how one component coagulates and dissolves, thickens, thins, refines, congeals and volatilizes another, as I have proven before, and in this way the right Quintessences are made and not the weak tinctures extracted by ⩔ .

This then is proof that coal is not feces but the more fixed tincture of everything. And if some coal is dissolved, it will keep on dissolving more and more, until the coal-body is completely turned into a *Liquor*. Preceding volatile parts must again dissolve and volatilize the remaining more fixed ones.

Another proof that coal is no feces: Let *Sal tartari* flow and add coal dust, no matter which, as much as the *Sal tartari* accepts. Then the salt of tartar will become very dark blue and green with the tincture. Now pour it out, powder it quickly and

pour highly rectified ♆ upon it. It will be colored in a few hours and attract the tincture. By this one can see what alchemists generally reject.

After this, take the blue *Sal tartari*, boil it thoroughly with spring water, filter it and ⏝‾ with ⩒ or ✟, or ⏜ ⊕ or other sour things, that precipitate the 🜄 down, and you will find a 🜄 at the bottom, the color of which is in no way inferior to that of ☉, ♂, ♀, ☿, 🜄 and which shows itself in ♎ more and more as bright yellow as gold. By this one can see what is contained in coal.

Now I must here note an error into which in general all alchemists fall superstitiously and stubbornly, namely, that the *Tinctura* ⊖♀ is derived from the salt of tartar itself (in which they believe as persistently and assuredly as in Doomsday). That this is a big mistake, however, is easy to prove, as will follow. But before I will note that they ascribe a great effect to that

tincture ⊖*lis* ♀*ri*, by which they should learn that a penetrating power is contained in coal and its 🜄, when the latter is dissolved. The proof is as follows:

When the *Sal tartari* is flowing with and through the coal-fire, every laboratory worker sees that coal produces various colors, red, green, blue, etc. These flames are nothing but the coal sulphur which, being an *Acidum*, likes to adhere to an *Alkali*. The, *Alkali*, on the other hand, eagerly absorbs the ☩ into itself, and one attracts the other like a magnet. Now, however, if burning coal is an *Acidum* and *Sal tartari* an *Alkali*, it is clear and evident that the alkaline ⊖ absorbs the ☩ of the coal sulphur and thereby produces a color (gestalt) or form for itself. But because those flames are dissipated into very fine small parts, one has to melt the salt of tartar for a long time before the salt of tartar will get colored.

But when one uses coal that jumps, crackles and throws about fragments, and due to the laboratory worker's inadvertence a small part of them falls into the *Sal tartari* in the crucible, it turns blue as soon as it meets but a little of the coal powder or dust, and the same happens to those who wish to catch "air-gold" or solar sulphur from the air in broad daylight. Here they see what they catch. If the *Sal tartari* flows too long, it loses its blue color again and becomes white as before. The cause of it is: Like produces like. The salt of tartar consumes the coal and turns it, together with itself, into salt by the most violent incineration, and thus the treasure takes on the form (gestalt) of salt.

Here I wish to show a trick to the inclined reader, how he can not only make tincture of salt of tartar in large quantity and more economically, but

377

also how to prepare potent tinctures from every fixed salt of every animal, plant or mineral body, with their own and not an alien ⊖, that is, from the extracted *Alkali* of every individual. For example, from wine:

Rc. Tartar or grapes, 6 lbs. Put 4 lbs thereof into an unglazed pot, not closed and not covered. Into another pot put the other 2 lbs and close and lute this one. Now send both to a potter, let them well calcine and anneal together. Then the open vessel will look white, the covered one black. Boil the white mass to a lye, filter it, coagulate it, then let it melt in a crucible. Afterwards, take the black mass, powder it and gradually add some of it to the Sal tartari, until it flows quite thick and very dark blue. Now quickly pour it out into a brass mortar, quickly powder it, put it into an alembic, pour on it some highly rectified ⩔, thereupon let it stand day and night in gentle warmth and it will draw out the tincture. Pour that slowly off the remaining substance, and you will have true tincture of salt of tartar. Also:

Rc. An animal or plant, as much as you wish. Divide it as before and burn it together in the stove, one covered, the other uncovered. In that way, they will both be burnt at the same time. Then leach one out, let it melt and add the coal-black mass until it is completely colored by it. After that, extract the tincture with ⩔ or its own volatile ᭜, then you have the true tincture of every individual.

In regard to mineral bodies, however, change
the mineral or metal back into ⊕, calcine it by
leaving one open as in a potter's stove, but so that
it does not melt back into a metallic body but
remains porous and spongey like ☠ ⊕li and take
⊖ Alkali from ☠. Add as much of the ☠ from the
⊕ or metal as the salt will absorb. Then extract
the alkaline ⊖, let it flow and add as much
mineral or metal as it will absorb, but only so much
that the salt continues to melt. Then the salt will
take color. Pour it off, powder it, pour ℣ upon
it, and you will obtain an extract or tincture like
the above.

Now you have made from all things a tincture of
tartar which is certainly a hundred times more
potent than all apothecary's waters. But if you wish
to know how much tincture your colored ℣ contains,
or how much 🜍, extract it (the ℣) in B.M. You
will be left with a very small quantity of powder,
which is the so potently effective 🜍 and coal.

Now you see, you alchemists, what you are
throwing away - a tincture which has such a great
effect in so small a portion that a certain writer
sold it as ☉ potable astrale and ascribed immense
potency to it. He believed that he had caught the
solar sulphur from the air in hot days - which

nonetheless was nothing but some coal or coal dust which had jumped into the melting ⊕.

If then sulphur does this in such a small amount and while it has not yet been made volatile in a *Liquor* but has only been subtilized and extracted by ♉ in its more fixed form, what will it do when it is turned into a *Liquor* by its own components, as I have taught before? The above-mentioned author called his extract "potable gold." What name shall I give to this one, when the thing to be dissolved stays together with the dissolvent and the fixed and the volatile are inseparably joined? Now the great potency of the rejected coal has been proven.

If they burn the coal to ash, however, and leach the salt from the latter, they believe that they have done the right thing and that they have separated the fixed. That salt is fixed, they know themselves, but the ash causes them doubts. But just visit a glass factory! There you will at once see what ash is and for what purpose it is used there, and what becomes of ash, namely, an incorruptible, eternal, permanent body: glass. If then it becomes glass, it cannot be dregs. Indeed, everybody can see that it is a solid body, fireproof to the highest degree, yes, a reborn glorious body like a precious stone. From this everybody can conclude, and intelligent men can judge, what they have thrown away, namely, the most fixed part, the more fixed subject, and the most durable fixing body.

O you alchemists! It is indeed your goal to make your tincture take on a glassy, precious-stone-like nature or redness, otherwise you do not think anything of it. But if you throw away the glass-making substance, how are you going to make such a fixed and fire-proof tincture? You do not see that salts may well flow in the fire but that they always also evaporate and become less. Oil has no stability at all, ☩ is volatile of itself. So now you see what you always overlook and do not heed. Therefore many say that you take the hulls and throw away the kernel. If you wish to fix, look first for the fixed body as the basis of fire resistance, like an architect who first puts the most stable stones on the ground. Afterwards, he builds all kinds of things on the ground. In the same way, you must also take the fixed substance and afterwards congeal its own volatile upon it, according to the natural order and law. Then you will obtain a wholesome medicine from all things.

Now, however, each and all alchemists say that animals and plants have nothing fixed within them. Yet no one has paid attention to the ash, which is such a fire resistant substance. The *Faex* and feces, or *Terra damnata*, has turned their minds, so that they throw the best and purest, the most transparent and most fixed parts of all plants, animals, also very often of minerals, upon the dung heap. That is why they have been unable to make anything fireproof, unless they borrowed it from the mineral kingdom. But if they had considered the animal and mineral hermaphroditic coal sulphur, which is both fixed and not fixed, and how it can quickly be made fixed and also volatile, they would have judged differently. What else is ash but the fixed, congealed plant and animal sulphur? But mixed with

dust and other impurities from hearths and stoves,
it cannot prove its ivory whiteness.

If one were to take coal, however, and let it
glow in the test to the highest degree in an
unglazed pot, in an open flame-fire, and let it turn
into ash, one would see its lunar whiteness and
greatest stability. Nevertheless, such ash or
sulphur made of coal is not as good as when it
appears in its cinnamon color, as shown above, which
it gets through its own or another *Alkali*. Nor is
this (ash or sulphur) by far as potent as that which
changes into a ruby-red *Liquor* with its own oil.

From the above anyone can see how ☩ is
transformed into oil, oil into coal, coal however
into salt and ash, and the longer a salt or *Alkali*
is melted, the more earthy it becomes, the more it
leaves a very pure, virgin sulphuric earth in its
calcination, dissolution, and filtration. It is very
easy to congeal the different components of that
earth and to transform them together with itself
into a glassy and yet soluble stone, which is the
perfect Quintessence and *Magisterium*. Any salty ash
can quickly be turned into a very subtle and snow-
white ▽ , which happens when ash is added to a
flowing *Alkali*. The latter very subtly and quickly
causes the ash to flow snow-white, and thus the
artist does not need to evaporate the salts by
lengthy melting. He can thus make a large amount in
one go, and has enough matter for congealing. But if
he does not like to do it, such work is not
necessary either, and the coal is adequate for
gradually congealing its volatile parts.

Since this is found *universaliter* and *particulariter* in each and all things in the whole world, let now someone prove to me that some feces are present in the total nature of things, and let him show me those, and I will give him a conqueror's crown. If someone speaks to me about earth, I will point out vitrification to him. Glasses show him that they of all things retain the glory of permanence. Do take note, however, that no earth, no matter which, can be made into glass without salt. Either some salt must already be innate in it or some must be added from outside. And when it has some salt, it becomes more volatile and more easily fusible. The longer it flows in the fire, the more the excessive moisture evaporates. This combination does not retain more moisture than it requires to become glass. The glass, in turn, immediately retains the salt, so that almost no element can rob it of anything.

From this anyone can gain the brightest insight. If he does not know how to change a salty tincture into a *Vitrum* (glass), let him add such a pure white prepared ▽ (i.e. the pure ash), in its specific weight, melt them together for several days and nights in the glass-oven, in a closed crucible, and they will merge and turn into a liquid, glassy *Corpus*. To have no doubts at all, let him add some prepared animal earth to his animal tincture, and a plant or mineral ▽ to the plant or mineral tinctures respectively, such as metallic bodies offer after the sulphur has been separated. When the ♁ is mostly or altogether out, the *Corpus* becomes an *Electrum* or metallic glass.

From this now anyone can see that ash is precisely that which remains after its ⌒, vinegar, oil, salt and coal are drawn off.

It is a fixed vinegar, a fixed oil, a fixed coal and salt, and the ash differs only in the added fixity and not in the substance itself. Consequently, a most fixed tincture can be made from animals as well as from plants and all minerals. Animals and plants must therefore not be accused of corruptibility. Although they are not as fireproof as minerals, they can nevertheless achieve a fireproof state through the artist's intelligence and thereby prove that they, like minerals carry incorruptibility in their center.[160]

But so that the reader may finally clearly see that every dispute about these or those things is only *de lana caprina*, let him just consider, as I have often said in this tractate, that animals, plants and minerals by no means differ in their essence and primary matter but only accidentally, that is in regard to their greater or lesser volatility or fixity, their more or less density or coarseness, their higher or lower moisture content or degree of dissolution, and their dryness or degree of coagulation, in which they find themselves. In regard to their origin and the primordial water, however, they are one and precisely the same thing, and animals are volatile plants, plants are volatile minerals. Likewise, minerals are fixed plants, whereas plants are fixed animals.

---

[160] In the freemasonic *Versammlungsreden der Gold - und Rosenkreuzer,* No. XI, P. 281, a passage pertaining to the above by P. J. Faber is quoted, which is incomparable and deserves to be read.

Now I have proven that there are no feces in Nature. Whoever can produce better proof against this proof, let him refute it. He is at liberty to do so. Meanwhile, I stick to my opinion and experience. What I see with my eyes and manipulate with my hands, nobody will take from me.

Furthermore, the fact that I interrupt coagulation, remove the *Corpus* from the retort, grind it, water it, draw it off, incinerate it, let the fire go out again, and again grind it, etc. Herein I again follow Nature and thereby shorten all my works. What Nature dries up, withers and macrocosmically reverberates by day through the sun, she moistens and waters at night through the cold of the moon or also with a cool, moist rain by day. Then She dries, coagulates and reverberates it again through the sun from above and the central heat from below, continually and so to speak, *ad infinitum*.

Mark well, O artists! Nature does not keep in vain to her definite alternations of things; therefore, do likewise! There is indeed no advantage in taking a long way when I can reach my goal sooner by a shorter path. I leave others free to follow other philosophers. Whoever does not wish to follow me, I let pass on without hindrance. Only, let him go one way according to the prescription of others, and the other way according to mine, and watch then what progress there is on both sides. Furthermore, that the work of the philosophers is done in one vessel, is right. I myself have no more than one alembic, and for the sake of speed, at times a retort to lift the more fixed parts properly, as they do not easily rise so high.

On the whole: This tractate is not intended to invalidate the authors, but to present an elucidation given by Nature Herself. Whoever gets some advantage from it, let him give thanks to eternal God. And because these chapters about the destruction and generation of things require elaborate instructions and interpretations, the reader will approve of my adding - mostly briefly - the physical causes of every dubious point. This will be somewhat extensive in regard to the main parts of this second tractate, although there will not be as many as in the first.

I have mentioned that my practice with the Chaotic Water is tedious and tiresome, and have promised to teach some shorter and more amusing ways. They will now follow: The first being according to the Art, the second according to Nature herself, the third according to the method of those who are used to separating the feces. Let the artist choose any of these he wishes, it is up to him. As it happens here, so it also happens in all creatures of the animal, plant and mineral kingdoms.

First Way

SINE SEPARATIONE FAECUM CRUDI

Rc. Putrefied rainwater, stir it well, put it into an alembic, distil the more subtle ⌒, and you have the Volatile. Keep that separate. Then continue distilling, and you will get a coarse *Phlegma*. Continue distilling it to a still rather moist Liquor; keep the drawn off *Phlegma*. Put the remaining *Liquor* from the alembic into a retort and distill a sour *Phlegma* and oil in ashes or sand. The

coal or ☠ stays at the bottom of the retort. Take
that out, pulverize it and stir all the O O into it.
Put it into a retort, having poured its oil and
Acidum upon it, set it in B.M., distil in a high
retort with a head whatever will go over.

Digest it for four or five days and nights,
then pour its above preserved volatile ⏝ upon it,
let it digest in B.M. through the first grade for
two days and two nights. After this, gently distil
per gradus whatever will go over. When nothing will
go any more, set it in ash, coagulate and
reverberate it in ash through the second and third
grades until it gets a color at the bottom. Now take
it out, powder it, pour its *Liquor* which was drawn
off in B.M. and in ashes back upon it, and set it in
B.M. for two days and two nights.

Then distil everything off that goes, and keep
that as before for future imbibing. When then
everything is distilled ex B.M., set it in ash and
distil the moisture well off until it is dry, but
very slowly, *per gradus*, so that you do not awaken
the more fixed spirits. Afterwards, when it is dry,
reverberate it as before, then take it out again,
pulverize, imbibe, digest, distil, coagulate,
reverberate, and do this till it is altogether of
one single color. Subsequently, congeal it through
all grades in ash, then in sand, as I explained in
detail in the first work. Now you have the
Quintessence and *Magisterium macrocosmi*, and it is
as good as from the following ways.

Second Way

VIA NATURAE IPSISSIMA

Rc. Putrefied rainwater, distil all *Humidum* out of the copper alembic to the consistency of a thick *Liquor*. Put that into another retort with a head and a receiver, and distil in B.M. everything that will go, and the ▽ remains at the bottom. Put it into a retort in ash with a head and a receiver, and dry it up very gently *per gradus*, so that you do not burn it or awaken its vinegar or oil but only draw off the excessive moisture. When you notice a sour vapor through the spout of the head, let everything immediately cool down, because its vinegar is rising, which should not be, and the vinegar would be immediately followed by the oil. This would be a violent operation and not according to Nature which does everything nicely and slowly, until she has made a stone of ▽. In a natural way She does not easily, and very rarely, turn things into coal because she does not burn any, and if she does, unlike that made by the Art, except for lightning, when she burns trees, and that is neither *generatio, nec corruptio, nec generatio naturalist sed violenta destructio Vulcani superioris.*

When now the *Humidum* has been gently drawn off in ash, reverberate the earth gently through the second degree. Then take it out, pour its drawn-off *Humidum* upon it, as much as to turn everything together into a thinly melted honey, set it in B.M. to dissolve, distil from the B.M., then afterwards from ashes, and repeat this reverberating, exsiccating, imbibing, digesting, distilling, coagulation etc. until your earth is altogether of one single color. When it is all one color, reverberate, strengthen and imbibe it again, digest, distil, coagulate, exsiccate, and repeat this until it is again of one single color, for it will always change from brown to red. When it has several times gone through the colors, reverberate it strongly and congeal it in ashes, then in sand, as above, and you have the Fifth Essence.

Third Way

VIA FACEUM SEPARATORIA BREVISSIMA

1. SEPARATION

Rc. Putrefied rainwater, distil the volatile spiritual part out of the alembic, keep it separate and mark it with "A". Then distil the phlegmatic part off to the thickness of thinly melted honey. Keep this *Phlegma* also separate and mark it with "B". Remove the honey-thickness from the alembic, put it in a retort, set it in sand and draw off first a coarser *Phlegma* then a ✝, and after that the oil *per gradus*. The ⚕ stays at the bottom.

Separate the (latter) coarser *Phlegma* and vinegar from the oil by decanting it through a glass funnel, and mark it with "C". Put the oil separate and mark it with "D". Put the *Phlegma* with the ✝ in the B.M. in a low retort with a head and a receiver, draw the *Phlegma* off from the ✝, and the *Acidum* remains at the bottom. Add the drawn off phlegma to the above "B", and you have now separated all the parts. These you must now rectify.

Rectify the volatile ◠ (A) in the B.M. out of a high retort, so subtly as you like it yourself, and you have rectified the volatile spirit A. Now take the ✝ (C) and drive it gently over in ash through the retort, and it is also rectified.

Rectify the oil "D" as follows: Take the ☠ from the retort; of it take two parts; of oil "D" take three parts. Stir these together, put it into a retort and distil in ash or ∴, and the oil "D" is also rectified.

Now take the ☺ and calcine it in an open flame-fire, turning it into ashes. Extract this ash with the Phlegma "B", filter and coagulate it, and you have a brown salt. Set this salt to glow, dissolve it again in its *Phlegma*, filter and coagulate it. Repeat this flowing, dissolving, filtering and coagulating until the salt is snow-white, and then all parts are rectified.

2. CONJUNCTION

Rc. Of the salt two parts, of vinegar three parts, of the volatile spirit A six parts. Pour the volatile spirit upon its salt in a retort then add the vinegar. Put a head and a receiver on, and distil in B.M. to an O O. Set this oil in a cellar, let it crystallize, and it will precipitate refined crystals. Take these crystals out and dry them. The Volatile (residuum of the crystals), however, draw off again to half the amount or to the consistency of oil by B.M., and let it sprout again. Repeat this until there are no more crystals. Now take all the crystals together, dry them gently in the sun or at a warm stove, and you have the Fifth Essence of the macrocosm and of the great *Ilech*. Enjoy it now as you please.

If you wish to make a stone with it, however,

take the crystals, dry them to powder and seal them pulverized in a retort. Set it in sand, give fire *per gradus* for three hours, and they will flow into a salty stone like butter and wax, without smoke.

3. AND CONGEAL

If now you wish to coagulate the oil "D" of this stone, pulverize the stone and take three parts of it and two parts of the oil. Stir them well together in a glass dish, put that into a retort, set the retort in gentle ashes *per gradus* for four days and four nights, and the oil will also become fixed. Then stir once more two parts of oil into it, and congeal it again and so long until at last it flows together into a stone, and you have finished your work.

Now we will clarify some points in these three works. In the first work, the reason why I did not make any dephlegmation and rectification is that like to go through the work quickly, because I know that the more fixed $\triangledown$, that is the coal, does not retain anything of the *Phlegma* but only the essential parts, and because they are all identical things, I do not suspect that any other vexatious things might be there. Again, the reason why I pour very little water on, or all of it at once, is because I know that the earth does not absorb more than it requires, and it willingly lets go of the rest.

That I do not burn coal to ash, however, is because I know that the essential embryonic sulphur is contained in it, and I desire to lose that as little as I do the other parts.

392

THE OTHER WORK

Many people will have much hesitation and wonder where Nature might work as this one here (the author) believes. Then I say, everywhere. True, everybody readily admits that Nature passes through putrefaction in the dissolution of things, as may be clearly seen with our eyes in plants. Such a growth withers and, wetted by rain, finally turns into slime, mold and mud, as farmers and gardeners continually learn from their compost heaps piled together from fir trees and other trees and grass that these wetted by rain in the woods at last become quite black and turn into a fat mud and earth. Such is the natural calcination. In that mud or earth there is an essential nitrous salt, a fattiness or O O which is burnt to coal by closed calcination. In a glowing △, however, the essential ⊖ becomes an *Alkali*, and that is done by our strong fire.

At first, however, Nature never undertakes such a tremendous incineration on the surface of the earth, only, a gentle reverberation by the suns rays, as if she did not burn the O O and essential salt but were only reverberating it to make it wish to attract some moisture, namely, rain and dew, from which plants take their nourishment and grow up in the air. But if the air is taken from such an essential salt and it is yet always watered, as laboratory workers do, imbibing and abstracting in glasses, the plant growth is impeded and receded into a mineral nature, namely, on account of the continual imbibing, abstracting and reverberating, and it becomes ever more fixed, earthy and stony, which is what we want.

That stoniness, however, is not like a stone from which the salty radical-moisture has been thus removed to the utmost, but we demand of our medicine a salty purity; a balsamic saltiness, which alone refreshes our bodies, keeps them from putrefaction, Salts, embalms and preserves them. Therefore, whoever would now wish to go this way of Nature, let him follow Nature, and he will not fail. But if he knows still better ways, let him follow those. The third way need not be elaborated on, because those who separate impurities will themselves consider this more pleasant than the previous.

CHAPTER VI

WHAT IS FINALLY TO BE CONCLUDED FROM THE PRECEDING LONG CHAPTER

The preceding chapter deals only in general with the destruction and dissection, and also the regeneration, of all natural things, in particular, however, of the regenerated Chaotic Water, according to which rule, all creatures of the plant, animal and mineral kingdoms must sustain themselves and necessarily follow this path (i.e. modus of regeneration), because all of them have sprung from this natural origin and primordial mother. Just as the artist proceeded in the preceding separation of the universal water, by separating one volatile part after another from the more fixed parts, he must deal with his *specifices* and *individuis*, animals, plants and minerals. He must separate them in the same order and join them again as they had been separated from each other, turning them into a fifth essence.

But, as Nature Herself proceeds, without putrefaction no-one can achieve separation and regeneration. Putrefaction can be natural or artificial, that is, a natural slow way or through artificial fast manipulation of the work - as an artist wishes or is able to do - for the faster he rushes putrefaction the faster the work is done, for which sufficient directives have been given in Part I. Also no separation of the volatile parts from the fixed ones can be undertaken and made perfectly without distillation, although many kinds of separation exist. Those, however, are not required here, except the one which Nature Herself shows and does all the time: First, preparation, then

putrefaction or dissolution, after that,
distillation or rectification; then conjunction,
coagulation and fixation; further, imbibing,
liquifying to a water, multiplication, fermentation
and application. Nature proceeds by steps, as these
are taught in many ways in this and the preceding
tractates.

Therefore, when the artist undertakes a
separation, he must at all times consider the
volatile parts as the uppermost, heaven and air, but
the fixed ones as water and earth or, to speak in
the parlance of the Alchemical Art, he must separate
the parts into a Volatile, an Acid and an Alkali,
into ☿, 🜔 and ⊖, into soul, spirit and body, or
into the four Elements, according to the teachings
of the Aristotelians: into Fire, Air, Water and
Earth, as the artist wishes to do the separation,
with the difference that he must not mix up his
constituent parts and change and confuse them during
coagulation, thereby producing a heterogeneous
factor. It is a question of terminology, and matters
not what name is given to the child. When he has
turned his separated things into four or three, he
can again undertake a more subtle separation of each
of these four or three by rectification. That is he
can subdivide them into their parts, as I taught
about rainwater in the preceding chapter: into the
subtlest, the subtler and the subtle; likewise, into
the thickest, the thicker and the thick;
furthermore, into the most volatile, the more
volatile and the volatile; into the most fixed, the
more fixed and the fixed. In this way, he can give
each its name. When he has done the separation, he
can immediately proceed with the conjunction,
coagulation and fixation, which does not take as

long as putrefaction and dissolution or separation.
If an artist but recognizes its advantage, he can
himself, with some reflection, more quickly shorten
the work than I could describe it to him.

As far as the volatile parts are concerned, he
should consider them a volatile seed, but as to the
Acetum or the *Acidum*, he should consider it the
intermediary or half-fixed and half-volatile seed:
the salt or *nitrosum in universalibus*, in
specificatis, the Sal resolutum essentiale nitrosum,
or the fixed, is the fixed seed. Likewise the O O:
because oil is a coagulated or thickened and
concentrated *Acidum*, and *Acidum* a dissolved O O. He
should consider coal the more fixed part, an earthy
or coagulated oil. But when it is changed into ash
or an alkaline salt, he should consider it the most
fixed part, a ⏝ted, alkalized, fixed salt,
because coal can be made as fixed and fireproof as
ash in a violent, fast calcination fire.

If oil and coal are triturated and their
moisture is removed from them by means of a high
necked vial in ashes, it also turns into coal. But
if they are fiercely distilled, the oil is distilled
into a sour *Liquor* or vinegar. If the coal is put
into an open fire, however, it will finally turn
into ash and salt. An artist must know such first
principles as these before everything else,
especially when dealing with this tractate. If an
alchemist does not know what is a volatile, an
Acidum, or coal, or ash and *Alkali*, how does he know
what to do when mistakes occur and what reasonable
decisions to take?

Consequently, from the preceding chapter the general purpose of all separation, coagulation and fixation is to be noted, which must and should be followed by everything else according to a specific natural law. Just as people follow their king, everything else must follow the universal seed or *Chaos pluviali aquoso*. And as has been logically explained in the preceding chapter, the reader should proceed thus - he should reason, reflect and do practical work, if he wishes to do something useful and reach his goal. Books are published so that the reader should first grasp the subject matter in his mind, and understand the author's intention regarding this or that point. He must ponder it well a hundred times before beginning manual work, so that he does not need to have regrets later when mistakes are made and does not revile the author who had the best intentions. Since the author does not know everybody and cannot show the manipulations involved, he publishes his writings, so that his practical experience can benefit his fellow Christians and they can profit by it.

That is why I wrote first about rainwater, being the universal chaotic water, and its separation and coagulation, because everything grows and originates out of it, so that the lover of the Art should have an example and a rule from this on the general purpose with which all other *specifica* and *individua* must comply, and he should judge them and deal with them in the same way.

One thing is certain: the whole of Nature had been water in the beginning. Everything was born of water, and precisely through water and through this chaotic regenerated water everything must again be destroyed. Where there is no water, there is no separation in our Art, enabling us to separate the subtle from the gross.

Therefore, just as Nature gives birth, sustains and destroys everything through the regenerated chaotic water, and again begets and destroys everything with Herself and through Herself alone out of the destroyed matter, we must also follow in her footsteps and example. Like water which is of the same nature as, and has affinity with all and every thing in the world, we must destroy again all coagulated regenerated things, and give birth to them again in a more noble nature, that is, in an everlasting Quintessence.

As this water has been divided into its parts - into a volatile and a fixed nature, we must resolve, prepare, putrefy and corrupt the volatile things, animals and plants, with the volatile chaotic water or its like, as they are not as closed, compact, and hardened as minerals.

With the more fixed parts of the water, however, nitrous and the *salalkali* ⊖, we must dissolve and destroy the minerals and other harder, coagulated bodies, because they are composed of and generated from these more fixed parts. It must necessarily be a penetrating sharpness that is to release the stone-hard bodies from their bonds. But just as every single thing carries its *solvens &* *coagulans* with and in itself, though it is not powerful enough to destroy its body, we help it with its first mother, the universal chaotic water, rainwater, snow water or dew water, to its destruction, decomposition or putrefaction, and thereby arouse the coagulated sleeping spirit, that which is its own homogeneous inciting agent of its destruction so that then after suffering the pain of the fire of purification, putrefaction and separation, it can be glorified and acquire the immortal glorification of a fifth essence through its coagulation and fixation. Although there exist some subjects which contain an excess of their own destructive moisture by nature, and carry their own adequate destroyer or death on their own backs, thus not requiring the help of the chaotic or rainwater, there are nevertheless many which are in great need of its help.

Animals and plants are of a very juicy, moist, liquid nature. After they have been cut up small or bruised, they immediately go into putrefaction, corrupting and fermenting through their own juice. Should they be lacking some moisture, some volatile from putrefied rainwater can be added to help them. The more moisture they get, the sooner putrefaction sets in.

Instead, stones, metals and minerals, the *flumidum* of which is too much dried up and too strongly coagulated, will not only obey this volatile water, but when the fixed and half-fixed chaotic water, that is, ⦶ and ⊖, are turned into the same nature as that from which the minerals themselves were born, the doors of hell are broken and the indwellers released.

Above, in the first tractate about the generation of minerals, I said that minerals are generated of a resolved, salty, spiritual ✝ which is a ⦶ and ⊖. In the great belly of the earth, it is acidified by strong fermentation, rises and congeals on the inner surfaces of mountains by the central heat in the form of a spiritual vapor, and there it brings forth various kinds of minerals.

These vapors are universal seed and a dissolved ⦶ and ⊖ and salty nitrous spirits which have the same general nature as all minerals. With or by such a spiritually-made ⊖ or ⌒ ⦶ and ⊖ the coagulated and exsiccated minerals must return, in order to arouse to action their own mineral coagulated and exsiccated *Acidum* and to destroy their own body. Thereby they become that which they had been in the beginning of their coagulation,

namely, a salty, mineral, metallic spirit, a ⦿ and this, by turning back, a spirit, and this spirit, through regeneration, a regenerated, penetrating medicinal balsamic body, each after its kind. When they have been transformed in this way, only then can they be changed and raised into a further spirituality or pleasantness, volatile and fixed, by means of the volatile universal water or seed and chaotic △ or through plants and animals, as one wishes - plant or animal or even universal. Every creature is changeable into any other, because everything is born of one primordial substance.

Animals are extended plants, plants contracted animals. Again, plants are extended minerals; minerals, instead, concentrated plants; plants, concentrated animals; and all these are a concentrated universal seed or chaos. Because

animalness is volatile, and the more the volatile is
contracted, however, the more it becomes a plant
☩; the more the plant acidity is contracted,
however, the more it becomes mineral; and likewise *e
contra*: the more the mineral is extended, the more
volatile it becomes, so that it is by degrees
transformed into a plant and animal state.

Having mentioned this beforehand, we will now
undertake the destruction of the animal kingdom and
search for its fifth essence.

CHAPTER VII
DISSECTION OF ANIMALS

No strong water can be obtained by separation and distillation without putrefaction, both in the animal and the plant kingdoms, except for the plant or animal smell. On the contrary, all powers are unlocked by a preceding putrefaction, in order to obtain a urinous volatile salt from the animal kingdom and a volatile [161] burning from the plant kingdom.

As we intend to describe a true dissection of things, we rightly begin to proceed according to the laws of Nature and finish according to her grades of preparation, putrefaction or dissolution, distillation and rectification, of the union or conjunction, coagulation and fixation. As all *Subjecta* and *Individua* of the whole wide world differ one from another, it is here in the animal kingdom, where there are whole bodies with blood, flesh, yams, marrow, bones and skin, urine and excrements. All together they are taken to produce the medicine. Again, one can take the different parts of a body by themselves for the medicine, especially blood, urine, excrements, the legs, skin, hair, horn, etc. and make a special medicine from each part.

How to dissect these pieces each and all, also how to turn them into a fifth essence, we will now describe, first in regard to all liquid things, then

---

[161] There is a large space here in the R.A.M.S. edition – missing symbol? -PNW

in regard to the dry ones.

More than the others, this realm has the most horrible, atrocious *Praxis* on account of its stench during putrefaction. On the other hand, because of its highly penetrating *Salis volatilis*, it has a much mightier and swifter power to act than the other kingdoms.

I advise alchemists, however, not to work with blood obtained directly from an animal, that is, with warm blood, because the following happened to me: When I tried to distil the more fixed parts through the alembic, the *Evstrum* both of human beings and of animals appeared in the recipient in a very hideous way. That of human beings started a rumbling noise in the alembic as if a poltergeist were in it, which was very horrible to see and hear - although it does not happen each time.[162]

If, on the other hand, blood and flesh are allowed to putrefy, they cause a horrible stench. Therefore one should take, provided one can get them, the excrements of every animal, such as urine, feces, which are best, and contain the full power of the animal; after that, horns, bones, nails, hair, scales, etc. But we will describe the work on all of these, so that nothing is lacking.

Now Rc. The blood, juice or urine of an animal and what is liquid, either one of these or all together, it does not matter. As they are all made of one substance, although one is more volatile or fixed than another, they have nevertheless one substance, coming from one subject. Put it into a

---

[162] Here the R.A.M.S. edition refers to chapter endnote (a), but it does not exist. -PNW

covered vessel and set it in a lukewarm place to putrefy. But whoever wishes to avoid the stench must put it into an alembic, well closed with a head and a recipient, set it into B.M. of the first degree, let it stand for 14 days and nights, and he need not fear the stench. When then it has thus stood for as long or longer, depending on circumstances, distil everything off in B.M. according to grades, and preserve that. If you wish to rectify it and separate the *Phlegma* from it. Thus you will obtain a very penetrating urinous spirit and volatile salt. The *Acidum* does not rise from the B.M. through the head. When this is done, remove the alembic, put the residual matter into a retort in the sand and drive it once more through gentle grades. A *Phlegma* will come first, followed by a penetrating ⌒ that

attacks the tongue, which is the ☩ *animale*. This

it is followed by a stinking, thick ⚬ ⚬. After this, a matter burnt to coal stays behind, which is the alkaline part.

Now you have separated the volatile, the acid, the oil and the alkaline coal. This then is the substance and inner essence of the animal, and these are the parts it consists of. In order to put these back into one, you must here once again heed the philosophical saying: *Non transiri posse ab uno extremo ad alterum absque medio*: one cannot go from one extreme to another without an intermediate thing.

The volatile ⌒ and coal are the two extremes, they will never unite without their intermediate nature. Their intermediate nature, however, is their water or *Phlegma*, their penetrating spirit or *Acidum*, and ⭘ ⭘. In turn, these will not unite if they are not taken in reverse order, or else so slowly that your efforts and work will annoy you: If you wished to unite the volatile ⏝ and the oil, or vinegar and coal, and overwhelm the vinegar in the first instance and the oil in the second, you could not do it. Instead, they will readily and instantaneously combine when cohobated in their proper order.

Therefore, this order must be followed. As one went before the other during the separation, so they must combine in the same order during their conjunction. Then they will immediately coagulate through moderate and appropriate degrees of heat.

Whoever should wish to rectify such parts, may well do it in a more subtle way but not faster or better. Therefore, whoever wishes to work faster, let him take the volatile with the *Phlegma* (or separate the *Phlegma* from it) and pour the spirit to the Azoth, then both are joined. Thereafter take the oil, mix it with the coal or ☠, put it into an alembic, pour the vinegar and the volatile part over it, set it in B.M. for two days and two nights to digest moderately. Then distil through gentle fire grades, and the volatile spirit will rise quite weakly, together with the *Phlegma*, while the most volatile and acid stays at the bottom. Take that out of the B.M. and set it in ash to dry out, to coagulate and reverberate, as was taught in the fifth chapter. When it has been reverberated, imbibe it again with its distilled volatile, set it in B.M. to digest, distil, then to coagulate, dry out and afterwards congeal, as has been taught in detail with rainwater, as it must also happen here in the correct order. Then the Quintessence of the animal kingdom is ready.

After this, someone might ask and say: My! Why does he say to leave the *Phlegma* with the work or to separate them? Is it of no use? To answer this: To begin with, I leave the *Phlegma* with the rest, because, although it remains and is not driven away from the volatile through rectification, the essential congealing part does not absorb it but lets go of it each time during distillation.

Let someone take note but of this, as I said above, that *Phlegma* is still an unripe and not salty seed and therefore a guide and tool of the active and passive universal spirit, by means of which the natural indwelling, coagulated and slumbering spirit in a body forges everything, or has forged everything, changes everything or has changed everything. For as long as the *Phlegma* is still there, it arouses the spirit again, so that it begins once more to act and work and to produce a constant change via this *Phlegma*.

To confirm this, take a Quintessence, made in such a way, when everything volatile has been coagulated and concentrated, put it in an alembic which the *Phlegma* may fill completely to the brim, pour its own *Phlegma* upon it, set it in a warm spot, and watch:

You will see a wonderful sight, because the spirit or *Evestrum* will represent the shape of the animal as it had been previously during life. Of it, the *Subjectum animale* had been taken. But if it is then put in the cold, it will disappear again. From this an artist may now infer to what purpose the *Phlegma* is taken, because it arouses the implanted spirit to action.

Besides, this *Phlegma* should not be rejected, because it is altogether filled and impregnated with the spirit and the spiritual power of its Subject, like all distilled apothecary waters. Instead of taking the Quintessence in another vehiculum, I am taking it in its own separated *Phlegma*.[163]

In addition, this *Phlegma* serves to produce putrefaction in a fresh subject, instead of using other *Species*, such as spring water, rainwater or leaven etc., although spring water or rainwater are actually of the same nature. This has been said about the liquid parts of animals. Now we will proceed with the dry or dryer parts.

For this purpose take the flesh, bones, horn, hair, claws, skin and whatever is hard in animals. Pound, crush, cut up, grate, file and chop one up, whichever you wish and as well as you can. Then put it into an alembic and pour on it some blood, putrefied urine or juice of the same animal from which your subject has been taken, or, for want of all these, putrefied rainwater or ⊡ *microcosmi*. That is, of a human being, because he is the concentrated center of the entire animal kingdom, in which all the powers and virtues of the other animals have converged, just as in wine all powers and virtues of the other plant growths, and in ☉ and its vitriolic Guhr all powers and virtues of the

---

[163] The R.A.M.S. edition indicated footnote (b) here, but that footnote does not exist. -PNW

minerals have converged. Pour, I say, some of those on your crushed or powdered subject, set it in B.M. or B. vaporis, or digestion apparatus, etc., let it putrefy, then separate all parts from it through B.M. and ash with the alembic and retort, as was said before. Then, if you wish, rectify each part according to the abovementioned instruction and unite them, also coagulate and congeal them just as the aforementioned.

But because the hair of an animal are first of all an almost pure fat, coagulated substance, fat oily things, and oily things are largely balsamic and do not easily go into putrefaction, and if so, very slowly, just like the bones and horns, enough to scare a lover of the Art to undertake such tedious labors. I must show him two shorter knacks by which he can quickly get on.

When, therefore, you have ground, grated and filed hair, bones, horns and claws, etc., boil them in the urine of the same species of animal or in human urine, or putrefied rainwater, or in saltwater, until you have a gelatin or a jelly. You must thoroughly boil them for 24 hours, or two or three days without interruption, until the bones and horns also turn into a jelly, although some take less time, according to whether the compound is hard or soft. When they have turned into jelly, they will start putrefying in a few days and nights with the addition of more rainwater or urine which must be foul and bad-smelling. Thereafter, the separation and conjunction is in everything as above: First, you must drive the volatile parts out of the B.M. through the head, then the more fixed parts out of the sand or ash in a retort; after the rectification, you must conjoin, coagulate and congeal.

The other knack is the following, but it does not give as much substance as putrefaction. Nevertheless, it will be interesting. Take the horns, claws, legs, hair, excrements and skin, grind them like hazelnuts, the hair cut small, put them in a retort with its receiver, and distil with a gentle fire everything that will go over. When the separation has been done, unite the parts again in the same order as they parted from each other. But here the artist requires no volatile but only a gross *Phlegma*, *Acidum*, ⚬ ⚬ and coal, because the volatile in those hard parts has partly evaporated through coagulation and exsiccation, and been partly transformed into Acidum or ♁ *animale*.

This then is the artificial separation and conjunction without separation of the superfluities, in which all parts have been concentrated and congealed except the excessive waters and the *Phlegma*.

I must here remind the reader and anticipate. Because I often refer to something and repeat it, he must not think that it is superfluous, but I do it so that he should infer further and have an opportunity with every word to see ever deeper into Nature. Therefore, many a man will say that I try always to follow Nature, and that I yet have many violent ways that are counter to Nature. To those I have added the way of Nature which does not destroy a thing completely and burn it to coal, or it happens but very seldom. An artist, however, must consider the final purpose of Nature and the Art. Nature does not desire to destroy a plant or animal body to the utmost, because it is enough for her to dissolve those bodies into a salty, essential, spermatic, salty watery or guhric juice, as food and seed, to generate another thing of the same nature. But she does not have the power to make a quintessential and eternally glorified body, as the Art does - a body which is never corruptible in itself. Likewise, all glassy bodies are the most durable - more even than ☉ and ☽. One never hears, or very rarely, that glass and precious stones have decayed, unless the artist destroys them into the first matter with a great deal of work. In a natural way, however, it does not happen easily. Instead, one can see in mines that ☉ and ☽ are awakened again by the arsenical vapors and destroyed to such an extent that they leave nothing behind but an empty shell, given its structure by preexisting crystals etc.

   Here I will make known two more ways. One a quite natural way as Nature herself uses, the other an artificial one. With these, an artist may kindle a great light in his mind and select one or the other, as he wishes. In the first way Nature operates as follows: Nature softens the dead animals and the tender plants with dew and rain or other kinds of water and moisture. She causes them to putrefy, then she distils one subtle and volatile part after another into the air through the solar or central heat, but the weak heat cannot lift the ⊥ and the oil etc. In today's common laboratories this left over *Residium* is the animal or plant, essential

411

⊖, which I call a ⊕ *animalem & vegetabilem*, precisely because it candies in such a way and is

afflicted with a precipitable earth. This ⊖ or ⊕ gives off an acid spirit during distillation, whose acidity has a mineral taste, namely, the vitriolic

acidity, which is followed by a thick O O and later on the coal. Nature does not separate these three in the plant and animal kingdoms. In the mineral kingdom, however, she congeals them even more, and concentrates them, so that they become ever sharper

and more corrosive, as may be seen with ⌒ and O O

⊕ *li*.

When now Nature has thus turned animals and

plants into an essential ⊖ or ⊕, she waters them all the time with the volatile parts, rain and dew, etc. Consequently, the artist can do a similar work

by turning animals into an essential ⊖ substance or gelatine, thereafter distilling it with a volatile ⌒ of the parts of the same animals, imbibing therewith, coagulating it and congealing it into a fifth essence by frequent repetition. However, should he not have any volatile part from that animal, he should take the volatile spirit from human urine or from rainwater, dew, etc. If Nature imbibes frequently, the essential grows up into the air and turns and is transformed into a plant or a tree, since she does not raise anything to a fifth essence. The artist, however, changes it into a quintessence, which Nature does not yet achieve.

But to throw more light on this subject, I will

describe the process clearly. Take an animal (the same applies to a plant): Turn it into jelly or gelatin by its own putrefied or human urine or by putrefied rainwater. Let it putrefy and ferment, then pour off the clear part, filter it, and distil ex B.M. everything volatile from the filtered part, to one-third or the oil. Preserve the volatile.

Remove the O O or the settled (clarified) *Liquor* and set it in the cellar to crystallize, or to turn into a jelly. This is the essential ⊖ of the animal or the 🜨 *animale*. Now take these crystals or jelly, set them in ash to coagulate gently, so that they become dry but do not burn to coal. Now Nature stops but the Art begins. Let this cool down, pour upon it as much of its *Volatile* that it stands but two or three to four fingers above it, and digest it again in B.M. Let it rise and go over whatever will go, because in the B.M. nothing burns to coal or ash. Afterwards, when nothing rises any longer in the B.M., set it again in ash and coagulate it till dry, reverberate it somewhat more strongly, then remove it, powder it and imbibe it once more with its volatile. Set it again in the B.M., distil it again, coagulate it in ash, and repeat the procedure with imbibing, coagulating, reverberating, congealing until it has gone through all colors, as said above, and the quintessence will appear.

In this manner the volatile will be congealed, as it must be, and finally an insipid *Phlegma* will rise which has left behind in concentrated form all its essential parts. This is nothing else but a mineral fixed nature, resisting all fire. This then is the simplest way, such as Nature uses herself in her work. The other way is all pure as it does not suffer any *feces* (as the scrupulating alchemists imagine), but a quintessence purified to the highest degree.

After you have separated the volatile acid or oily parts of an animal or plant, rectify the volatile and acid of all Phlegma in the best possible way and as is in general described by almost all authors. Then, as follows:

Rc. Take the oil. Mix it with two parts of coal and distil it likewise over through the alembic in ash or sand. Or if you do not wish to have the oiliness, mix the oil with its coal, put it on a cupel and set it in a baking or pastry oven. When the oven is hot, the flame will reverberate the coal from above and the oil will turn into ash and salt. But you must put the cupel in a spot where no wood or coal can fall into it and only the flame can go over it. Then, when it has turned into ash, lixiviate it with its own *Phlegma*, filter it, coagulate it, and you have the *Sal alkali*. Put that again in a cupel and let it reverberate once more in such an oven, and calcine. After this, dissolve it again in its *Phlegma* or distilled rainwater, filter and coagulate it. Continue this reverberating, annealing, dissolving, filtering and coagulating until the salt is beautiful, clear and white. Then the three parts, the volatile, the acid and the *Alkali* are cleansed in the best possible way. Now the conjunction takes place.

Before, however, I must recommend the *reverberationem per flammam*, because laboratory workers in general reverberate with coal under the

414

rnuffel, which is by far not done as subtly as by the flames of the wood: They penetrate much faster and more sharply than coal, because the flame has a pure and very penetrating *volatile*, while coal is a very strong *Acidum* and corrosive. Yet anyone is free to choose any of these two. I consider the method with the flames a better one, as experience has taught me.

CONJUNCTIO

Take two parts of the rectified *Alkali*, put it in an alembic, pour four parts of its volatile upon it and then three parts of its acidum, and they will become united and congealed before your eyes, so that they flow constantly together in the fire as an incombustible O O. In the air, however, they stand like ice, and nothing is now required except that you put it into B.M. with a head and a receiver and draw its *Phlegma* off to the point of oiliness. Set that in a cold spot to crystallize, and the quintessence will crystallize.

Remove that, again draw off the moisture to the oil, or let it evaporate to a skin, and crystallize once more. Do this until it crystallizes no more, and you have the quintessence. Dry that gently, put it into an alembic, set it in sand, heat through the four degrees of fire, and it will flow together into a stone. It is so clear that a light can be seen through it when molten and stands like oil, and when the fire goes down and is out, it is a stone. Now smash the alembic and take the fifth essence out, put it into a boxwood can and carry it thus dry with you through the world. If you wish to use it, take a few grains of it and some appropriate water or wine from the nearest pharmacy, throw them into it and they will melt like sugar and ice. Then administer it and note its powers.

However, you will find (although you have separated all your parts as much as possible from all *Phlegma* or excess during coagulation) much and more *Phlegma* than quintessence. But aside from this, you will also see how quickly the homogeneous parts unite, coagulate, grasp each other and stick together so firmly that they sooner penetrate the crucible or glass below before they will part again. That is how fast they congeal. Even if they are driven fast through the alembic in a volatile state, they yet participate in each other and none can be recognized from another.

Now the artist knows several kinds of work and manipulation, so that he can concentrate and turn the whole substance of every single thing, except the excessive water or *Phlegma*, into a dry, fixed fusible form and carry it with him through all lands without danger. A grain of it is more potent than a great amount of common distilled water.

But someone will ask: Why does he burn the oil, which is such an essential part? I wished to do it so as to work faster and so that the artist should notice that the Art transforms the oil into salt, and that the ⊖ or *Alkali* is a congealed and reversed O O, which one can see by its tincture when the *Acidum* and *Volatile* are poured over it, when it shows either a deepest ruby-red or a golden-yellow or other colored tincture. But whoever wishes to keep the oil and take the fixed salt from the

calcined coal alone, may do so. When the
quintessence has taken the form of a stone, he can
add the oil, mix it with the stone, add his
distilled *Phlegma*, boil them in B.M. and distil them
to the point of dryness gradually through the
degrees of heat, afterwards coagulate them in ash
and sand, congeal them and melt them to the stone,
as I have clearly described the method in connection
with rainwater.

Here someone might complain and say: Yes, this
way might well be good if one could produce the
stone in quantity, and still better if the
apothecaries could give it for little money, so that
all, poor and rich, could enjoy it.

This is indeed easy, and if someone were to
reflect just a little, his intelligence would inform
him itself, thus: Let an apothecary take three
basketfuls of some herb, such as balm-mint, or let
him take the blood, urine or flesh of an animal, and
set it to putrefy in a big alembic. Then he should
take the bones, horns, claws, hair, etc. Meanwhile,
while the volatile parts are putrefying, he should
put the crushed dry parts into a retort filling half
of it with them. The *Acidum* and the ⚬⚬ must be
distilled there from to leave the coal. Thus he will
get the ✝ and ⚬⚬ also a large quantity of coal.
He should give the other half of the dry parts to a
potter, to be calcined in an open pot. From this ash
he must leach the *Sal fixum*, no matter how much
there is of it. Of the above volatile part which had
been standing in putrefaction he must distil a large
quantity of the *Volatile*. He can also calcine the
rest and leach the salt, and he has the constituent

parts in quantity. After this, he must do nothing but the conjunction and coagulation, and he will have the quintessence in superabundance and to sell for reasonable price.

I must here note that animals do not yield much fixed salt but give much saltless earth. But how will someone proceed to obtain much fixed salt in order to congeal the volatile parts? That man must run back to Nature, where Nature Herself often forges a universal *Alkali*. This Universal Alkali is homogeneous with all creatures. There exist indeed entire salt mountains, and common table salt is indeed the best balsam for all animals, especially for men. To specify it, however, for each *Subjectum quint-essentificandum* is quite easy if he takes the dry parts of the animal, which he sends to the potter to be calcined. He crushes them small and mixes a third or fourth part of table salt with them. Then the salt burns and specifies along with them and turns into a specified animal *Alkali*. Thus the artist should not complain that he cannot separate the quintessence out of all things in a large quantity, and an apothecary could stock up his whole drugstore with nothing but quintessences which, if he but had a supply of them, would never corrupt and become moldy like his waters, oils, ointments etc. If he manufactured the quintessence but once every three years, he could make of every subject several pounds which would not spoil, and which he could sell cheaply to his neighbor. He does not sell the quintessence by pounds, ounces or half ounce lots, but by grains and scruples (1/20 grain) which acts powerfully and more quickly in a *Dosis*.

Consequently, the time he has spent in making

oil and water can pay double and fill his purse as well and even better than before. Plants, however, can be made even more easily, as will be taught in the following chapter. He must take a large quantity of plants: three basketfuls. He sets one to ferment and putrefy, and gently dries the other two in the shade, so that they dry thoroughly. Of these he burns one to ash in an oven or in a potter's kiln. From the other he distills the vinegar and the oil. From the putrefied matter he distils the volatile; from the ash the salt, and when he has rectified everything, he unites them, and has thus the fifth essence in quantity.

From what has been said an artist can see that Nature can very well be united and also separated through middle natures, provided one studies diligently. She Herself reveals all means, and sets the ☩ between the volatile and the *Alkali*, which is to be found in all subjects and without which there can be no lasting union. It is neither fixed nor volatile but an intermediate, a true hermaphrodite and Janus which sees in front and behind. If it meets the volatile, it is pleasant; if it meets the *Alkali*, it is its equal; with the volatile it becomes volatile, with the fixed, fixed. No author has described this point. Therefore, consider it a very precious secret and thank the one who disclosed it to you.

After having dealt with this (the dissection of the products of the animal kingdom), we follow the order and go to the hermaphroditic realm of plants, whose head is next to the animal kingdom, and whose root next to the mineral realm in order also to extract their inner essence. Therefore, it now follows.

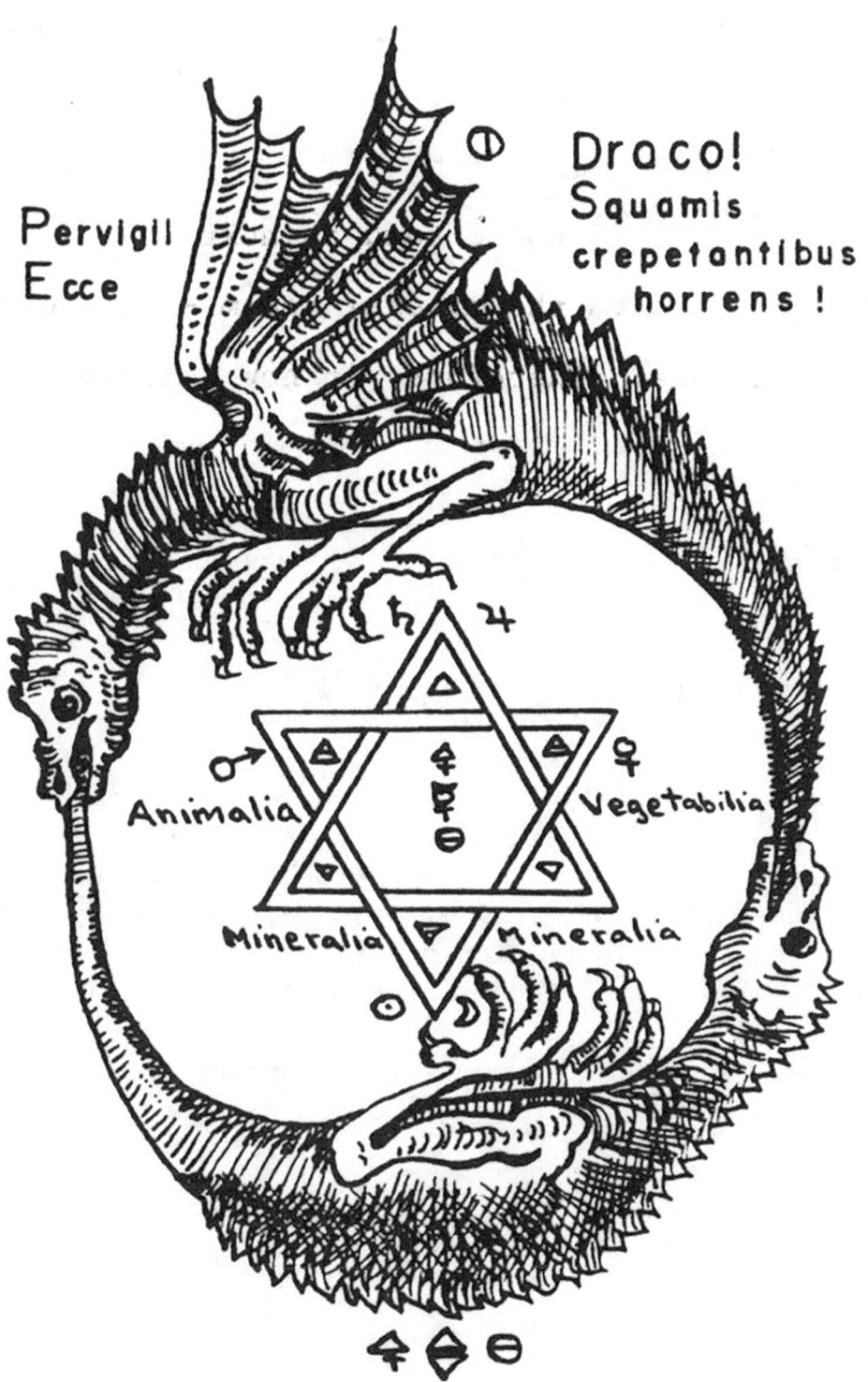

Pervigil
Ecce

Draco!
Squamis
crepetantibus
horrens !

Animalia

Vegetabilia

Mineralia

Mineralia

Whenever the Dragon meets an Enemy, they fight. The Volatil must become fixt, Vapor and ▽ must become ∀, △ must become corporeal, or no life can enter into the ∀. The Superius must become Inferius, and Vice Versa. The fixt becomes volatile. The ∀ becomes ▽ vapor, △ and △, whilst △ returns to the Centre of the Earth. Heaven, i.e. △ must be converted into a fixt ∀. The Dragon with Wings kills the Dragon without Wings, and the Latter destroys the former. Thus is manifested the Quintessence and its power.

About the Dragons Illustration
by Philip N. Wheeler

The illustration appeared in the 1762 Latin edition of The Golden Chain of Homer and in the 1781 edition. It is also said to be found in "Das Hermes Trimegists," Leipsig, 1782. Its exact origin is unknown. The illustration contains the following Latin phrases and alchemical symbology:

Pervigil Ecce Draco Squamis crepetantibus horrens! (Ever watchful, the dreadful scaly dragons burst asunder!)

⊖   Niter, Saltpeter, Potassium Nitrate, Soda

♄   Lead, Saturn, Plumbum

♃   Tin, Jupiter, Stannum

♀   Copper, Venus, Cuprum

♂   Iron, Mars, Ferrum

☽   Silver, Moon, Luna, Argentum

☉   Gold, The Sun, Sol

△   Fire

🜁   Air

▽   Water

🜃   Earth

🜍   Sulphur

☿   Mercury, Argentum vivum, Quicksilver

🜔   Salt, Sal, Sodium Chloride

Animalia (Animals)
Vegetabilia (Plants)
Mineralia (Minerals)

CHAPTER VIII
ANATOMIA VEGETABILIUM

Concerning separation and coagulation, this kingdom resembles the animal realm, except that it differs somewhat in the quantity of its components. The animal kingdom volatile keeps its bad-smelling urinary salt, while the plant kingdom has its bad-smelling burning spirit, although it is more pleasant than musk and amber to many an alcoholic. This realm is also different in its subjects, like the animal kingdom, for it has partly juicy herbs, stems, fruits, juice, gum, resin, oil, seed, wood and roots, and partly hard dry parts, stems, roots, wood and seed. Therefore, we must describe slightly different approaches for the benefit of the lover of the Arts, namely:

Grind, cut up, squash everything green as well as you can, and if it has not enough juice by its own nature, pour as much putrefied rainwater, or wine, or salt water upon it that it turns into a thin paste. Or, if you wish, press the juice out of it and let it ferment, like wine, or the way farmers make cider from pears or apples. Every soft and juicy plant can be treated this way, as can also the hard ones when they have been ground fine and a sufficient quantity of moisture has been added to them. Or, keep all the plants turned thus in a paste together. Put the paste in a lukewarm spot and let it thus soak until it gives off a sour or foul smell - about fourteen days and nights, or three weeks. Then put it into an alembic and slowly distil the subtle volatile with its soft *Phlegma*. Remove what is left over, dry it completely and put it into a retort in sand and distil by degrees. In this way, you will first obtain a grosser *Phlegma*, then a ☩,

then the thick O O; and then a lump, burnt to coal will stay at the bottom.

Now the plant has been separated. You should note however, that as plants are not one like another, they are also unequal in regard to their component parts, for one has a great deal of

volatile, the other more ☩ or oil, according to whether it has specified, coagulated and congealed much universal seed in itself. In accordance with these constituent parts, their virtues and powers are also likewise distributed and are to be assessed and then applied. For a fragrant plant, if it has much *Volatile*, that is, burning spirit, has the power to strengthen and heal not only the natural vital or animal spirit but also the *mentalem*. If it has no pleasant fragrance, it invigorates the natural animal and vital spirit, although it does not always depend upon the outer noble fragrance but rather on the inner, through which, distilled by the *Archeus*, often quickly comforts and heals the wounded organ. If the plant contains a great deal of *Acidum*, it is specified to heal the tighter limbs, such as muscles, sinews, bones, cartilages, thicker and viscid vessels, etc. Likewise the oil: the thicker the essential parts are, the thicker and more coagulated parts of the body they strengthen or destroy depending on how they are applied.

Any physician knows that a highly volatile thing cannot serve as food for the fixed bones. When such a volatile essence reaches the body, it is immediately driven out by the warmth to the outer

surfaces of the limbs, and finally completely out
through the pores of the skin in the form of
perspiration. On the contrary, an *acidum* is not
expelled this way, but it excites either the urine
or the evacuation of the bowels, or mostly a grosser
perspiration. We can see, when we hold something
fragrant under the nose of a highly melancholic
person, that he feels relief and comfort for his
saddened heart, the moment he smells it, although
such quickly passing fragrance does not achieve
lasting relief, especially when he has committed a
mortal sin or the like, or if the excesses of his
debts or his wicked wife are making him sick.
Nevertheless, one can see that he admits that it is
a pleasant and comforting fragrance and spirit for
his mind and heart. Instead, if one makes a stench
under his nose out of wickedness, he will get yet
sadder, sicker and angrier that very moment.
Likewise, the warm spirit of a plant or an animal
warms a cold melancholicus, whereas a cold soporific
or painkilling spirit cools a *cholericus*. This *en
passant*. A physician knows anyhow how to cure the
specific qualities.

When now the plant has been thus separated, the
conjunction is done in just the same order and
manner as in the case of rainwater and animals, and
may be treated in everything like them, in
accordance with all above-disclosed different ways
and manners or processes, as has been taught
concerning the rain - ▽ and the animal kingdom. It
is not necessary to repeat it, in order to avoid
verbosity.

I must make a present to the lover of the Art
of yet another method, namely: Many alchemists have

426

tried to obtain the volatile salt of a plant, but have not met with success, which is nevertheless such an easy thing to do. When you have set the plant to macerate and putrefy, let it stand until maggots and worms grow in it, which happens soon. When you have this sign, distil with a high alembic in the B.M., and an animal-urinous ⌒ will rise over and the volatile ⊖ attaches itself to the head. This is the reason and visible cause that the plant has turned into an animal, which animal kingdom is full of volatile salts. Let the reader take careful note of this. Through reflection he will obtain many other things, which he endeavored to seek and find for a long time.

In regard to harder growths, such as wooden plants and roots, wood, etc.: One deals with these as with the hard animals bony parts, and grates, saws, files, pounds and squashes them small, as one can, and one pours putrefied rainwater or wine or saltwater or saltpeter water upon the matter, lets it soak, or boils it until it is soft, and then putrefies them. Or one distils them, cuts them into thin small pieces, dries them in a retort, as has been taught in connection with the products of the animal kingdom, and when they are separated, the conjunction is done as described there.

So that finally the reader might not be beset with doubts that wood (when it is not rotten but is distilled dry) has no *Volatile* like animals, etc., I will impress upon him to the point of disgust, and work with it, namely, that which is homogeneous with every single thing in the world. All he has to do is take putrefied rainwater or snow water and distil the *Volatile* or the delicate volatile spirit of rain

427

or snow water, etc. Then he has a *Volatile* for all
things which have none. Likewise, if he has no
Acidum or *Alkali*, he should take the saltpeter or
its spirit. The *Alkali* is the salt and its alkaline
spirit (in fixed form). But if he believes that the
⦶ and the ⊖ the are too strong or corrosive, he
should distil all the *Volatile Phlegma* from the
rainwater. What is left over is to be driven *per*

Retortum, and he will get a ✝ while the *Alkali*
stays behind in the sediment which will become
visible during the reverberation. Then he has
everything he may wish to have and he now does not
lack anything.

    An artist should take note that a universal
subject can be specified for all particular things.

Supposing I had no *Volatile* but an ✝, ○ ○ and
Alkali. I add to them the volatile part of the
rainwater as a general essence. It becomes specified
with the other components and takes on the quality
and specification of the same *Acidum* to which it is
added. It is said, *a potiori fit denominatio*. The

Acidum, ○ ○, and *Alkali* are preponderant as far as
quantity is concerned. They can therefore easily
master the volatile and transform it into their
nature.

    Likewise it is with ✝ and *Alkali*. If there
were a subject in the nature of things that had no
✝ or *Alkali* but were only volatile, where would I
now take a ✝ or *Alkali* to congeal this volatile
and concentrate it into a stone? I look about in

Nature for something homogeneous. If I do not find any suitable to its kingdom (as there exist sufficiently however), I resort to the universal, to rainwater, snow, saltpeter and salt. Here I already have something homogeneous, as I require. If I have but one part of the specified product, the universals at once take on and act according to its quality and specification.

Everybody can see with his own eyes how the universal subjects, hardly born, immediately take on again the *Specificum* and transform themselves therein. When rain etc. in falling down adheres to the animal, plant and mineral creatures, it becomes the same as them. All one has to do is boil a plant, animal or mineral with saltpeter and salt, either in liquid or dry, and they will immediately partake of their qualities.

It is, however, not necessary to rely on the universal realm in all cases, since God has given a principal subject to each kingdom. The principal subject of each kingdom comprises in itself generally all subjects like itself and can be compared with every single *individuo* of its realm according to the observations of all philosophers, in its components the volatile, acid and alkali. In the animal species it is the human being, man and woman, with all their parts, urine, feces, flesh, skin and bones, etc. In the plant kingdom it is the wine and corn or wheat. In the mineral it is the *Sal commune* and *Nitrum*, which assist all hot and cold subjects and can fill their deficient parts or components, the volatile, the acid and Alkali.

There might finally occur a small hesitation concerning the separation of the animal and plant kingdoms, and someone might say: He has indicated most of the components and he is yet lacking one or the other. Because in the distillation of animals and plants, in view of the volatile, there also goes over in the B.M. or in the alembic, a delicate oil together with the volatile ⌒. That he has omitted and has not said anything about it.

Answer: I have said above that the more something is opened and subtilized, the tenderer and more volatile it becomes. What precisely is every burning ⌒ but an O O extended to the highest degree, or a highly volatilized, dissolved saltpeter or seed of a burning ⦵? I have indeed sufficiently proven that the *Volatile* and *Fixum*, the *Acidum* and *Alkali*, together with the volatile are in no way different in regard to their essence, but are only accidently different, according to whether one has been made very volatile or very fixed. After this they absorb the terminum *distinctionis* or *distinctionem termini*, otherwise they are *materialiter idem omnia et universa*.

Let nobody worry about such things. Even if the volatile part were to go over first, he should throw it again upon its fixed part in the conjunction, namely, to rectify and coagulate by means of the fixed. It is precisely such scruples that have prevented many a man from reaching the central point, so that they thought: Hello! This is surely a heterogeneous thing or the part rejected by Nature, and it does not belong to the work, etc. In this way they rejected the best, retaining the tripe in hand,

like the brandy distillers who keep the ⚗ and give
the larger quantity and better, more fixed parts to
the pigs. I am telling you, however, that everything
Nature has put together, be it poison or theriak is
all good. The artist can bring everything to
perfection: He will not transform theriak into
poison, but poison into theriak, and make it turn
out well. Whatever Nature has left unripe, poisonous
and raw, the artist must ripen.

Now then, it is known that all mineral, plant
and animal poisons are mostly volatile, raw and
unripe. When these have been made fixed, however,
they are no poison but an antidote and the best
tonics for the heart. Therefore, what Nature has
started and not perfected, is left for man to
complete, to contemplate and admire the million
wonderful different works of God, and not to destroy
himself because of his stupidity, but to thank the
God of gods that he has let him recognize his
wonders, see them with his eyes and touch and grasp
them with his hands. With this, this part is also
completed. Let us now turn to the mineral kingdom,
being the subject most in demand at this time.

CHAPTER IX
ANATOMIA MINERALIUM

In its outer appearance, this kingdom is completely different from the animal and plant ones. Although they are inwardly one, except that minerals are longer and more strongly fermented, digested, coagulated and congealed, and because they have chased away the excessive *Aquam recolaceam* or *Rumidum superfluum*, together with the most volatile seeds or volatile spirits, by the stronger degree of heat, also because they are of a dryer and stonier nature, they seem to be the contrary of the preceding. But whoever considers this realm thoroughly, has no doubts, as has often been experienced.

I said above that plants and animals are born of the more volatile universal seed, and that they are changed back and reborn into their first essence, precisely through that volatile seed. Minerals, however, have sprung from the more fixed parts of the universal seed, saltpeter and salt, that is, out of the strongly fermented and spiritualized (volatilized), corrosive vapors of both; in a word, out of ⌒ ⨁ and ⊖, both mixed together. They violently attack, gnaw, corrode the earth, changing, dissolving and transforming it with themselves into a vitriolic or aluminous Guhr.

Just as these were born of the more fixed spiritual universal seed or ⌒ ⨁ and ⊖, so they must also be dissolved and turned back into an essential salt or ⊕, according to each appropriate stage, by just that seed or ⌒ ⊕ and

⊖. The essential salt or ♏ must return to a vapor or corrosive water as its origin, according to the saying: *Ex quo aliquid fit, in illud rursus resolvitur, & per quod aliquid fit, per illud ipsum resolvi necesse est: sed fiunt amnia mineralia & metalia ex apermate universali fixioni spirituali seu* ⌒ ⊕ & ⊖ *ut eaepe probatum; engo necesse est, ut per sperma fixum spirituale resolvantur in primum* — "That out of which a thing becomes, in just that, it is again dissolved. And that by which a thing is made, by that it must necessarily again be dissolved." Now, however, all ores and metals originate in the more fixed spiritual universal seed or ⌒ ⊕ and ⊖, as is proven in many places. Consequently, it is necessary that they are again dissolved into their first essence by the fixed spiritual seed.

But this kingdom also has various *Subjecta*, such as ♏, alum, volatile and fixed △, Arsenic, marcasite, metal, rock, etc. Therefore, according to such grades of coagulation the grade of dissolution must also be arranged. But so that a man may not go wrong in this natural science and turn to other subjects, which come from, and out of Nature as from the breasts of the grandmother, he must also take natural subjects, or those that come straight put of the mines and have not yet been processed by the Art. For things that have got into the hands of men are greatly changed by fire and various additions and eliminations. With these the natural *Praxis* must be done somewhat differently than with the previous ones, thereby also driving them back to their first origin.

Well then, it is a basic rule of this anatomy that saltpeter or its ⌒ does not attack the alkalized or congealed minerals as much as those which are still full of acidity. On the other hand, salt and its ⌒ falls short in all acid things. The reason is:

If the *Acidum* comes in contact with an alkaline matter, it is either corroded to death, or it does not attack it at all and congeals thereby instead of dissolving. Similarly, if an alkalized subject or a *Menstruum alkaline* comes in contact with an *Acidum*, it is also corroded to death, or does not attack it, or congeals thereby instead of dissolving. Instead, like dissolves like, such as one acid another, one *Alkali* another. But what has been driven together hermaphroditically by Nature and is united (that is, where Nature has yet worked too little, or congealed or alkalized too little, where the *Alkali* is forming but is still in equilibrium with the acidity), it attacks and dissolves both the *Acidum* and the *Alkali*, and both get satiated thereby, as will be said later on.

I have indeed said that the ⌒ ⦶ and ⊖ are universal-*Menstrua*, or *Spermata mundi fixiora*, which associate intimately not only with minerals but also with the fixed animals and plants. Whoever considers this point and heeds it carefully, is again nearer his goal of accomplishing many otherwise lengthy operations.

I have indeed also said that the *Spiritus specificatus individuatus*, in so far as it has

itself not sufficient excessive *Humidum* to revert to
its primary essence, must be awakened with the help
and addition of the universal spirit to operate in
itself. This applies first of all to the minerals,
which are mostly dry bodies and which have almost
totally chased their *Humidum* away. Those and similar
dry bodies must be helped with the universal *Acidum*
or *Alkali* because of deficiency in their own acid
vitriolic or aluminous moisture, so that their
incorporated vitriolic or aluminous be aroused to
act in its own bodies and to transform them into its
primary essence.

However, it is well known to all natural
scientists that in the mineral nature various
juices, *Liquores* and water are certainly found which
are suitable to dissolve different subjects, such as
petroleum, naphta, alum, salt, and saltpeter waters, [164]
water, acidic mineral water, sulphurated baths, etc.
But because these are far too weak in their natural
condition to attack a rightly congealed metal or
rock, let alone bring it into its primary essence,
we must look to the right origin and beginning of
all metals and minerals, by which all minerals
revert into their Species by means of various
digestions. We have now and then sufficiently
pointed at it, that is, that the universal seed, [165],
when it dissolves the earth in the bowels of the
mountains, turns it into a vitriolic and aluminous
Guhr and is thus further generated in various
subjects through the varying degrees of inner
warmth.

---

[164] The R.A.M.S. edition appears to be missing a symbol here. -PNW
[165] The R.A.M.S. edition is missing symbols here; possibly "⌒ ①," followed by "and ⊖." -
PNW

Therefore, since the first beginning and the original essence of all minerals is a vitriolic or aluminous acidity, we must also use that as our chief means for transforming minerals and metals grown out of it, into their original state, namely, into a vitriolic and aluminous essence, which must later on turn totally into a corrosive mineral vapor by further reversing.

Now then, the vitriolic or aluminous acidity looks to the mineral nature with the root, but to the plant nature with the head. With and through plants it can become a plant, and finally an animal through animals, or, however, once more a mineral or regenerated metal, with and through minerals. The lover of the Art again sees that Nature or the Art operates with nothing but *media mediata homogenea*, and from one origin to another - which he should consider carefully.

Now someone will say: Has he no other *Menstruum* than ⌒ ☉ and ⊖ such as ⩔̵, ⩔̵, ⌒ or ∴ ⊕, ☿ and ○. The world has long known that these are the most harmful corrosives and have let go of them a long time ago. There is no need to smear the paper, etc. Answer: Do not reject my wares before I exhibit and offer them.

Have you not heard out of what basic nature minerals grown, namely, out of these two or only a fermented ⊖ and ☉, i.e., their vaporous acid and alkaline spirit? If you know that, you also know the origin of minerals. And when you know that, follow Nature, how she generates, coagulates, congeals and interlinks minerals. With this and by this undo the knot and go back as they went before, and again make the fixed volatile in this way. For Nature congeals the spiritual corrosive vapors through the earth and dries their *Humidum* up, and when that is gone, they become dry. The drier they become, the more they are turned into earth, coagulated and congealed. But if they remained liquid, they would always remain liquid. Therefore, because they are thus dried up, give them a consubstantial moisture and an excessive *Humidum*, and you will again awaken the locked up, chained up and congealed spiritual *Sperma*, and you will again change it into what it had been in the beginning, namely, into a vitriol or alum, or a similar Guhr.

What then shall I say to you, you who shy so much away from those corrosive solvents as I recommend and advise, and instead desire to get the Alcahest, otherwise also called *Acetum acernimum*, without any corrosive, so that it should be sweet and without sharpness like sugar? Don't you know any more Latin, so as to know what is meant by acernimum?[166] Tell me, if you understand what is *Alcahest*! Why and what are the reasons that the philosophers add spirit of wine to the *Alcahest*? Or what are the reasons that they use spirit of wine when they intend to dissolve their subjects with the Alcahest? Don't you yourselves say, although without understanding it, that spirit of wine improves all corrosives?

By this and the following you should understand and judge according to the laws of Nature, whether my way is better or the (sweet) *Alcahest*. If you do

---

[166] acernimum = sharp. PNW

not wish to recognize this in Nature, study until you recognize it, what the *Alcahest* and its *Menstruum* actually are. In one word, many laboratory workers produce the *Alcahest* themselves, but because they do not know how to apply it, they reject the master-key to the fortress.

Now we will start by dissolving our minerals with the *generali*, i.e., the more fixed ⌢ mundi ⊕ and ⊖, which is the *Homogeneum vehiculum generale* for all *individuis*. Just as with animals and plants we took either their own juice, if it was abundantly present, or in the absence of it the chaotic putrefied rainwater. So we must do the same here if the mineral moisture is lacking or we cannot have enough of it. We then replace and increase it with the universal moisture. With it and by it we strengthen and awaken the mineral *Humidum*, i.e., the vitriolic or aluminous coagulated mineral spirit to operate within itself or in its own coagulated subject, to destroy and reduce it, and to loosen the interlinked bonds in its *proprio coagulato*.

Because minerals are such fixed, strongly combined, desiccated bodies, they also require a stronger penetrating solvent than animals and plants. Therefore we take the more fixed universal seed, ⌢ ⊕ and ⊖, and what saltpeter cannot achieve, the salt will, or both together.

This is, however, not to be understood to the effect that one has necessarily always to add the universal seed, but only when the mineral *Humidum* is insufficient or too weak. Then we strengthen it with the very subtle and penetrating universal medium, the ⌒ ①.

    NB. Otherwise, however, the ⌒ ⊕ and ◯
nis are at all times to be made in quantity. They
are the moisture of the minerals for all red and
white *Astris*. But because the old sages wrongly

added saltpeter to ⊕ by which ①, the ⊕ was to
be sharpened so as to better penetration through
minerals, they distilled a universal *Menstruum* from

① and ⊕ for the mineral kingdom. But because of
longstanding ignorance it had not been applied
correctly; they used it as a separation water or

▽ and could not use it for anything else,
although metals always become more volatile through
long digestion and finally they could distil a good
part of their color over. They did not take note of
this but rejected it as useless, because it was done
by a corrosive. They had especially great doubts

that metal can always be ⌣ed again into an earthy
powder out of the corrosive, because they did not
understand that the mineral kingdom is corrosive
from its first birth; secondly, that although it is
dissolved by a corrosive and made volatile, it
nevertheless always aims at all times at becoming
earthy again, because its goal is to be earthy.
Therefore, they easily revert to their former nature

through ⌣tion. But if they knew what the volatile
thing is, that keeps metals always volatile and mild
and in liquid form, irrespective of all additions,
they would know that metal would not resume a

metallic nature but rather unite with the ⌣ing
means and turn into a third factor.

They should take note that it is not really to
be found in the mineral kingdom but must be sought
elsewhere in Nature, and recognize precisely this
main point: Through it, the mineral acquires a
nobler alteration, so that not only the mineral
nature but also all others can enjoy it without the
slightest danger. One can indeed see how plant

spirits, water, oil and ☩ retain their volatility
longer and more stably than minerals, and love the
volatility of animals even more, although each and
all try to become earthy due to their innate and
acquired nature. They are seeking their places of
rest, without which they can never be mobile. Just

as one can see that all ☩ dry up and become a
mother, that all oils turn into a thick gum like

nature, all waters leave an earth, and when the
issimus but finds a subject to which it can adhere,
it also turns earthy immediately. The purpose proper
of the whole of medicinal alchemy in the mineral
kingdom is to revert the mineral through its own
basic moisture; and because such a reverted product
is purely mineral and corrosive, it is heteregeneous
with the plant and mineral nature. Consequently, the
purpose is to improve and sweeten that corrosive
nature and transform it from the mineral into a
plant, and from this into an animal nature. This is
the purpose, and otherwise the mineral is a
Heterogeneum and stays that way until it is
improved.

So many hundred *Menstrua radicalia solventia*
have been described by alchemists, and each
recognized his own as the best, although many a man
obtained a bad result thereby. Yet everyone would
have had a much nearer way if he had considered the

foundation and law of Nature. Many have made such *Radical-Menstrua*, and one can also buy them everywhere, but it all depends on their correct application.

In general, a *Menstruum* is described - which they call by the common name ▽̶ or ▽R - which is made of two parts of ♁, and one part of saltpeter, or in āā ; or also with alum; after they have calcined the ♁ somewhat. They mix it with raw saltpeter and distil therefrom a ▽̶, which also produces an effect the way they commonly use it - but which is not the right way. The reason is as follows: When the saltpeter meets the ♁ in the heat, the ♁ has a burning sulphur which is adverse to the saltpeter and chases its spirit away too quickly before it can attack and dissolve the ♁. Consequently, the ‿ ☉ goes over into the recipient, taking with it a little bit of the volatile sulphur of the ♁, of which the ▽̶ also smells, as the difference between ▽̶ and ‿ ☉ also shows when the latter is prepared by using "glue." But what stays behind is the congealed ♁ as much as the saltpeter and the fire are able to do, because it has been more congealed than dissolved by the ☉, being alarmed (frightened), flowing and sweating in the fire.

The right way however, is the following: One makes ⏑ in the normal way, or a ∿ ☉ prepared with "glue." Of this 1 lb. pour it over 1 lb. of pure ⊕ calcined white, put it into a retort, and distil the ⏑ off with a gentle fire so that the ⊕ be not calcined, but only slowly in sand to the third degree. If you distil the ⏑ too strongly from the ⊕, you will congeal the ⊕ more than dissolve it. When then the ⏑ has gone over, pour 1 lb. of fresh ⏑ upon it, and pour everything together upon the ⊕ and back into the retort. Let it dissolve and digest together for one day and one night. Then distil it again slowly until one third remains. In this way the ⊕ will lie like butter, quite greasy like another oil. This then is a reborn spiritual Guhr, which must now be changed and dissolved further into a liquid or moist vapor, if it is also to turn its like or other subjects into such a nature.

To this end, take the distilled ⏑ and add again 1 lb. of fresh ⏑ so that 3 lbs of ⏑ are added to 1 lb of ⊕. Let it dissolve and digest for one day and one night, afterwards distil again with a low grade of fire, and most of the ⊕ will rise over quite spiritually with the ⏑; and if all of it did not rise, one has to cohobate it until it has all gone over without leaving any sediments. When everything has gone over, it has to be driven over again once, twice or three times *per se*, and then the right *Radical-Menstruum* is ready. It will turn and reduce all red *Astra in primam materiam* and make them equal to itself.[167] Do likewise with ⏜ ⊖, if you wish - although it is not necessary. The previous one dissolves all *acid* and *alkalica subjecta*, as experience will show.

If one wishes to make a distinction, however, between the red and the white *Astris* - although it is not necessary - one should take the *Menstruum* of the ⊕ for the red and the *Menstruum* of alum for the white, and make of it a *Menstruum* with the ∇ or ⏜ ⊕ and ◯, in the above-mentioned manner, as with the vitriol.

Here I have again discovered a technique which many omit and of which they have known nothing nor paid attention to. In this kingdom I have written about *minorem*. But you, diligent and thoroughly taught artist, will know how to argue a *minori ad majus* and draw the conclusions thereof, else I cannot help you further. At least you have here the means to render fixed things volatile. If you understand me quite clearly, keep it a secret and be discreet, because many will read this, as it is so open that anyone can enter the door. Nevertheless,

---

[167] The virtues and qualities of this excellent radical solvent have been described in detail by *Verbum Electri,* in his second discourse, which is the eighth in our *Versammlungsreden.* Only, an artist will hardly attain his purpose by following this general description, unless he be in possession of one of our special instructions, which we give with pleasure to all our practicing worthy Brothers. This our Brother Homer, now reposing in God, seems to have presupposed, for he writes chiefly for the Sons of Sages, since he was our worthy relative.

it will appear difficult to those who are not chosen. The door is already open, enter the room, friend! *Aperta jam porta, intra in conclave, amice*! Mark, however, that I have only given you the key to open all locks, but one lock is not like another. Although they can be opened with one means, there must yet be varied and frequent trial and error, so that many will think that the key does not fit all locks. Now you have the key and you have the hands to unlock the doors, and feet to enter. Or shall I carry you like dogs to the hunt?

Well! Sit down upon my back. I will carry you to the bed of the naked queen through all doors. Be careful, however, when the Nature king comes. The danger is hanging over your head, for He will be enraged if your mind is full of vices. Then that is the end of you. Therefore, walk about with pious, decent gestures, so that you are not chastised by bites of your conscience loaded with vices. Be careful, I am telling you, for the king is such a Lord *qui acrutatur corda & renes, D E U S !* who "searches the heart and the kidneys," namely, God.

Now we will mention a small provisional distinction before preceeding further, so that this Art be understood all the more easily; namely, by Alkalized subjects I understand all mineral, embryonized, metallic sulphurs, congealed to the

highest degree, such as the minerals ☉ and ♂,

♃, talc, emery, bloodstone and countless similar ones which are not yet known, in which Nature strongly reverberates the *Acidum*, or coagulates it and congeals it, and has turned them into such an alkaline fixed nature that they do not easily revert to their first essence, even using an alkaline essence.

By acid subjects I understand all those which are still dominated by acidity and are also easily dissolved by acidity, which are not so strongly alkalized and congealed, such as ♄, ☽, bismuth and other white sulphurs and arsenics which in practice, during dissolution, show themselves of what quality they are, as I have indicated in the chapter on the generation of minerals. Consider that of an acid nature which the *Acidum* can attack, and what the *Alkali* attacks, consider as alkaline, and what attacks both, consider of a dual nature.

Among the hermaphroditic nature I count all those minerals and metals where the acidity has begun to congeal but has remained half and half on account of a weak digestion, such as ♀, ♂, ☿ etc. Such subjects can be dissolved and treated, not only singly but also together equally well in an acid spirit as in an alkaline one.

This sentence, however, is not to be understood as applying to the aforementioned *Menstruum*, but only if one wishes to treat such subjects solely with the common solvents, such as ▽̶, ⌒ ⊕ and ⊖, because with one or another subject there will be some delay on account of its subtle universality.

If they are specified with their own mineral ♁ - or ○ - acidity, however, one can dispense with such a precaution.

445

Therefore, we divide the following in regard to the red or the white Menstruum (from ♁ or ☉), into (a) red and white minerals ♄i, ♃is, ♂is, ☉is, ♀is, or ☽ae; (b) and then the marcasite minerals, such as ☿ii, ♂̇, bismuth, zink and all other marcasites (of metals), ☉is, ☽ae, ♂is, ♀is, ♄i, ♃is, or ☿ii. (c) Further, into the fixed embryonized sulphurs, such as bloodstone, emery, bolus, red stone, talc, magnet, blende, Galmey, Tutia, etc. (d) Afterwards into the volatile embryonized sulphurs, such as ♀, bismuth, arsenic, ♁, pyrites, and all volatile marcasites and quartz sand.

We will now teach how to dissolve these four *Species in genere* and turn them into a fifth essence. Only mark, if you wish, and dissolve the red Aetna in the vitriolic and the white ones in the alum menstruum. Therefore, take some ore, whichever you wish. Let it get red-hot according to its fixity, because one requires more heat than another. First you must powder it, then let it become red-hot in the crucible, and when it is red-hot, pour some 🜔 (common), stir it well with an iron wire until the sulphur is burnt out. In this way the mineral is prepared to be dissolved by the *Menstruo*.

Better, however, does the man who takes the mineral out of, or from the mountain, no matter which, powders it finely and washes the mountain or rock from the ore on the "Saxen" as ores are usually turned into a slurry. Thereafter, one lets it burn red-hot according to its degree of fusibility and together with the sulphur, and it is also prepared.

Now take such prepared ore, put one part into an alembic, pour the above-mentioned Menstruo upon it, of vitriol for the red, alum for the white, three parts, set it in the sand to digest and dissolve. What is dissolved, pour off clearly and gently, and pour again Menstruum upon that which is not dissolved, three times its weight. Set it again to dissolve until all is dissolved, and a clear *Liquor*. Thus the ore has again been turned into its first essence, because if you distil this Liquor to one-third in the sand in a retort or an alembic, and then let it cool down and crystallize in the cellar,

it is a ⊕ and the first matter of this distant

mineral. But if you again dissolve this ⊕ in three parts of fresh *Menstruum*, distil it through the retort and cohobate it until all has risen over, it is a vaporous primeval *Liquor* which cannot be reversed further without infringing on its mineral nature. If it is changed back further, there would occur a transformation and specification into something else, either into a plant or an animal or a universal. As long as it remains a corrosive vapor, it remains in the primordial essence of minerals and stands with its root in the mineral realm. With its head, however, it reaches into the plant kingdom, and is now very easily transformed into an animal *per vegetabile*.

Here you have again the whole of the mineral

with all its components, for neither its ⚥ nor its arsenic, nor marcasite or metal has been taken from it, but all its vital spirits and innate parts are together in a *Liquor*. If you take, for instance, the congealed malleable ☉ and ☽; ♄, ♃, ♂, copper, etc., you have but one part. During melting and congealing, its vital spirit has vanished, which is the vitriolic acidity, the ⚥, arsenic, marcasite, i.e. all that the metal has lost in the violent fire. Here these are preserved and used for the best, and it loses nothing of its natal parts.

Well, there is but one single *Modus* by which all and sundry ores can be treated, as experience will teach you. But if you wish to coagulate and congeal such a distilled Liquor or mineral ⚬⚬⚬, nothing else need be done except digest it in a not too low alembic through the grades, for three days and nights, in the B.M., under the head with its recipient, by means of gentle boiling, and distil the excessive *Humidum*. When nothing will rise any more in the B.M., set it in ash and subtly draw off all *Phlegma* or weak spirit through the first, second and third grades. Remove the sediment, put it into a phial, set it in ash to coagulate, and it will turn into a salty stone in the fire liquid as oil, and in the air it is like ice. You must not close the phial, for it does not easily rise. Thus you also have the *quintam essentiam* of minerals, only very corrosive and harmful to human nature, because it is still mineral at this time. If it is to be useful to man, it must be transformed into a vegetable and animal through plants and animals, for plants and animals are man's food and not minerals, as will be

448

explained subsequently. But in regard to minerals gone through fire, such as common 🜍, melted 🜩, bismuth, fine ☉, ☽, ♂, ♀, ♃, ♄, we must again transform them back by means of consubstantial components and must add back what has been taken from them in the fire. Now then, the sulphurated and arsenical acid spirit, together with its stony mother, have been taken from the raw 🜩, by means of which acidity the 🜩 could easily have been turned back into its first essence with the help of the universal or mineral-vitriolic acidity. The common 🜍, which is made from pyrites, has been robbed of its sulphurous ⌒ and ° °, also the coppery essence (out of which the ꝏ is leached). Similar parts have also been taken from ☉, ☽ and other metals.

But so that a lover of the Art with little understanding may see what has been taken from every metal and mineral, and how it is prepared in order to reduce it again to its prime matter by the addition of what has been removed, we will add a small table. Gold is calcined with sulphur, arsenic, 🜩. This calx is afterwards easily dissolved by the above *Menstruum*. ☽, ♀, ♄, are easily calcined with sulphur, as is also ♃ ore, and afterwards dissolved by the above *Menstruum*. Likewise, if 🜩 is mixed together with sulphur in the fire, until the

449

sulphur is burnt up, it is easily dissolved by the above solvent. ☿ can be ⚖ed with 🜍 and common salt. 🜹 is also immediately dissolved.

Sulphur, however, is a dry oil, and no oil is like a ⊖ or salty *Menstruum*. Therefore Nature herself has shown an easy and homogeneous *Menstruum*, that is, petroleum, which is a liquid, dissolved sulphur. Sulphur is a coagulated petroleum, and with it one can cook some fragrant liver (i.e. liver of 🜍) which does not make such a bad smell as rape-oil, linseed-oil and tree-oil (turpentine). Such "liver" is afterwards dissolved into a vitriolic ⊖ or *Liquor* in the above-mentioned *Menstruum*.

When now the reader has turned all metal and mineral into a 🜹 in the above-indicated manner, and has turned that into a *Liquor*, and dried the *Liquor* again into salt or a salty stone, it is prepared and ready to take on the plant or animal transformation, as will follow.

Indeed, I have said that the corrosive quality is innate in the mineral kingdom and that it is by nature less disagreeable and repugnant to the plant than to the animal kingdom. I have also said that no corrosive is useful to man but is a poison. The artist must know how to improve that poison into a counter poison or *Alexipharmacum*. This, however, cannot be done except through sweetening, and sweetening can only be done with plants and animals. Such a sweetening, however, is a transmutation or specification into something else, so that a mineral turns into a plant or animal, or into a homogeneous medicine for the plant and animal kingdoms. This is the reason why minerals are generally improved and sweetened by plant products, such as ℞ and ✝.[168]

Because all alchemists cry *dulce dulce* and yet so few of them understand what *dulce dulce*[169] means, I will disclose that extremely secret technique in a Christian manner. Very few have known it until today, and the volatile ℞ was supposed to accomplish everything. What kind of an effect they get of it, however, they experience daily. When a corrosive is not tasted on the tongue, it is called *dulce*. You alchemists, do you understand Latin? Does then the word *dulce* mean that everything is sweet, even when it is not sweet? No - but the word *dulce* must not be changed in the alchemical understanding but rather be and remain *nomine, re & actu dulce*. Sweet is sweet and must be sweet like honey and sugar, if it is to be called sweet. Consequently, you should make your medicines so that they are truly sweet and not comparatively so.

But, am I to give the pointed knife into the hands of all physicians and theriak pharmacists as I would do with children? Shall I reveal that which the fathers have kept from their children? How the

---

[168] The sweetening is quite remarkable and fundamentally in accord with our brotherly concordance. *Verbum Electri,* in the eighth of our Versammlungsreden, No. 8, P. 207 ff., deals, although briefly yet fundamentally, with this matter and shows why the gentlemen *Professores Pharmaciae* do not reach their goal, although they are using the same means. Our author is quite right in saying that his extremely secret knack, disclosed in a Christian way, had been known to very few. Besides Basilius Valentinus, in his tractate on the fifth essence, printed at Erfurt, I would not know anyone who had dealt with this subject.
[169] Literally "sweet sweet" in Latin. -PNW

physicians and *Doctores*, who have spent so much money and time, complain about me, let alone the sons of the secret science! If I could only speak into the ears of each one of them, I would like to entrust this in secret and not publicly, especially as those people are in general quacks and chatterboxes who care little about referring to philosophical books and *Autores*, so that they do not even understand the terminology of the Art, let alone the *Praxis*. Therefore, and because the illumination must mostly come from above, it will not be everyone's work who wishes, but *miserentis illius folius, cujus pleni sunt coeli & terra majestare* - "his alone on whom he takes pity, of whose glory heaven and earth are full." If He does not enlighten a man's intelligence, that man will be left sitting in empty talk and darkness, in spite of all his trouble and zeal, like all peripatetics. Love of our poor fellowman is that *qui vincit amor proximu*m. The rich man finds his medicine and comfort in the contemplation of his gold coins.

To help a man understand rightly the significance of the sweetening of minerals, for they are not sweet by nature but mostly bitter, and consequently adverse and contrary to all vegetable and animal creatures. Yes, so that this adverse nature also becomes homogeneous to the other kingdoms, alchemists sweeten the minerals through and with plants and animals. When they are sweet, they are no longer mineral but transmuted into a plant, and if this plant is eaten by an animal or a growing plant, it is transformed in the stomach or digestive tracts, transplanted and transformed into a plant or an animal by the subject's own *Archeus*, and thus it is sweetened consubstantially.

Therefore, we will here reproduce an *arborem dulcificationis & harmoniae*, whereby the animal is to be joined agreeably with the plant and this or both to the mineral, whereby such sweetening will help us and the mineral can be absorbed without any harm.

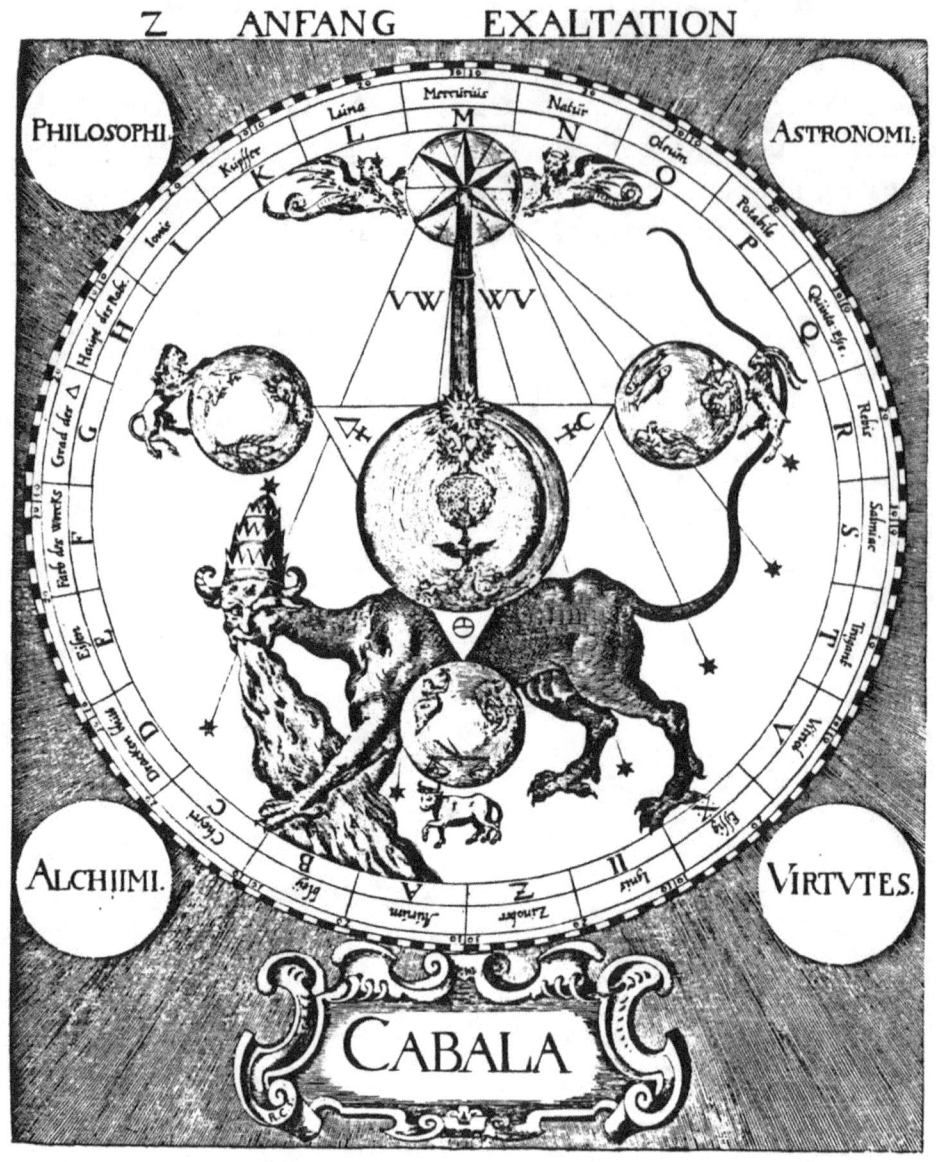

A wheeling Zodiac of good and bad signs encircling the composite image of the universal cow mother and evil dragon.

CHAPTER X
ARBOR DULCIFICATIONIS

So that a lover of the Art may see that I try to keep to the principles of Nature at all times, to go along with her and to follow her in her steps, let him contemplate Nature herself with me, how she sweetens the minerals and makes them homogeneous with the human and plant natures. Every researcher will indeed see that Nature sends up mineral corrosive vapors from the center of the earth. Because they are highly corrosive from their first origin, they deposit their stronger corrosive in the bowels of the earth, as they attack stones and earth, eat them up, dissolve them and thereby coagulate themselves. Every distillant knows that no mineral-corrosive ‿ rises as high as the milder plant and animal vapors, otherwise he would not need such low vessels as the retort, neither the highest grade of △.

When thus the strongest corrosive has been deposited in the earth, the remaining vapors rise up higher to the roots of the plants driven by the central heat. When they reach the plant kingdom that which is still sharp is intercepted, attracted and absorbed by their roots and transformed into their nature. What the plant kingdom has not retained rises still higher into the region of the lower air of the animal kingdom. Here then animals absorb the now mild vapors through the breath and transform them into their food and finally into their specific animal nature. Here is the *Arbor dulcificationis*.

From this every sage can conclude how it works in this perfect order in Nature Herself, and we must confess that the corrosive mineral is first changed into a plant nature before an animal should, and can take it as food. Nature Herself intends to say, so to speak:

If you wish to eat or digest a mineral, turn it first into a plant product, else it will be horrible and disgusting to you. Thus mineral is first turned into a plant in the natural order itself and then only when an animal eats it does it become homogeneous with the animal nature. In the same way, the animal decayed in the earth first becomes a *vegetativum* or *Sal essentiale nitrosum* and when this runs to the center of the earth owing to its dissolution, it goes into a mineral association.

In this way every artist can see how Nature, as the precursor and teacher, at all times takes to the middle course. She does not immediately jump from the mineral nature to the animal but first to the plant nature, and when she has been changed into that, only then does she willingly, eagerly and without disgust assume the animal creatures, which she first causes to putrefy on the surface of the earth, in the vegetable realm, turning them into a soluble salty nature and adapting them to the vegetation. When they now stand at the threshold of this quality, dissolved by water and conducted by it, to its center by means of cracks and fissures of the earth, then they stand at the threshold of the mineral nature, to be made mineral there. Here the mineral nature is in excess, so they are transformed into a mineral by the weight of that which exceeds. In the bowels of the earth everything is fermented together and again driven upward by the heat to the inner vaults of the earth, just as the minerals smaller Quantum becomes plant through the greater quantity of the plant components, and the latter in turn becomes animal through the exceeding quantity of the animal components. A human being will never consume as many plants as he is tall and heavy. Consequently, he exceeds the plans in quantity or mass, and this exceeding quantity causes a change in the lesser to the extent that the plant *per excessum alterius sicci animalis* also turns animal, as we can daily witness.

When enemies of equal strength meet, there can be no victory. But when one exceeds the other in strength or mass the weaker must be defeated and obey the stronger. In this way Nature proceeds and we must also do in the Art, if we wish to accomplish something. But if someone were to object: Yes! If I pour a bucketful of animal or vegetable *Liquidum* over half an ounce of mineral, it will of course become animal or vegetable, for the quantity will disperse the mineral corrosive so far that one can no longer notice the corrosive at all - this would look to me just as if I said that I will put one drop of o o ⊖ into the whole sea. I do of course not believe that the fish would die of it or, that this drop would be able to corrode all ships and fish. This objection is put forward by a simpleton and not by a philosopher.

Nature and all natural things have within their means of composition and destruction in their natural circumstances, in their weight, measure and order, and none of these factors can be exaggerated, or else it would result in a contrariety. Indeed, every artist can see it and notice it without any difficulty, he can see it with his eyes and grasp it

with his hands, feel it with his tongue and all his
senses, whether Nature has enough or not, whether
She has received too little or too much. If She has
got too much of the sweetening, the excess will be
discarded through distillation. If there has been
too little, the taste will show whether it is still
too sharp or not. Thus each can help and counsel
himself. I hereby describe the art of sweetening in
the order as follows:

Whoever wishes to make the animal and vegetable
realms homogeneous with the mineral and soften its
corrosives, must each time bear in mind the
philosophical *Axioma* and heed it well: *non transiri
posse ab uno extreme ad alterum sine medio*, that is,
one cannot pass from one extreme to another without
an intermediate. Minerals do not jump directly to
the animal quality without being harmed, likewise
can the animals not pass to the mineral quality, as
both, being heterogeneous, rather spoil each other
than that they should produce a tasty fruit.

Therefore, one has to use the intermediate and
advance the minerals NB. to the animal nature by
means of the plant. Instead, transform and congeal
the animals by means of the plant — then they will
easily unite through the levels established in
Nature. For example, if I wished to unite the three
volatile or the three acid spirits of the three
kingdoms and took first the animal and the mineral,
poured them together, they would fight each other
like two fires and oppose one the other. Instead, if
I follow the order of Nature, I first pour the
volatile from the animal kingdom to the volatile
from the plant kingdom, and a homogeneous union
immediately occurs without the least resistance.

Only then do I add the volatile spirit from the mineral kingdom, and when I then distil, they will go over inseparably or remain behind, all three together.

Rc. ⌓ ⊡ *volatilis* and 𝒮 a̅a̅ , pour together, then add one part of the volatile acid ⊕ *Phlegma*, and they will all unite without opposition. Likewise, take one part each of the animal acidity and the vegetable acidity, pour them together, then add one part of the ⌓ ⊕ li, and they will also easily unite, because the plant is the *Copulator* which associates as much with the animal as the mineral kingdom.

The mineral kingdom is easily sweetened provided one deals with it according to the levels of Nature, but not otherwise. And so as to serve a lover of the Art rightly, Rc. Putrefied urine, distil its volatile spirit of urine and salt from it, rectify it of its grosser *Phlegma* in a phial, so that it becomes quite pure, crystalline and clear, and preserve it, and the volatile has been prepared from urine.

NB. Distil the sediment which stayed in the B.M. to a honey-like *Liquor*, and the grossest Phlegma is thus separated. Remove the latter, which is of honey thickness and mix it with leached ashes, so that it becomes almost dry and you can form a ball of it. Put it into a retort and distil all that will go over in the sand. Then you get the † of the animal kingdom together with a thick stinking oil. Separate the oil through the *Tritorium* or a

459

glass funnel. Filter the Acidum and the volatile
salt, which have gone over with it and distil once
more gently through the ash, and it is also
prepared.

Now take a good old wine, make of it a ⟨S⟩ that
ignites powder, as is taught in many books. When you
have distilled the wine out of the vessel, take what
is left over and let it evaporate to honey thickness
in a copper kettle. But take care, when an acid
vapor rises into your nose, stop. Now take that
substance free of phlegm but containing its vinegar
or the honey thick materials, mix it with coal dust
or leached ashes, distil it through the retort, and
first a gross *Phlegma*, then the wine-vinegar, later
a thick stinking oil will go over. Separate the oil
from the vinegar *per tritorium* or through a funnel,
and distil and dephlegmize it two or three times
through rectification - and it is then ready.

Now you have prepared everything necessary for
the sweetening of all corrosives, and you will
experience that this sweetening is as different from
the common one as heaven from earth. I need not
praise it any more - *Praxis* will indeed show you.

MODUS DULCIFCANDI

If you now wish to sweeten, follow the steps of
Nature, else you will confuse everything. Therefore,
Rc. ⟨S⟩ and ⌒ *volatilem* ⊡ pour them together,
put them into a high alembic, distil ex B.M. &
cinere, and wait until a gross and empty *Phlegma*
stays behind, and it is ready. Now Rc. also the

Acidum ⊡ and the acidity of the wine, pour them together and draw them over together through the retort, and this is also ready.

Now take a corrosive, whichever you wish, be it in liquid or dry form. Take one part of the corrosive, pour upon it three parts of the prepared *Acidum*, put it in B.M. and in a low alembic, draw over a *Phlegma* to the stage of oiliness. When you stop distilling, try the remaining oil to see if you find it sweet enough or not to the taste. If it is sweet enough and right, let it be. But if it is not yet sweet enough, pour once more three parts of *Acidum* upon it, and do as before. Then it will become ever sweeter, and you can sweeten it until it pleases you.

If it is now sweet enough, pour upon it three parts of the prepared spirit of wine, set it in B.M., distil it again to an ⊙ ⊙, and it will become still sweeter and ever more homogeneous with human nature. Pour once more, or for the third time, each time three parts of ⩔ , and it will each time become sweeter and pleasanter. Each time you pour fresh spirit of wine upon it, distil it each time in B.M. to the oil. Then the ⩔ , as well as the *Acidum*, will always go over weak and watery, because the volatile salt stays with the corrosive in the process of sweetening, which is as it should, otherwise it would not be able to transform the corrosive.

When now you have thus sweetened the corrosive and distilled it to the oil, put it into a retort

461

and distil it into a sweet, pleasant oil, which all animals and plants can consume without the least danger. And this is the Quintessence, Magisterium, Arcanum minerale, from which you have taken it.

If you wish to coagulate this oil into a salty stone, fixed and liquid like butter, put it into a high alembic with a head and recipient, set it in B.M., distil its excessive moisture by degrees, because the essence does not rise further in the B.M. Afterwards set it in ashes and draw the moisture which did not easily rise in the B.M. off by gentle degrees of heat. It will now become ever thicker and thicker, so that it flows in the fire like oil and stands in the air like ice. Then you have it in liquid and dry form. Therefore, thank God.

Now take note: The stronger your *Acidum* and ⚕ is, the quicker they sweeten; the weaker, the slower. The sharpness, however, consists in that the *Aqua recolacea*, or the *Phlegma*, be separated as much as possible, so that the ⚕ and the *Acidum* are concentrated as much as is feasible.

Again - let the lover of the Art take note - if he intends to use the mineral or the corrosive mineral essence only for plant work and not for the animal one, it is not necessary, although there is no harm in it, to add the ⌒ and ✚ ☐ to the ⚕ and plant vinegar, but only the spirit of wine and the plant acidity. But if he only desires to reach the mineral nature, sweetening is not necessary, if he does not wish to do it.

But because physicians are at all times eager to seek the well being of their sick fellow men and do not pay much attention to other transformations, they must sweeten the corrosive in the above-mentioned manner.[170]

Here now come the objections. For some will say: This process is almost counter to the opinion of all sages who order us to separate the ♃, ☿ and ⊖ from every mineral, actually the components confirmed for ages, but here he turns every mineral into a ⊖ or ♅ this into a corrosive o o, and he congeals that again into an ⊖. Where then is the sulphur and mercury in a dry and flowing form?

My dear friend! Whoever you are and demand such a way as you find described in all books, I freely admit to you that you have not yet gone far, let alone rightly examined the nature of minerals, even less understood the philosophers.

You will also have read in the philosophers, although it is here not to be understood to mean so much, because that is a more *advanced way than*

---

[170] From a godly impulse of true human kindness, to which we are also obliged by our holy vow, I cannot but advise the physicians by their soul and conscience, to introduce this magnificent art of sweetening in all pharmacies and to see to it with sincerity and strictness that it be manufactured according to the prescription, and they will be able to await not only honors and advantages, but also the reward thereof or here in the temporal world and there in eternity. In the laboratories the ☿ *dulcem* is manufactured, which is a medicine which stands on a good foundation. But if it is sweetened by this art until it shows the signs that it should have, that is, that it stands as a liquid o o, its effect is always unfailing, sure and certain.

explained in this book: Sal metallorum est lapis Philosophorum & basis totius artis, that is, the salt of the metals is the philosopher's stone and the basis of the whole Art, and concealed in it is the ☿ and the sulphur. If it is turned into oil, it is called sulphur and its internally working spirit is ☿, then ⊖, 🜍 and ☿ are together. When this oil is again coagulated and made fixed into salt, it will immediately coagulate through gentle distilling of the moisture. In heat it will flow quite steadily like wax or O O, in the cold, however, it will stand like ice, and in all moistures it will dissolve like sugar in water, without any precipitation. This is then a most effective medicine, capable of healing all sicknesses.[171]

Someone will object again and say: This

---

[171] Here the mechanical physicians will sneer again and say: "This one will again warm up the long-flogged teaching of the universal medicine, but we know it. Through the art of separation, we have seen with certainty and conviction that the body of man is not to be considered other than some clock, and that he must be healed in precisely this way. If the wheels and mainsprings of this clock are too strained, they cannot recover their elasticity and flexibility except by such means as will relax their overly strong driving impulse. If they are too slack and have lost their *Tonus,* the latter cannot be restored except by such means as can strengthen the fatigued parts. Consequently, two different kinds of sicknesses cannot be cured by one kind of medicine, and a universal medicine is nonsense. If these gentlemen were as thoroughly convinced as we are by the experience of several thousand years that there is but one single spirit which both destroys and restores unceasingly above and below in all realms of Nature, they would think differently and endeavor to get to know this universal spirit of Nature better. With this intention, they would also look deeper into our true writings on natural science as the only true ones, especially this Golden Chain. They would learn to release this spirit from its bonds, the chaotic water and its products, and to make it acceptable to human nature. Thereby they would finally obtain certainty in medicine and forget their armchair philosophy with pleasure. See *von Plumenof, Geoffenbarter Einfluss,* p. 31-40. But because they are no lovers of spiritual essences, both in the ethical and the physical sense, they stay with their old humdrum way, whereby they save themselves much trouble, but heal the fewer sicknesses.

operation is not only done with corrosives, but the corrosive is left with the work and is not separated.

Now, in order to help this man, I must again commit verbosity and even return to the origin. Accordingly, consider: God has created something visible and something invisible, as we can see daily with our eyes, that is, two things out of which everything has originated: the universal chaotic water as a body, vehicle and tool of the spirit or seed. Such is visible and palpable. Spirit, however, or the seed as the indweller is invisible until it becomes visible, palpable and corporeal through the successive *Gradus putrefactionis, separationis, conjunctionis, coagulationis & fixationis*, as we have explained sufficiently above.

Now look! Water is *recolaceum*, and with the seed and together with it, it does not become a body, except what it badly needs in making a body. The rest it drives away violently through fire and heat. Now mark well: The *Aqua recolacea* NB. is a tool and vehicle of the universal spirit or seed, by means of which spirit must accomplish its work of fixing or volatilizing itself heavenly or earthly. Without this water spirit would be dry and would, so to speak, be obliged to lie low asleep or dead, without any effect. For as long as this *Aqua recolacea* remains with the spirit, or the spirit with the water, it knows no rest, for it is forever being awakened to action. We can see this clearly in animals and plants, especially in those which are excessively moist, in which the *Aqua recolacea* is not separated. Therefore, as long as the animal lives and the plant is alive and green, this spirit

or seed wanders in the moist limbs or veins together
with the water and digests, putrefies, separates and
coagulates in order to distribute food for the
growth and preservation of the subject. If the
subject is headed for death and destruction by
withering or dies, the spirit reverses. Whereas it
previously helped the animal and plant grow and gave
them food, it here starts again *in puncto*. Because
the animal or plant has lost its balsamic vital
spirit, it brings about its putrefaction and
dissolution until later on occurs the regeneration
into something else, and it does all this through
and by means of the water, without which it cannot
act, as can be proven.

If a subject is reduced to its essence and
coagulated quite dry, so that all the *Aqua recolacea*
has gone and the spirit or seed is quite dry, the
spirit is as if it were dead or asleep, because it
has been deprived of its medium, tool or Aqua
recolacea. But when it obtains another one, either
from the universal effluences, such as air, dew,
rainwater or *Specificis*, that is, when it is
injected into the plant or animal subjects as a
medicine, it again acquires some excessive moisture
or a like specified tool of the water, which rouses
it again into action, when it will thereafter either
heal or destroy the plant or the animal, according
to how it is applied and manufactured.

Now I say, the more the universal spirit is
separated from its *Aqua recolacea*, the more fixed
and concentrated it becomes, and when this fixed and
concentrated spirit or seed is again made spiritual
by the excessive heat, it is a pure fire and winged
devouring dragon which consumes everything in its

mouth. Therefore, consider ⌒ ⊕, ⋎̄ , spirit of
copper chelate. Each is a pure consuming fire, and
as long as it is such a fire, it is harmful to all
single things, especially, however, to animals and
plants. But because various means have been shown to
mitigate its raging cruelty and to turn it into a
pleasing gentleness, a lover of the Art has nothing
to complain about and should not worry about it.

The artist should know that if the spirit or
seed did not have such a sharpness, how could it
dissolve hardened bodies and earth? If, however,
someone does not like this way, let him dissolve

stones and metals in a ⋎̃ or ⊡, with a plant or
animal acid, and then note and learn for himself the
difference. *Praxis* will probably finally make him
see reason.

Now it is said that I leave the universal seed
or spirit with the ⌒ ⊕, ⋎̄ etc., as the
solvens with the *solute*, and I am giving this
explanation. When the universal seed or spirit is
added to the specified seed, and thus the universal
spirit turns into a like *Specificum* through the
specified seed, the mother is added to the child, or
the child to the mother, and the mother feeds and
increases the child with its own food. Thus the
child also draws its nourishment from the mother,
through whose substance and blood it came originally
into being. Consequently, I do not know what harm
there could be if the child is fed by its right
mother and transforms that food into its nature and
growth, whereby it differs from the Mother. Neither
can it be harmful that the child has taken and
advanced its own growth thereby.

467

It is true that all *Universalia* make themselves homogeneous with the *Specificis*, assume their nature and actuality, and act with the predestined power of the *Specifici*. Not only are the virtue and power increased by the concentrated and thus sharpened universal seed, but they are also exalted and heightened in *quantitate & qualitate*. The sharper and more fiery the spirit is the more powerfully is its effect; the more concentrated it is, the stronger it is, and the smaller must its *Dosis* be.

Neither have I taught to add this sharp spirit to plants and animals before its sweetening. After the sweetening, however, let anyone prove to me that I have made a mistake. Whoever does not wish to believe this in theory will surely be taught by the praxis and be thoroughly convinced.

I will now add a very simple example by which every artist will immediately perceive the rapid transformation of the sharp or corrosive spirit or seed into sweetness. Now take a well dephlegmed ⌒ ⊕ or o°o ⊕, one part. Pour upon it a simple distilled wine vinegar, six parts. Distil it in ash to an O°O and a pure *Phlegma* or *Aqua recolacea* will go over through the first or second grade of fire, in a not too low alembic. Now pour once again six parts of fresh vinegar again distil it to the oil, and do this three times. Then taste the o°o ⊕li on your tongue, whether it has not already changed most of its sharpness into a sugary sweetness. To sweeten this even more, pour six parts of ⩔ upon it, and abstract to the oil in an alembic in B.M.,

468

in everything as with the ☩, but the ♆ in B.M.

Repeat this a third time. Then the ○ ○ ⊕li, especially if the *Acidum* and the ∿ vini were strong, will be sweet as sugar and so strongly sweet that everything one eats and drinks thereafter will also taste sweet. That is how strongly it penetrates the Pores of the tongue and fills one's taste. If

the Azoth and ♆ are sharp, the more and quicker they sweeten, and even more so, if the animal kingdom is added to them.

Here there will be yet another objection and someone will say: He asserts only two components, the *Aquam recolaceam* and the spirit or seed hidden in it. When then the *Aqua recolacea* is alone to be separated, the whole world, all mountains, stones, rocks, meadows, fields and soils are a pure coagulated spirit, seed and *Sperma*.

Answer: Whoever does not wish to believe that the *Punctum terrae* is a pure *Sperma*, let him take some earth, where and which he wishes whatever is available. Let him but leach the salt out, so that the spiritual corrosive seed does not devour itself to death, dry it and calcine (in glowing heat) it somewhat. Take note of its weight, pour upon it ∿ ⊕ or ⩒. If it does not attack, add some more ∿ ⊖ until it is completely dissolved. Then distil the spirit and you will find a salty white corrosive earth at the bottom, which earth has again reversed to its prime essence or salt through the *Primum* or primordial spirit. Now consider whether this earth is a *Terra damnata* or a leaven!

Here I must once again explain a point or problem to which many alchemists take exception, namely: When they handle 🜅, 🜆, ⌒⊕, etc., and try to dissolve minerals with them, and if these solvents, mostly however the 🜆, either do not attack or only very weakly, they say that it has gone bad and is of no use, while they themselves are spoiling the work. Let us assume that they wish to dissolve ☉. Then they take 8 Loth ✳ or ⌒⊖ to 1 lb of 🜅. If the 🜅 is strong and has not much water, it will indeed dissolve the gold. But if there is little 🜅 and too much water, it leaves the gold as is, or dissolves only a little. That then is the problem.

However, if someone wishes to dissolve a sulphurous ☉lar mineral, such as the ☉lar marcasites, gold pyrites, gold sulphurs, etc.: If the 🜆 had at first also been rather weak but had now been made very strong, it would dissolve barely half or one-third or one-eighth part, when it had previously completely dissolved gold. What is the reason? This: the 🜅 is an *Acidum*, the ⌒⊖, however, or ✳ is an *Alkali*. It is indeed known that where an *Acidum* and an *Alkali* meet, they kill, beat down, sweeten and congeal each other, and thus produce a third non-corrosive salt which, in liquid form, does not have the power to attack such a hard body. If it is weak in dry form, it congeals more than it dissolves. If then 1 lb of 🜅 and much

470

water are present, the ▽̄ eats, ⏝̄ ates and congeals itself to death with the eight Lots of ✳ or ⌒ ⊖, and hardly attacks anything. But if it is strong it may well attack strongly, but the *Alkali* is nevertheless too much. This may be seen when one tries to dissolve stony marcasites with it. True, it attacks gold more readily, because it is a finely finished body, separated from all stoniness, sulphurs and gangue; but not so the marcasites and pyrites, and even if they are washed ever so purely of gangue on the "Saxen," they still retain a stony mother, totally intermingled with them in their core, at which the *Acidum* eats or congeals itself to death, as with the △̟ of marcasite, so that there is no result in distilling or dissolving. The more a body dries up and is separated from all moisture, the less a Humidum can act within it, unless it be awakened again by a moisture of the same degree, so to speak. The means to do just that, the *praxis* will teach.

Take 1 lb of ▽̄ and eight Lots of ⌒ ⊖, pour them together, distil them gently in a retort in ash to a good oiliness. Set this in the cold and let it crystallize and it will turn into crystals These are a generated ⊕, because the ▽̄ is *Acidum nitrosum*. The *Spiritus salis* is a spiritual *Alkali*. Thus one can see that the *Acidum* is coagulated and congealed by the *Alkali*, and that thus the sharpness of the acid is broken, so that it can no longer attack so strongly.

Likewise with ✳ or common ⊖: From eight Lots of ✳ or common salt distil 1 lb of ⊽ through a retort in ash. Then remove the ☠. Then take some fresh ✳ in one hand, in the other the ☠. Try them against each other on the tongue and you will find that the ✳ has retained a great sharpness of the ⊽. As much sharpness as the ⊽ has lost to the ✳, as much weaker and broken the ⊽ has become, so that it can no longer attack strongly.

To prove once more that the ⊽ eats itself to death at the marcasites, dissolve the marcasite in ⊽ℛ, and when it no longer attacks, decant it. Pour spring-water on the bottom, set it in warmth, let it boil somewhat, then decant the water, filter it, coagulate it - it is quite dry, and you will get a salty earth, or a ⊕ produced by the ⊽ℛ and the marcasite. By this one can see that the ⊽ℛ has eaten itself to death at the marcasite, whereas it has dissolved little enough.

So that ⊽ℛ and similar *Menstrua* should dissolve more than normally, one must add them to alkalized subjects and sharpen them with an *Alkali*, but not so much that the *Acidum* eats itself to death at them. Supposing I take four Lots or even only two

472

Lots of ✳. Let it gently digest day and night in sand or ash, then either distil it over or use it immediately for dissolving. In this way I dissolve almost twice, three or four times as much as someone else with a weakened solvent.[172]

But someone will ask what is the reason that one must add ✳ or ⌒⊖ to the ⩔, in which the Aquafort is anyhow very strong? Following is the reason. I have said that all minerals are made of the universal seed. That *Acidum*, however, produces the lesser metals much sooner than the perfect ones, for the *Acidum* is not yet congealed or alkalized in the imperfect ones, and made as far earthy as in gold and ⊙*lar* subjects, so that the acid still prevails in part or wholly, although some subjects are more strongly congealed than others. Since these imperfect metals and minerals are still of an acid nature, an *Acidum* easily attacks such a one. Instead, it either eats itself to death at the alkalized or strongly congealed minerals or it does not attack those at all. But because these must also be attacked and dissolved, an *Alkali* is added to the *Aquafort*, so that it might arouse its like. And if the fixed *Alkali* is but once aroused, it separates its bonds itself with the help of the acid and is easily reversed into an *Acidum* through such an awakening. This is because everything volatile desires to become an *Acidum*, and that which is acid,

---

[172] This is a beautiful technique in which many err which our in God reposing worthy Brother Homer has thoroughly proven. A lover of the Art should repeat what has been said above, and in that way he can become convinced. Although we cannot use it in our high natural works of the Royal Art, it nevertheless serves perfectly to teach our young students the basic reason for the action of a natural thing and to let them discover an indispensible knack in their practical works.

wants to become alkaline or fixed. And thus *e contra* - what is alkaline wants to become acid again, and the latter desires to become volatile again, so that the uppermost become the lowest and the lowest uppermost, in a perpetual chain.

Just as the *Alkali* dissolves its like alkaline subjects, so it does not dissolve acid things or causes them only to swell. The reason is that the *Alkali* is not so penetratingly subtle, as it contains a fat earthiness which prevents the *Acidum* from passing through its tiny openings. Even if it attacks, it will eat itself to death, and solely corrode, so that it turns into dust or swells like a sponge.

Here someone will now say: If you have re-aroused the *Alkali*, being the ⊖, the ⌒ ⊖ ✳ci, etc., by the *Alkali*, there exists a contradiction, for in this way the *Alkali* would be strengthened and the *Acidum* would eat itself to death each time. I reply: By the term *Alkali* I do not only understand the volatilized and fixed alkaline salts, such as all volatile *Alkali* of animals, Sal ammoniac, salt and other fixed alkalis but also the volatilized and fixed alkaline earths and, as I have said, when the ▽ contains much volatile or fixed *Alkali*, it will eat itself all the more and sooner to death before it begins to dissolve. If the ▽ contains but little or less *Alkali*, it dissolves all the more.

Dissolving depends solely on the saturation of the Menstruum, meaning that its *Pori* are filled by

the extended and dissolved subject. The emptier the
Pori are, the more the Menstruum can absorb and
dissolve. The more filled they are on the contrary,
the less it absorbs. Since too much $\ast$ or $\ominus$ or
$\sim\!\ominus$ fill the $\underline{\vee}$ or its *Poros* too much with a
subtle alkaline earth, before it begins to dissolve
a marcasite, not many Pori can be empty, and as many
Pori as are still empty, as much it will still
absorb. From this one can clearly see the mistake
and difference in many *practicants*.[173]

In the lower kingdom, Nature does indeed make
an acid of the volatile, and an *Alkali* of the acid,
and even if something appears quite volatile, it yet
contains its acid part and *Alkali*, although the
volatile is preponderant, which prevents the acid
and *Alkali* from dominating. But if the *Acidum*
prevails, it associates again with its like and
readily absorbs the other *Acidum*. Consequently, when
the *Alkali* prevails, it loves its like, even if it
is already mixed with the volatile and the acid, and
it also wants to be treated by its like. The artist
must take note of this to avoid many mistakes.

---

[173] No solvent can dissolve more than its mass is capable of seizing its meanest little parts in
those empty inter-spaces which in their touching each other must necessarily leave in the
whole expanse of the water, by virtue of their natural impenetrability. This is the reason why
(1) the diaphaneity proper to it, is changed by the solution in its own way - although not if
the color is pure; (2) the dissolution can only last as long as and not longer until the empty
inter-spaces of that which has dissolved are filled, and (3) the precipitation of the thing
dissolved must necessarily follow as soon as something is dissolved that is pleasanter, that is,
naturally more appropriate to the hunger of the devouring solvent, which occupies its
interspaces and displaces the *Solutum* there from; (4) no precipitation or other separation
can be effected by those solutions which the pure components of one and the same product
(in the weight and way of Nature) constitute among themselves; but (5) the whole solution
either coagulates or congeals through the fire degrees and time suitable to its nature, when
its fixed component is preponderant, or, if its volatile component prevails, becomes totally
volatile without any separation.

From this each can draw his conclusions, and if I should perhaps have erred out of frailty, let him correct me with gentleness if he has a better reason in theory even though corroborated by true praxis. Everyone is free and unimpeded to expand this little natural science teaching further, to improve it, add to it, increase and enlarge it, etc.

Well, philosophers also say: (i) Our solvent and that which is to be dissolved must stay together, either both volatile or both fixed. (ii) The solvent must be homogeneous with the dissolved.[174] (iii) It must be a *Mercurial-Menstruum-ubiquoticum*, and this is to be known by its universality.

Now someone will again say: Assuming we agree that the ☉ and ☽ are *ubiquotica* and universals for all *Specificis*. ⊕, instead, a pure *Acidum* and mixed thing, which appears to be heterogeneous to the *Universal-Mercurial-Menstruo*, because the ⊕ has more 🜍 than ☿.

---

[174] It is an age-old axiom of hermetic philosophy, corroborated by experience: Heterogeneous or contrary things cannot be united. Accordingly, our students of the true and genuine alchemy must carefully note that they must unite in all their compounds not the distant but the nearer natures. To please them, I will quote a passage from *Sendivogius,* which can cause them to reflect a great deal. It is in his letters which *Rothscholz,* Nuremberg, published in 8vo in the German language, and is as follows: "Nature is not visible, although she works visibly, for she is a volatile spirit which performs her function in bodies; she has her seat and place in the will of God. Her location is of no use to us, except that we know the place and location which are most proper and agreeable to her, that is, we know how to unite one thing with another according to Nature, so that a man is not united with wood; or cattle, or some other animal, with metal, but so that each should work and operate in its like. Then Nature will also do her share." With universals, however, it is different. They appropriate all particulars products of Nature and specify themselves according to them, as the author has wisely noted in the immediately following paragraphs.

That ♁ is a *primum ens Mineralium* has been proven above; that it contains ☿, 🜂 and ⊖, is known to all artists; but that it is more sulphurous than mercurial cannot prevent - as has been proven above - that ☿ and all arsenical-mercurial subjects originate in sulphur.[175] All these scruples do not matter provided it does its effect.

Moreover, how many authors are there who assert that ♁ is the prime matter of metals as well as ☿. They have even recommended it as the substance of the philosopher's stone, according to the known philosophical saying: *Visitando interiora terrae* etc. If ♁ is the prime matter of metals, it must indeed have the power to turn metals into their prime essence after its dissolution, and must also be homogeneous with all minerals. If it is the substance of the stone, it must be the extract of the fifth essence of the whole mineral Nature.[176]

It is also known that saltpeter and salt are a universal subject which a great many also recommend *pro subjecto physico* in all dung-hills. They call it all in all because it can be found everywhere. Because it is a universal, this subject can also

---

[175] Here he says something which is only tangibly and understood with great experience.
[176] That the philosopher's stone is made out of the prime matter of metals, and that it has a vitriolic, that is, a salty quality according to the axiom: "The salt of metals is the philosopher's stone" - this is known in our school of wisdom as an indisputable truth. But one must make a well-considered difference between the common ♁ and the ♁ of the philosophers, to write more of which is inappropriate here.

indifferently assume every form or specification.

With the ⊕ they become specified, become one with it, stay with it volatile and fixed, and what they dissolve they again make volatile and fixed, and it stays with them inseparably, and whoever intends to separate it from them, separates the volatile part and the fixed part stays behind nevertheless. For one seed likes to stay with another, especially *Specificum & universale*, and they let go the *Aqua recolacea*.

Therefore it is a stupidity of many erring alchemists who think that they can separate the solvent by distilling or reverberating, or by burning off the ⊽ , or by digesting with it, etc. All they should do is taste the drawn off *Menstruum* to see if they do not find it weaker by half, and they should do this the sooner if they intend to dissolve with it other fresh subjects. Then it is too weak for them.

Just look at the dissolved bodies and weigh them before and after the dissolution, to observe what difference there is in their weight. Everything intended to be fixed adheres to some earth, such as all acid things. And everything intended to become volatile rises upward, and this can neither be denied nor hidden in spite of all controversies, no matter how much boasting there is in theory or practice.

I am telling you absolutely that when someone says or writes that he has a *Menstruum* out of dew or rainwater, or other insipid menstrual ▽ etc.:

478

these are self-praising empty fantasies, sweet talk, which lead a poor seeker into nothing but wrong ways, loss of time, wasting of his little remaining money. They fleece his purse in a very unchristian and unscrupulous way, whereby such a man often dies slowly and very sadly. One should examine the solvents, and divide them into four parts, the volatile, the acid, the alkaline, or those mixed or compounded out of them.

Well, it is known that all volatiles, such as dew and rain, ♅, ⌓ ⊡ae do not even attack a hard-coagulated body, or, even if they contained an *Acidum*, they would color and satiate themselves so little thereby that one would require a whole bucketful to dissolve but one lb.; and when it has dissolved, it is no dissolution but an extraction, because the volatile spirit flies off again through the distillation and leaves the body lying dry, dissipated into small parts, and it is no better than before, except that it is crushed more finely or smaller.

If one takes the *Azoth* or the vegetable or animal acidity, they will indeed attack more strongly than ♅ or ⊡ or a half-volatile. But what kinds of *Subjecta*? No stone or alkalized mineral. Only those which are anyhow full of acidity or strongly filled with it are easily dissolved by it. But they dissolve in a way that cries to heaven. With 10 lbs of distilled spirit of wine, I do not dissolve 1 lb of ♀ or ♂, which are quite open - but I can dissolve 1 lb of ☿ and still more ♂

with three lbs of ⚬ ☉ or ⊖, ⚬ ♁, ⁰ₒ⚬
♁, ⁰ₒ⚬ ⚨is, and turn what has been dissolved
immediately afterwards into the first matter, that

is, into a ♁. Instead, if I distil the *Acidum*, I
have verdigris or Crocus ♂ in the sediment, and
little enough of that. True, one can dissolve more
with a spiritualized *Alkali*; but without *Acidum* NB.
every solution is almost a totally lost cause.

    We will now compound the above-mentioned
waters, strengthen them and intermix. Perhaps they
will now dissolve more than otherwise and be better

than the mere sharp corrosives. Pour some ⚡ to the
☩, or a volatile subject to an acid one, or the ⌒
⊡ to its acid, or these four together. Then pour
this upon a stone that has been calcined as usually,
or upon another strongly (chemically) combined
mineral, but in sufficient quantity. Watch what they
will do. They won't do anything. But if you pour
them over a subject that is open or not so strongly

combined, such as ♁, ○, ♀, ♂, ♄, ☽, etc.,
they will immediately attack them and produce a

sugary sweet ♁. But how much of it from one lb.?
When you have poured on six lbs. of *Menstruum* it

will hardly dissolve 1 or 2 Lots from 1 lb of ♀ or

♂. But I don't say anything of ♁ or ○ because
these are pure, very easily soluble salts. Here now
you have your powerful uncorrosive *Menstruum*.

If you pour a mineral *Acidum*, such as Aquafort, ⌒ 🜍, etc. into wine vinegar or ⩗ you will indeed sharpen the wine vinegar etc., but you will sweeten the corrosive and kill it, so that it can no longer attack as powerfully as before, although it

will dissolve more than the mere *Azoth* and ⩗ .

If, however, you pour a ⌒ 🜔 or *Azoth* 🜔 into Aquafort, ⌒ 🜍 etc. into the corrosive, you will completely kill the corrosive and turn it into a third salt which dissolves very little or nothing at all. But what is the reason why these and similar solvents do not dissolve? This: the further and more the corrosives are extended, the weaker they become and the less they dissolve. On the contrary, the more concentrated they are, the more

sharply, the more violently they attack. ⩗ and *Azoth* are extended, diluted, dilated corrosives, completely filled with *Aqua recolacea*, and even if they are made quite fiery through rectification, yet one lb. of them does not do as much as four or even

two Lots of the dephlegmatized ⩚ . Praxis will prove it.

Rc. A very fiery ⩗ and a sharp, fiery rectified vinegar, three lbs of ☦, one lb of the

Acidum, and half a lb of *Salis tartari*. Pour the ⩗ over the *Sal tartari*, then pour the vinegar upon it, set it in B.M. or ash, distil gently, and you will obtain an extremely clear, insipid *Phlegma*, weigh also the remaining *Sal tartari* which has retained

481

some of the sharpness of the ☿ and ⚛ or volatile
⊖. Now you will yourself understand that several
Lots of sharpness or volatile ⊖ had been contained
in as much ☿ and ⚛. However, pour 1 lb of
Aquafort or dephlegmatized ∿ ⊕ over half a lb of
∿ ♀ri. You will find the *Sal tartari* increased
by half or at least one-quarter after the *Phlegma* is
distilled. Now consider the difference in the
solvents.

When someone affirms that he has an insipid
solvent, it is a dissolved salty spirit which has
been dissolved by its own or an outside acidity. I
consider this the same as if I let salt and
saltpeter flow together and dissolved it then in dew
or distilled rainwater, and filtered it afterwards.
Then it is just such a *Menstruum*. Now let someone
simply distil such a *Menstruum* in B.M. or ash, and
he will find a well clarified middle salt or a
killed *Acidum* as a ⊖. And if he were to repeatedly
distil the same distillate off the same ⊖ a
hundred times and does not concentrate it, so that
the acidity would get the upper hand, it is unfit
for dissolving metals etc. If it is poured over a
subject, it may well get colored by the things or
metals to be dissolves, but it extracts their
sulphur so little when the *Menstruum* is distilled,
that one gets sick and tired of the effort and work.

What has been distilled is called by them 🜍 solis
and lunae. Yes, it is a 🜍, and it is supposed to be

482

the greatest tonic for the heart and to rejuvenate
old women as a true aurum potable. In addition, so
some philosophers say, although perhaps only to pull
the wool over other people's eyes, it is the 🜍. But
the ⊖ and the ♁ are supposed to be likewise drawn
out of the lower part. But I am asking an honest,
learned, faithful and compassionate alchemist how
much time, how much expense, how many things
neglected, how much nuisance, waste, want and less
of various precious substances and waters used in
the process there have been, and how much coal they
burn in doing so, before they but separate the 🜍
and ⊖ (not to speak of the ♁) and transform them
into a volatile essence? It is an imagined
foolishness and presented to alchemists as the
greatest possible trickery and to keep them from
proper work.

I do not say that it is impossible to change a
metal into liquid mercury, but that it is an empty
and vain, expensive and lengthy labor and procedure.
I do not know why writers hit on liquid mercury made
from metals, or demand mercury from minerals and
metals so eagerly since no liquid ☿ is ever found
in any ores (except solely ore of mercury) but acid
⊕, ◯, 🜍, arsenic, marcasite, out of which
metals grow and consist step-by-step, and not out of
liquid ☿, as has also been mentioned above.[177]

---

[177] What has been said above in no way diminishes the value of a certain mineral work,
made in the dry way out of liquid ☿. Neither is this the opinion of our author, but he will only

483

I am telling you, alchemists, do not trouble to extract the sulphur. You are deceiving yourselves greatly, because it is only a specific part of the softened metal, no more. Everything must be dissolved, yes, the whole body of the metal, and turn into a *Liquidum*, go over and be a spiritual

sweet liquid O O or a spiritual salt. In order to act upon human health, it must not be fixed but volatile, so that it can immediately be turned into vapor and steam by the Archeus of the stomach and be able to permeate the blood in this form, along with all veins to the marrow and bones. In this way it must be a right medicine. If the medicine is fixed, the Archeus must first make it volatile and work it over. You must make it first volatile and homogeneous, if you wish to revive dying bodies breathing their last. Although I have spoken in almost the whole book about the fixation of medicines, it was because each and all cry "fixed, fixed," and they do not know or do not notice that animals themselves make everything volatile for their food and growth.

Do you understand, however, that I demand a medicine that is as highly volatile as ? No, but so volatile that it is not too volatile nor too fixed, half and half and in the middle, like all acid things. It should be an *Acidum* by name, though in quality like sugar, because Nature pulls all sweet things eagerly toward herself. Thus it should be, as I have also taught in this book in many ways.

---

show that one cannot find mercury in liquid form in metals and that it must first be extracted from them by specific manipulations.

When a medicine is highly volatile like $\dot\vee$, it passes, driven by the heat, too fast through all veins and pores of the skin - with a bad effect. If, however, it is fixed and the Archeus cannot dissolve it, it is again no good, because in this case it is eliminated in the stool. But if it is of an intermediate type, it will adhere to the blood and unite with it. Together with the blood it will pass through all veins, and usually drive away all

sicknesses through $\boxdot$ and perspiration.

Do not make your medicine in that way, but stick to your sulphurous extract. Then you have the "shadow before the body." Even if all the philosophers deny it, I remain with Nature which does not put together any heterogeneous things or dregs, as some believe and say: Extract the soul and leave the body, because it is dregs *scilicet* in your head which are the dregs, and not in the body.

I am telling you to take this soul together with the body, if you wish to heal the body together with the spirit. They do indeed contradict themselves. If the disease is in the blood or in the body fluids, it is cured by the spirit. Thus the body must again cure the body, one spirit another, and in the same way one body another.

Such and similar errors have crept into the Art, whereby many, I do not wish to say thousands, but countless artists have lost all they had, died and decayed. Who knows where their soul is now? If someone has learned a manipulation after ten years' work, which he might perhaps have done in a quarter of an hour if fate had not prevented it, he boasts about it as if he had concentrated heaven and earth. He shouts that there is no other way than his, and if an angel came down from heaven and taught otherwise, he would call it a lie - as if God had not a thousand ways to help. But they wish to alone be the Master, and interpret all similes and parables according to their work. They run for paper, pen and ink, and about one single thing, to which he attaches a whole philosophy, he smears entire folio-volumes. In it there are bound to be nothing but *Hieroglyphica, perlexa, transposita*, as the deepest secrets of which the whole world is not worth it - yet it is written for the world. To it will be added a few old recipes, quite obscure to boot, of the universal tincture and the philosopher's stone, to fathom which many a man risks his possessions and money, yes, body and soul. Seen by light, one often finds his *Arcanum* in an old manuscript sold publicly in second-hand store. There, then, the treasure lies open and is esteemed little or not at all.

But as I wish to do another favor to the lover of the Art, I will do here just as I prescribed various methods in connection with the animal and vegetable kingdoms. If he then wishes to practice one or another, I am telling it without circumlocution and briefly, without obscure talk, so that anyone can see and get a Christian impulse to communicate his experiences for the common good.

Of what use is speaking in similes and puzzles? I would rather leave it alone altogether, so that I do not take time, effort and expenses from the poor human beings who are anyhow pursued by the archenemy, Satan, as well as their hard earned

livelihood. Anyone who writes books should take heed
of this and either bring out what he wants to write
for the temptation of the world in a clear form or
not at all. I can well understand my own riddles,
but someone else cannot look into my head and know
what I mean. Therefore, each interprets and explains
it as he pleases, and because of various
interpretations there arises confusion and
bewilderment, from which the ruin and destruction of
all alchemists follows. I will do the lover of the
Art another favor regarding the universal medicine
or the philosopher's stone, and disclose to him this
secret without circumlocution, without obscure talk.

    Therefore, let the reader but mark this: that
he must bring metals and minerals back into their
primary matter, either by a *Menstruum*, any he
wishes, corrosive or not, mercurial, sulphurous,
saline - whichever he considers best and works
fastest. With it he must turn the mineral and metal
back into the primary salt matter, that is, he must
change the metal into a salty nature, which is then
vitriolic, aluminous, or a mineral salt which can

subsequently be dissolved into wine vinegar or ⊽R
and does not leave any undissolved earth. And even
if it did leave one, it would be a sign that it had
not had enough *Menstruum*. Dissolve that with fresh

Menstruum and turn it likewise into salt or ⊕,
alum etc. Now dissolve that vitriol, salt or alum in

the afore-taught sweetening acid, likewise in ⩔
and proceed in everything as has been taught before.

The more often you dissolve with fresh ⩔ and ⊹,

and again coagulate to O O, the sweeter and more

volatile they become and are easily distilled over,

quite oily with little veins like ♈ or ⌒.
Afterwards, it can be dephlegmatized, coagulated in
a gentle heat of ashes. In warmth, it congeals
liquid like wax, in the cold like ice, it melts like
sugar in all moistures and cannot be precipitated:
It is pleasant, sweet and agreeable, and penetrates
everywhere like smoke.

   Now and again, countless simple and compound
solvents are described[178] (i), which I do not heed
but remain, and go along with Nature. And I am
telling the reader that he may do what he wishes,
but without a corrosive he will hardly achieve a
good mineral solution. Even if he had the *Alcahest*
and other root mercurial *Menstrua*, they must be made
- and indeed are all made - of the root of the
corrosive, and it does not matter if someone were to

say: but it is sweetened with ♈ , etc. The
corrosive is master of the process and will remain
so as long as the world exists.

Cape, si capere potes, that is, understand it, if
you can understand.

---

[178] Above there are several excellent *Menstrua,* especially for the improvement of
antimonial medicines, which also come from the alchemical laboratory of our praiseworthy,
in God reposing Brother Homer. They cannot be improved, and the inquisitive public owes
us every possible gratitude for having communicated them.

CHAPTER XI
(The Last Chapter)
OF THE ALKAHEST OR: ALKA EST

In order to enable the reader to get an understanding of the *Alkahest circulati* ✝ *acerrimi*, I will report about it in a circumscriptive way, and in this way finish this book.

Not to be too wordy. The Philosophers, seeing that they do not achieve anything, or little, with the usual corrosives according to the above-mentioned method, have invented and found a means: If a corrosive dissolves an acidic metal, it does not dissolve an alkaline one, and the corrosive that dissolves alkaline products of Nature does not dissolve the acidic ones, because the *Acidum* and the *Alkali*, when they get together, devour each other to death and give birth to a third thing. Therefore they (the philosophers) looked around in Nature to see if there were not one subject that would dissolve both the acid and the alkaline products of Nature without distinction, that would dissolve one like the other and which effects all uniformly in the dissolution. After having searched through everything, they all saw that it had to be a hermaphroditic subject, comprising both natures, and associating with this as well as with that. Indeed, they found this among others in all mercurial subjects, such as *arsenical, marcasite, realgaria,* after the sulphur was separated and in all *Mercurius currentibus & coagulatis*. Such *Mercurius* they took and again made a selection from among them, according to the whim of each. Most of them, however, take a ☿ that is most suitable according to the metal substance in question and penetrates to it during the conjunction to the marrow, and likewise remains undestroyed after its separation,

and such a ☿ that is transmuted into no other metal than gold in its coagulation and fixation.

They saw that that ☿ was too thick and not sharp enough to reduce metals into their first essence and to turn them into a liquid state; that metals, if they are to become homogeneous with all creatures, must assume either a salty or oily or watery nature. They found that ☿ could not give metals such a salt-nature in its simple condition. They also saw that no mere water or earth could dissolve ☿ or metals or turn them into a salty nature, because they noticed that if they wished to reduce metals into salt, oil or water, ☿ would first have to be turned into salt or salty water, so that like would be generated by its like.

Therefore they took such ☿ and turned it in various ways into salt and water, as they succeeded according to their experience. The sharper an artist made his ☿, the better it dissolved; and they saw that ☿ could penetrate little or nothing without such a (sharp) nature. Therefore they were now

forced to turn ☿ into salt, subsequently into
water, to call to their aid all *acida* and *alkalia*
and *nolena volens* to increase the corrosiveness,

without which ☿ did not wish to fight. Just as one
had a better or worse method than another, so they

sharpened their ☿ by means of the *Salia* as well as
they could. They partly took *animalia, vegetabilia*
and *mineralia Salia* mixed together into one in order

to sharpen ☿, and partly took *universalia* and
mineralia, after they found a way that worked for
them. They then recommended this way so heartily as
if there were no other in Nature and they were the
only ones who had the key to retrograde Nature. When

they had turned ☿ into salt, they thought that
Nature uses water in all *generation* and *corruption*
and in every mixture, and that She made almost no
dry *composition* in which she did not require water.

        Therefore they turned this salty ☿ into water
by means of water, so that it should better
penetrate metals and minerals on account of such an
extension and penetrate them to their *centrum
anamae*. Then they took this and turned it into water
with water. The more penetrating the water was, the

better and faster ☿ attacked the metals. The weaker

the water, however, the slower their ☿ began to
dissolve. Therefore they forged it partly with
animal water, partly with plant, mineral or
universal water, or they forged a compositum of

these and with them drove ☿ to and fro until it turned into water with them.

If they made this water sharp and spiritual, the swifter the result was achieved. Instead, if they left the water gross and *crud* and even corporeal, so that ☿ did not become spirit together with them, they achieved a proportionately imperfect *operation*. When they had turned ☿ into such a spiritual water, they called it *acetum acerrimum, acidum metallicum Philosophorum, acheronticum, infernalem, alias etiam circulatum majus*.

There have, however, also been some who turned ☿ into water without any salt, solely *mediante igne*,[179] and because in that way it did not penetrate, they were forced again to call salty, penetrating and sharp water to aid, and sharpen it with animal or vegetable or mineral universal waters. But some were so afraid and doubtful when they used mineral-sharp waters that ☿ might become a corrosive, they sharpened it solely with animal and plant waters and completed their *operationes* in that way, and they succeeded. They feared that the *Spiritus* sharpened with corrosives might hatch all basilisks.

If then someone wishes to produce such a *Menstruum* let him study one from this book,

---

[179] This water about which many have talked extensively is well known and should not necessarily be discarded. But to be honest about it, I think more highly of the menstruum mentioned in Chapter 10 (Part II).

whichever he pleases. Such processes are now and
then publicly printed in various *Autoribus* together
with all techniques as aids to several methods. They
have only been hidden under another name. He can
then apply his intelligence to deal with them.

End of Part II, The Golden Chain of Homer.☐

ABBREVIATIONS

B.M.	Balneum Maria; cook in a double boiler
e.g.	for example
ff	a reference. Example: "Pg. 300 ff" starts at page 300 and continues forward
i.e.	in other words; that is
iij.	3 fluid drams
ij.	2 fluid drams. 1 dram = 1/8 ounce = 3 scruples = 60 grains.
lbij.	2 pounds
lbj.	1 pound
Lot	1 lot = 1/2 ounce
p.	page
P.ij.	2 parts
P.iv.	4 parts
Rc.	recipe
□	

References

Alchemical Symbols, Philip wheeler, Ed. Alchemical
    Manuscripts Vol. 2, ISBN 978-1434812377. 49
    pages, 2007.

Annulus Platonis (The Golden Chain of Homer) by Anton
    Josef Kirchweger, Berlin and Leipzig, 1781,
    German.

Aurea Catena Homeri by Anton Josef Kirchweger, Jena,
    ben Christian Henrich Cuno, 1757, German.

Aurea Catena Homeri by Anton Josef Kirchweger, Ludwig
    (Ludovico) Favrat M.D., 1762, Latin.

Opus Mago-cabbalisticum Et Theosophicum by Georg Von
    Welling, Joseph G. McVeigh and Lon Milo
    DuQuette, ISBN 978-1578633272, Weiser 2006.

Philosophia Salomonis by Johann Jacob Lotters,
    Augsburg 1753, German.

The Betty Story by Philip N. Wheeler. Alchemical
    Manuscripts Vol. 4, ISBN 978-1434813510. 238
    pages, 2007.

The Golden Chain of Homer by Anton Josef Kirchweger.
    Assembled by Hans W. Nintzel. R.A.M.S. 1984,
    English.

The New College Latin and English Dictionary by John
    Traupman, Ph.D. ISBN 978-0-533-59012-8, 2007.
  Webster's New World German Dictionary Peter Terrell
  & Horst Kopleck, Ed. ISBN 978-0-13-953621-2, Wiley
  1987.

A Word from the Publisher

Thank you for purchasing this small work from The R.A.M.S. Library of Alchemy. During his lifetime, Hans Nintzel was dedicated to the identification, acquisition, study, retyping and, when necessary, translation of what he considered to be the most important known works on Alchemy. Hans was assisted by his sparse network of fellow Alchemists, all members of the Restorers of Alchemical Manuscripts Society (R.A.M.S.). I was an active member of R.A.M.S.

My goal is to publish all of the works originally made available through R.A.M.S. as photocopies. To facilitate this, I have chosen to have the books professionally printed. I also have a few titles that I intend to add to the original R.A.M.S. Library, selected by strict criteria established by Hans.

The works from the original R.A.M.S. Library are republished by R.A.M.S. Publishing Company in the collection, "The R.A.M.S. Library of Alchemy," with permission of the Estate of Hans W. Nintzel.

If you have a work on Alchemy that you believe should be a part of the R.A.M.S. Library, please contact me through R.A.M.S. Publishing Company.

Philip N. Wheeler